Methods in Molecular Biology™

Series Editor
John M. Walker
School of Life Sciences
University of Hertfordshire
Hatfield, Hertfordshire, AL10 9AB, UK

For other titles published in this series, go to
www.springer.com/series/7651

3D Cell Culture

Methods and Protocols

Edited by

John W. Haycock

Department of Materials Science and Engineering, Kroto Research Institute, University of Sheffield, Sheffield, UK

Humana Press

Editor
John W. Haycock
Department of Materials Science and Engineering
Kroto Research Institute
University of Sheffield
Sheffield
UK
j.w.haycock@sheffield.ac.uk

ISBN 978-1-4939-5727-9 ISBN 978-1-60761-984-0 (eBook)
DOI 10.1007/978-1-60761-984-0
Springer New York Dordrecht Heidelberg London

Printed on acid-free paper

Humana Press is part of Springer Science+Business Media (www.springer.com)

Preface

Since the advent of routine eukaryotic cell culture more than 40 years ago, the most common substrates for supporting cell growth have been made from polystyrene or glass and have taken the form of a flat two-dimensional (2D) surface. A tremendous number of studies have been performed using cells in 2D culture, and a frequently raised point is exactly how relevant are such studies for interpreting the information gained when compared to the complexities of real tissue physiology? In answer to this, a number of three-dimensional (3D) models have now been developed for a range of tissues where the culture environment takes into account the spatial organization of the cells therein. A common goal of many of these studies is to bridge the gap between in vivo studies at one extreme with that of simple cell monolayers at the other.

For this, it is therefore necessary to create a growth environment that mimics the native tissue as closely as possible. To achieve this, 3D culture models rely inherently on a number of key areas, in particular materials science, cell biology, and bioreactor design. The integration of these approaches is more important than ever, given the practical and applied directions of such work; we frequently hear of tissue engineering and regenerative medicine in the news, with the promise to treat conditions associated with an aging population. For this to become a reality, accurate and relevant 3D culture models are essential and underpin the development of such technologies, be it growing nerves for treating neuronal injuries or skin for burns patients.

While the ultimate goal might be to create an identical tissue ex vivo, many strategies have made tremendous gains by focussing on a single aspect such as biomaterial design, an appropriate cell source, or the bioreactor environment. For tissue engineering, a more common approach has not always been to make an exact copy of living tissue but quite often to generate a "nucleating environment" in which nascent 3D structures have sufficient information for permitting cellular adhesion, proliferation, and differentiation into a functional tissue construct.

3D culture models can be grouped into the study of whole animals and organotypic explant cultures (including embryos), cell spheroids, microcarrier cultures, and tissue-engineered models. While not all 3D culture models require a scaffold, the use of scaffolds for 3D models has increased considerably in the past 10 years, especially due to advances in biomaterial science and processing technologies. Methods for accurately shaping and forming polymers such as microstereolithography hold considerable promise for creating scaffolds quickly with micrometer resolution. Of particular interest is the "re-birth" of traditional processing techniques such as electrospinning and wet spinning, which are seeing a rapid increase in their use for generating 3D culture scaffolds.

The promise of stem cell technology is also highly relevant when practicing 3D culture, not only for understanding the basic processes of differentiation but also for therapeutic purposes. The importance of being able to isolate cells with the capacity to renew, mitotically divide, and differentiate into a diverse range of cell types with control is self-evident. However, many stem cell studies are conducted using 2D environments, and so the relevance of a 3D environment is imperative to understand in detail how one differentiates a stem cell toward a desired lineage, especially if intended for therapeutic purposes.

3D Cell Culture: Methods and Protocols contains a number of basic and applied methodologies taken from a breadth of scientific and engineering disciplines. Many of the topics deal with direct applications of 3D culture models, most notably in the formation of tissues for clinical purpose. I hope that this book will serve as a basic manual for laboratory-based scientists who not only need to have a comprehensive range of techniques contained within a single text but also require techniques described using a standard format. The chapters have all been written by practicing scientists and engineers who provide careful detail for the reproduction of a variety of 3D cell culture models.

The book starts with two review chapters which give an overview of the biological and materials scaffold requirements for successfully creating 3D models. Thereafter, there are 18 chapters which cover key areas for the construction of 3D culture models. These include general scaffold design and fabrication techniques, models for bone, skin, cartilage, nerve, bladder, and hair follicles, and chapters on bioreactor design, imaging, and stem cells. Topics include tissue engineering, where reconstruction in 3D is primarily for clinical purposes, the use of 3D cultures for in vitro models, where work is intended to have an impact in areas such as drug screening, and on tools and technologies, for underpinning the successful development of 3D models. There is also a focus on the design and use of bioreactors.

I hope that readers of *3D Cell Culture: Methods and Protocols* will find this book a valuable reference manual for their day-to-day use in the laboratory. I am personally indebted to all of the international experts who have very kindly contributed chapters and taken enormous trouble to carefully prepare their contributions for this volume of *Methods in Molecular Biology*.

Sheffield, UK ***John W. Haycock***

Contents

Contributors

OTO-OLA AKANJI • *School of Engineering and Materials Science, Queen Mary University of London, London, UK*

HELENA S. AZEVEDO • *Biomaterials, Biodegradables and Biomimetics, University of Minho, Headquarters of the European Institute of Excellence on Tissue Engineering and Regenerative Medicine; Guimarães, Portugal; Institute for Biotechnology and Bioengineering, PTGovernment Associated Laboratory, Guimarães, Portugal*

MICHAEL BARKER • *School of Medicine and Biomedical Sciences, University of Sheffield, Sheffield, UK*

GIUSEPPE BATTAGLIA • *Department of Biomedical Science, University of Sheffield, Sheffield, UK*

ROBERT BIELBY • *Centre for Immunology and Infection, University of York, York, UK*

MANUS J. P. BIGGS • *Department of Applied Physics and Applied Mathematics, Nanotechnology Center for Mechanics in Regenerative Medicine, Columbia University, NY, USA*

JUNE BRAND • *Queens Medical Research Institute, Edinburgh University, Edinburgh, UK*

ROBERT BROWN • *University College London (UCL-TREC), Institute of Orthopaedics and Muscoloskeletal Sciences, London, UK*

LEE BUTTERY • *School of Pharmacy, Centre for Biomolecular Science, University of Nottingham, Nottingham, UK*

RANIERI CANCEDDA • *Dipartimento di Biologia, Oncologia e Genetica, Universita di Genoval & Istituto Nazionale per la Ricerca sul Cancro, Genova, Italy*

ELEONORA CARLETTI • *Department of Materials Engineering and Industrial Technologies and BIOTech Research Centre, University of Trento, Trento, Italy*

ROSS CARNACHAN • *School of Biological and Biomedical Science, University of Durham, UK*

SARAH H. CARTMELL • *School of Materials, University of Manchester, Manchester, UK*

TINA T. CHOWDHURY • *School of Engineering and Materials Science, Queen Mary University of London, London, UK*

FREDERIK CLAEYSSENS • *Department of Materials Science and Engineering, Kroto Research Institute, University of Sheffield, Sheffield, UK*

TERESA COUGHLAN • *Faculty of Medicine & Health Sciences, University of Nottingham, Nottingham, UK*

AILEEN CRAWFORD • *School of Clinical Dentistry, University of Sheffield, Sheffield, UK*

MATTHEW J. DALBY • *Centre for Cell Engineering, Faculty of Biomedical and Life Science, University of Glasgow, Glasgow, UK*

Claudio Gentili • *Dipartimento di Biologia, Oncologia e Genetica, Universita di Genoval & Istituto Nazionale per la Ricerca sul Cancro, Genova, Italy*
Andrew A. Gill • *Department of Materials Science and Engineering, Kroto Research Institute, University of Sheffield, Sheffield, UK*
Sven Halstenberg • *Institute of Pathology, Johannes Gutenberg University, Mainz, Germany*
Paul Hatton • *School of Clinical Dentistry, University of Sheffield, Sheffield, UK*
John W. Haycock • *Department of Materials Science and Engineering, Kroto Research Institute, University of Sheffield, Sheffield, UK*
Vanessa Hearnden • *Department of Materials Science and Engineering, Kroto Research Institute, University of Sheffield, Sheffield, UK*
L. Araida Hidalgo-Bastida • *Institute of Science and Technology in Medicine, Guy Hilton Research Centre, University of Keele, Stoke-on-Trent, UK*
Daniel Howard • *School of Pharmacy, Centre for Biomolecular Science, University of Nottingham, Nottingham, UK*
Gemma Jones • *Institute of Science and Technology in Medicine, Guy Hilton Research Centre, University of Keele, Stoke-on-Trent, UK*
Paul J. Kingham • *Department of Integrative Medical Biology, Section of Anatomy, Umeå University, Umeå, Sweden; Blond McIndoe Laboratories, School of Clinical and Laboratory Sciences, University of Manchester, Manchester, UK*
C. James Kirkpatrick • *Institute of Pathology, Johannes Gutenberg University, Mainz, Germany*
Eleanor Knight • *School of Biological and Biomedical Science, University of Durham, UK*
David A. Lee • *School of Engineering and Materials Science, Queen Mary University of London, London, UK*
Sheila MacNeil • *Department of Materials Science and Engineering, Kroto Research Institute, University of Sheffield, Sheffield, UK*
Cristina Mantovani • *Blond McIndoe Laboratories, School of Clinical and Laboratory Sciences, University of Manchester, Manchester, UK*
Stephen J. Matcher • *Department of Materials Science and Engineering, Kroto Research Institute, University of Sheffield, Sheffield, UK*
Sally L. McArthur • *IRIS, Faculty of Engineering and Industrial Sciences, Swinburne University of Technology, Victoria, Australia*
Rob McKean • *Department of Chemistry, University of Sheffield, Sheffield, UK*
Claudio Migliaresi • *Department of Materials Engineering and Industrial Technologies and BIOTech Research Centre, University of Trento, Trento, Italy*
Antonella Motta • *Department of Materials Engineering and Industrial Technologies and BIOTech Research Centre, University of Trento, Trento, Italy*
Bridgid Murray • *School of Biological and Biomedical Science, University of Durham, UK*
Celia Murray-Dunning • *Department of Materials Science and Engineering, Kroto Research Institute, University of Sheffield, Sheffield, UK*

RUI C. PEREIRA • *Biomaterials, Biodegradables and Biomimetics, University of Minho, Headquarters of the European Institute of Excellence on Tissue Engineering and Regenerative Medicine, Guimarães, Portugal; PTGovernment Associated Laboratory, Institute for Biotechnology and Bioengineering, Guimarães, Portugal; Dipartimento di Biologia, Oncologia e Genetica, Universita di Genoval & Istituto Nazionale per la Ricerca sul Cancro, Genova, Italy*
JAMES B. PHILLIPS • *Department of Life Sciences, The Open University, Milton Keynes, UK*
STEFAN PRZYBORSKI • *Reinnervate Limited, Sedgefield, UK; School of Biological and Biomedical Science, University of Durham, UK*
SARAH RATHBONE • *Institute of Science and Technology in Medicine, Guy Hilton Research Centre, University of Keele, Stoke-on-Trent, UK*
RUI L. REIS • *Biomaterials, Biodegradables and Biomimetics, University of Minho, Headquarters of the European Institute of Excellence on Tissue Engineering and Regenerative Medicine, Guimarães, Portugal; PTGovernment Associated Laboratory, Institute for Biotechnology and Bioengineering, Guimarães, Portugal*
R. GEOFF RICHARDS • *AO Research Institute Davos, Davos Platz, Switzerland*
ANTHONY J. RYAN • *Department of Chemistry, University of Sheffield, Sheffield, UK*
DONALD SALTER • *Queens Medical Research Institute, Edinburgh University, Edinburgh, UK*
ANNE SARTORIS • *Institute of Pathology, Johannes Gutenberg University, Mainz, Germany*
KEVIN SHAKESHEFF • *School of Pharmacy, Centre for Biomolecular Science, University of Nottingham, Nottingham, UK*
JOANNA SHEPHERD • *Department of Materials Science and Engineering, Kroto Research Institute, University of Sheffield, Sheffield, UK*
LOUISE SMITH • *Department of Materials Science and Engineering, Kroto Research Institute, University of Sheffield, Sheffield, UK*
JENNIFER SOUTHGATE • *Jack Birch Unit of Molecular Carcinogenesis, Department of Biology, University of York, York, UK*
TAO SUN • *Centre for Cell Engineering, Faculty of Biomedical & Life Sciences, University of Glasgow, Glasgow, UK*
GIORGIO TERENGHI • *Blond McIndoe Laboratories, School of Clinical and Laboratory Sciences, University of Manchester, Manchester, UK*
NICOLA TIRELLI • *School of Pharmacy and Pharmaceutical Sciences, The University of Manchester, Manchester, UK*
DESMOND J. TOBIN • *Centre for Skin Sciences, School of Life Sciences, University of Bradford, Bradford, UK*
RONALD E. UNGER • *Institute of Pathology, Johannes Gutenberg University, Mainz, Germany*
CLAIRE L. VARLEY • *Jack Birch Unit of Molecular Carcinogenesis, Department of Biology, University of York, York, UK*
NOHA M. ZAKI • *Department of Pharmaceutics, Faculty of Pharmacy, Ain Shams University, Cairo, Egypt*

Chapter 1

3D Cell Culture: A Review of Current Approaches and Techniques

John W. Haycock

Abstract

Cell culture in two dimensions has been routinely and diligently undertaken in thousands of laboratories worldwide for the past four decades. However, the culture of cells in two dimensions is arguably primitive and does not reproduce the anatomy or physiology of a tissue for informative or useful study. Creating a third dimension for cell culture is clearly more relevant, but requires a multidisciplinary approach and multidisciplinary expertise. When entering the third dimension, investigators need to consider the design of scaffolds for supporting the organisation of cells or the use of bioreactors for controlling nutrient and waste product exchange. As 3D culture systems become more mature and relevant to human and animal physiology, the ability to design and develop co-cultures becomes possible as does the ability to integrate stem cells. The primary objectives for developing 3D cell culture systems vary widely – and range from engineering tissues for clinical delivery through to the development of models for drug screening. The intention of this review is to provide a general overview of the common approaches and techniques for designing 3D culture models.

Key words: Cell culture, Bioreactor, Biomaterials, Tissue engineering, Imaging, Scaffold, Stem cells

1. Introduction

Since the advent of routine eukaryotic cell culture more than 40 years ago, the most common substrates for supporting cell growth have been made from polystyrene or glass and have taken the form of a flat two-dimensional surface (1). Thousands of published studies ranging from cancer drug screening through to developmental biology have relied on this format for the growth of adherent cells. A major criticism of these studies, however, is an assumption that animal physiology can be accurately reproduced using a cellular monolayer. Clearly, the presentation of a eukaryotic cell to a two-dimensional glass or polystyrene

John W. Haycock (ed.), *3D Cell Culture: Methods and Protocols*, Methods in Molecular Biology, vol. 695,
DOI 10.1007/978-1-60761-984-0_1, © Springer Science+Business Media, LLC 2011

substrate is not an accurate representation of the extracellular matrix found in native tissue. As a result, many complex biological responses arising such as receptor expression, transcriptional expression, cellular migration, and apoptosis are known to differ quite significantly from that of the original organ or tissue in which they arise.

The role of a normal cell from division, through proliferation to migration and apoptosis, is an accurately controlled series of events that inherently relies on the principles of spatial and temporal organisation. The culture of cells in two dimensions is arguably far too simple and overlooks many parameters known to be important for accurately reproducing cell and tissue physiology. These include mechanical cues, communication between the cell and its matrix, and communication between adjacent cells. On the point of intercellular communication, many two-dimensional culture experiments fail to consider the interplay between different cell types, with the vast majority of cultures being of a single cell type. 2D co-cultures overcome some of these shortfalls, but are some way off in accurately reproducing cellular function observed within a tissue.

2. Three-Dimensional Cell Culture

In answer to these problems, a number of three-dimensional methods have been developed for a range of tissues where the culture environment takes into account the spatial organisation of the cell (2–5). A common goal for many of these studies is to bridge the gap between the use of whole animals at one end of the spectrum, with cellular monolayers at the other. It is therefore necessary to create a growth environment that mimics the native tissue as closely as possible, and a simple starting point is the introduction of cells into a porous biocompatible scaffold. However, the complexity of 3D systems then becomes apparent with a number of parameters to consider. Important criteria include the choice of material for the scaffold, the source of cells, and the actual methods of culture, which in practice varies considerably according to the tissue of study. A number of common approaches exist, but so too does the opinion of investigators – from the precise design of scaffolds through to the sourcing of a particular cell type. For example, does one use naturally derived or synthetic materials for a scaffold? Does one use autologous or adult-derived stem cells? Does one invest time and money fabricating an accurate nanostructured scaffold, or produce a microstructured scaffold with an approximate geometry for maintaining cell growth?

3. Three-Dimensional Culture Models

Three-dimensional culture models can be grouped into the study of whole animals and organotypic explant cultures (including embryos), cell spheroids, microcarrier cultures, and tissue-engineered models (6). Not all three-dimensional culture models require a scaffold; however, the use of scaffolds for 3D models has certainly increased considerably in the past ten years. Whole animal and organotypic explants are principally used in studies where an absolute requirement for tissue-specific information is needed (7). These models enable data where the cell is physically located within its native environment. Examples include *drosophila melanogaster* (fruitfly) and the use of zebrafish and mouse embryos. Experimental versatility in terms of environmental conditions is permissible for non-mammalian models such as the fruitfly and zebrafish, but maintaining cellular viability for mouse embryos is an absolute necessity, and so culture conditions such as pH, temperature, and O_2 levels must be very carefully controlled for these models (6, 8). Organ explantation for culture has largely been pioneered in the areas of brain and neural physiology. Here, explants can be maintained in vitro in gels or on semi-permeable membranes in the presence of an isotonic or nutrient medium. Advantages include the maintenance of tissue architecture and importantly the presence of differentiated cells within the tissue (6). Technical demands for these models include the time available for maintaining specimen integrity and the need to image deeply into samples.

Cellular spheroids are simple three-dimensional models that can be generated from a wide range of cell types and form due to the tendency of adherent cells to aggregate. They are typically created from single culture or co-culture techniques such as hanging drop, rotating culture, or concave plate methods (6, 9, 10). Spheroids do not require scaffolds and can readily be imaged by light, fluorescence, and confocal microscopy. Consequently, spheroids have seen a use in modelling solid tumour growth and metastasis studies and are also used in a multitude of therapeutic studies, e.g. for high throughput screening (11). An analogous approach is in the development of epithelial tissues to form polarised sheets, such as the epidermis of skin (12). Normal human keratinocytes can be isolated from skin and cultured on supports such as collagen gels, synthetic polymer membranes, microfibre meshes, or de-epidermalised human dermis (DED) (12). The use of DED involves removing the dermis of its original cellular components, but, importantly for 3D cell culture, it maintains many of the native basement membrane proteins (e.g. collagen type IV). The presence of these proteins in the matrix is an absolute necessity for the reconstructive adhesion and growth of keratinocytes thereafter (13).

4. Biomaterial Scaffolds for Fabricating Structure and Shape

As one increases the size and complexity of a three-dimensional model, the need for a scaffold becomes apparent. Cellular aggregates require the careful exchange of nutrients and gases in addition to spatial control, and problems with cell death arise when aggregate thicknesses of 1–2 mm arise through a lack of mass transfer, principally through a limited exchange of nutrients and waste metabolites (6, 14). This has been addressed by the use of highly porous scaffolds where basic designs consider shape, cell adhesion sites and the flow of gases, nutrients, and metabolites (4). Different cell types are embedded within matrices possessing distinctly different properties and shapes. For example, if engineering peripheral nerve, one must consider the native structure where axons are surrounded by a soft uniaxially aligned lipoprotein myelin sheath. In contrast, osteoblasts adhere to a hard surface of bone within cuboidal sheets. Consequently, the design of scaffolds must reflect the tissue of interest and a tremendous diversity exists in the design of scaffolds for the engineering of tissues (4). An important consideration is the intended application and use. Clinical work that requires a functioning implant may require just a temporary biodegradable scaffold, which after implantation is remodelled by the body and replaced by native tissue to restore original function. In this instance, the scaffold must support cell growth and differentiation, and a physical match must exist between the size of the scaffold and that of the defect. Furthermore, the scaffold should break down into metabolites without a toxic or immunogenic response. Alternatively, scaffolds may be intended as a 3D in vitro model, e.g. to further our understanding in a fundamental aspect of tissue biology or to generate systems for drug and cosmetics screening (15). Here, there is a need to accurately reproduce the native tissue structure containing cells at a given stage of differentiation, and arguably there is a greater need to image these models for cell function and response. The absolute size of the scaffold for these models and the need for hydrolysis or degradation may not be quite so important.

The choice of bulk materials to be used for scaffold fabrication includes metals, glasses, polymers, and ceramics (4). Polymers are commonly used due to an ability to control their chemical and structural properties, in combination with methods for fabrication. They are typically grouped into synthetic and natural derivatives (4). Synthetic polymers include materials such as poly glycolic acid (PGA) and poly lactic acid (PLA), whereas natural polymers include materials such as chitosan and collagen. A general requirement for all biomaterial scaffolds is to reproduce an extracellular matrix environment for supporting cell growth outside of the body.

The bulk chemical composition of a biomaterial must therefore be the first consideration when designing a scaffold, with biocompatibility being a priority for implantation (3). In particular, a material must be selected that avoids triggering an immune response or the development of a fibrous capsule. A degradable scaffold should ideally be used for clinical purposes and most degradable synthetic scaffolds such as PGA undergo hydrolysis in situ. Consequently, the body must be able to metabolise the monomeric products released during breakdown without a toxic or inflammatory response. For systems such as PGA/PLA, degradation rates can be readily tuned by the composition of PGA versus PLA, where a higher PGA content degrades faster. Natural scaffolds such as collagen are degraded by enzymolysis and consequently less control is possible on tailoring the breakdown rate. However, natural scaffolds tend to exhibit better biocompatible properties over synthetic materials – but their clinical use is concerned with potential disease transmission (4, 12), a situation avoided by the use of synthetic scaffolds.

The surface chemical properties of a biomaterial are fundamental for dictating the adhesion and spreading of living cells (16). Such properties are not necessarily governed by the bulk chemistry, in particular due to surface modification with soluble proteins derived either from the growth medium or from the cells themselves (17). Surface chemistry is predominantly controlled by charge and polarity, which in general terms control the attractiveness of proteins in solution to diffuse and adsorb at the surface. The rate at which this happens is determined by the Vroman effect, whereby highly mobile proteins in a heterogeneous mixture will reach a surface quickly, but in time may be replaced by more slowly moving proteins with a higher affinity (18, 19). This arises in particular for serum proteins, where fibrin will adsorb to a polymer surface rapidly leading to fibronectin depletion in vivo. A relationship exists between the extent of charge at a surface and the proportion of proteins that are adsorbed. This is known to correlate with the tendency of cells to adhere to a biomaterial, where the cell interacts via an adsorbed protein layer, rather than directly to the biomaterial surface (16).

Optimising the surface chemistry of biomaterials can therefore be controlled either to increase or decrease protein adsorption and in turn cellular attachment. A good example of an approach for increasing cellular attachment is given in chapter 10, where Schwann cell adherence to aligned PLA microfibres is improved by the deposition of a plasma acrylic acid layer (20). Here an increase in the negative surface charge of acid groups is associated with an increase in cell attachment and proliferation. Conversely, the deposition of allyl amine serves to prevent Schwann cell attachment. Although both layers contain surface charges, the surface chemical groups must also dictate not just the extent of protein adsorption, but the folding conformation of the protein. At a molecular level, the interaction between acid or amine groups in the plasma polymer with amino acids containing polar, non-polar, and charged groups will dictate how a protein interacts and folds at the surface. This determines whether the adsorbed protein presents adhesive ligands permissive for binding to receptors such as integrins. A number of similar studies using plasma deposition are reported in the literature where the aim was to optimise cell adhesion and growth, e.g. the adhesion of human keratinocytes to polymer sheets for clinical delivery (21). Conversely, non-fouling surfaces such as polyethylene glycol serve to minimise protein adsorption (or fouling) and in turn cellular adhesion. The theory as to why PEG surfaces are non-fouling is highly complex – indeed the mechanisms are still being investigated. Predominant reasons suggest that chain mobility and a steric stabilising force are important, with protein-resistant properties arising through both a mixing interaction and excluded volume component (22). Thus, when a protein approaches a PEG layer, the available volume per glycol unit is decreased resulting in a repulsive force, due to a decrease in conformational entropy. In addition, the compressive force of a protein into a PEG layer reduces the total number of confirmations originally available to the chain, which creates an osmotically repulsive force, effectively pushing the protein away from the PEG layer (22).

Cell adhesion can also be controlled by integrating precise structural motifs into a biomaterial. Original work from Massia and Hubbell in 1991 reported that the alpha-V-beta-3 integrin adhesion ligand RGD, when covalently attached to a surface with a critical spacing of 440 nm, was permissive for the attachment of fibroblast cells in vitro (colloquially known as the "Hubbell limit") (23). If the separation distance between ligands was decreased to 140 nm, then fibroblast stress fibre and focal contact formation was observed. This has led many investigators to conjugate RGD-like ligands for attachment into and onto biomaterial surfaces for controlling cellular adhesion (4). However, many direct conjugation methods, while elegant, are confined predominantly to cell culture in 2D. In contrast, surface modification techniques such as plasma vapour phase deposition have proved to be effective for influencing cell adhesion in 3D scaffolds (20, 21). For example, Barry et al. report on the use of an allyl amine plasma polymer specifically for encouraging fibroblast cell attachment, morphology, and metabolic activity into 3D P(DL)LA porous scaffolds without changing the bulk characteristics of the scaffold (24). A major advantage of this approach is in the rapidity, reproducibility, and chemical control possible for modifying 3D scaffolds.

5. Scaffolds and Length Scales

The ultimate aim of a scaffold is to produce features found naturally within the extracellular matrix required for native cell function. Consequently, design criteria must consider lengthscales which span the macro-, micro-, and nanoscale. Macroscale structures are important for determining the overall size and shape of a scaffold. If constructs are for clinical purposes, then a range of shapes may be needed for implantation at a defect site. This has implications for the tissue engineering of constructs and whether they should be required as "off the shelf" products or alternatively be tailored to individual patient requirements. Bespoke products can be manufactured by modelling a defect site, e.g. a bone lesion can be characterised using computerised tomography or magnetic resonance imaging for producing a 3D macrostructure. This can be followed by computer-aided design and fabrication techniques such as stereolithography for generating a bespoke scaffold (25).

Micron length scales must be considered when reproducing tissue architectures. For example, organised parallel fibres are important for reconstructing peripheral nerve (26), while random non-woven networks may be more relevant for dermal replacement (27, 28). Irrespective of the tissue under study, particular attention should be given to parameters that control pore size, connectivity, and geometry (4). Microstructural features in general terms are important for ensuring cell adhesion, as the size of many adherent somatic cell types typically spans a distance of 10–150 μm. However, nanoscale features are being argued as more important for scaffold design (29), discussed below. Microscale features must permit a balance between scaffold porosity and the volume occupied when introducing cells for maintaining effective mass transport and nutrient exchange (30). The microstructure of a scaffold is also important for determining the overall mechanical properties. This is not only important for reproducing the properties of the native tissue, but also for withstanding experimental procedures in the generation and culture of the construct. Mechanical properties are also known to influence the function of cells contained within a construct. For example, it is well known that mechanical forces affect bone remodelling and repair. This knowledge has been applied to the exertion of mechanical forces on 3D bone scaffolds for tissue engineering (31). More recent information also reports on the direct effect of scaffold mechanics on adherent cells, where a stiffer versus a softer material can dictate the differentiation of stem cells in the absence of any externally applied force (32).

The fabrication of scaffolds with micron scale control has been possible for many years. Techniques such as electrospinning, wet spinning and sponge-like fabrication methods, such as freeze drying and gas foaming, are reasonably commonplace (4). A number of these methods also enable nanostructured features to be made, e.g. the electrospinning of fibres can produce diameters ranging from tens of nanometres to tens of micrometres (33). However, while many of these techniques allow the creation of microstructural features and have controllable process conditions, they do not allow for ultimate control where the exact positioning of microstructural features is possible for copying the extracellular matrix. This raises an interesting point as the designs of many scaffolds for 3D cell culture are not an exact mimic, especially if the intended purpose is for implantation and where the scaffold is ultimately biodegradable. One question is therefore whether an intricately designed scaffold should be made if cell adherence, growth, and the restoration of tissue function can be achieved using a more approximately designed scaffold? Clearly, a balance needs to be established between investing considerable resource making a perfect scaffold versus the manufacture of a device that fulfils a number of basic criteria. A practical approach must be taken where the intended endpoint of the work dictates the 3D method used for getting there. This is particularly relevant if the 3D model is intended for clinical implantation.

Nanoscale features are important in determining how cells physically interact with a substrate and how they respond to it. Interactions between cell integrin receptors and adhesive ligands in native tissue arise when inter-ligand distances vary from tens to hundreds of nanometres (23). For example, collagen fibrils are typically organised across a length scale of 50–200 nm and enable the adhesive interaction of fibroblast cells (34). The precise control of these structures is determined by the primary sequence of amino acids and secondary structure of component proteins. The importance of nanoscale substrates for cell function has previously been studied in depth using model systems such as the spatial control of RGD ligands and their interaction with integrin receptors for determining cell adhesion, morphology, differentiation, and apoptosis (23, 34). Therefore, nanoscale features for scaffold design may be important. An increasing emphasis is being seen on methods for fabricating nanoscale structures (known as nanoengineering (29)) and includes the use of peptide hydrogels (35), the control of process conditions such as thermal-induced phase separation (36), the use of post-fabrication modifications (21, 24), and the incorporation of nanostructures into a matrix (37). This subject is dealt with in detail elsewhere (4), but irrespective of technique, one could argue that the ultimate scaffold will have controllability of desired features across the nanoscale, microscale, and macroscale. This is a major future challenge and not only requires an in depth understanding of the bulk and surface physicochemical properties of the scaffold, but also tremendous control and versatility over the methods for fabrication (29).

6. Bioreactors for 3D Constructs

An important consideration when moving from cells in culture as a 2D layer to a 3D construct is the maintenance of mass transport (6, 30). A limiting factor for survival concerns not only the ability to supply nutrients and oxygen, but also the simultaneous removal of waste products and metabolites. This is most readily seen for spheroids where diameters greater than 1 mm are associated with hypoxic centres containing necrotic cells, surrounded by an outer shell of living cells (38). This is known to arise directly due to nutrient starvation and metabolite toxicity. Information on spheroid models is useful in directing the design of more complex cell-scaffold constructs. As spheroids are entirely cellular, one could predict in general terms that the maximum depth for a given cellular mass within a larger nutrient-maintained construct will be of a similar order. This is therefore a major consideration in the design of all 3D culture systems whether for clinical purposes or for in vitro models. Early simple 3D culture models were based on static methods; however, the design and use of bioreactors is increasingly being integrated together with 3D culture systems and tissue engineered constructs (30). Bioreactors enable the precise and reproducible control over many environmental conditions required for cell culture. These include temperature, pH, medium flow rate, oxygen, nutrient supply, and waste metabolite removal. In addition, increasingly complex systems are being designed for the simultaneous control of seeding cells into scaffolds, and where relevant, the application of external forces to encourage differentiation and maturation. Common to many advanced systems now is the ability to maintain and monitor the environment during growth (30).

Several designs of bioreactor exist, but broadly, these can be grouped into rotating wall vessels, direct perfusion systems, hollow fibres, spinner flasks, and mechanical force systems (30). Rotating wall vessels provide continuously moving culture conditions where cell constructs are grown under low shear stress forces and enable high rates of mass transfer (39). The speed of rotation is such that forces exerted on the construct by rotation of the bioreactor ensure that constructs are in continuous free-flow. Direct perfusion systems allow the culture medium to pass through the construct (26, 40). A major advantage here is the ability to seed cells directly into the scaffold under flow conditions, which usually allows for a high seeding efficiency. The control of medium flow thereafter enables cell adhesion and growth, where a high mass transfer rate is typically achieved throughout the entire construct. Hollow fibre systems are used for cells that have a high metabolic rate (41). Cells are usually seeded within a matrix or scaffold contained within porous fibres. The medium is then perfused externally over the fibres to increase mass transfer. A consideration when designing these systems is the porosity of the scaffold and whether the entire scaffold experiences metabolite exchange or just the periphery. Spinner flasks can be used to seed cells into constructs and also culture them thereafter (38). Seeding is conducted by the introduction of cells into the medium and their perturbation by the spinner, generating convection currents. Mass transfer for subsequent culture is maintained by the spinner mechanism. Mechanical force systems exploit the mechanism by which tissues respond to force during growth (42). Cells such as osteoblasts are known to be mechanoreceptive and respond to force with the activation of intracellular signal transduction pathways (43). Secondary messenger signals arising can control gene expression and determine the expression of differentiation genes thereafter, and consequently enhance conditions for 3D construct maturation (43). Bioreactors can therefore be exploited by using physiological loading regimes for determining the optimum conditions for exerting and detecting forces on a construct (42).

A future direction of bioreactor design is in the reproducible and automated production of tissues, where temperature, pH and oxygen levels are monitored and controlled simultaneously. Taking this one step further, monitoring a developing tissue by non-invasive methods such as 2-photon microscopy, MRI, or CT scanning could assess the extent of cellular growth and differentiation, allowing for flexibility in variations expected during development. While some way off, state-of-the art bioreactors such as the Advanced Clinical Tissue Engineering System (ACTES™) are in development (reviewed in (38)). ACTES™ systems are intended to be based within hospitals whereby an automated closed loop system takes a patient biopsy, isolates and expands the cells, seeds them on to a scaffold, and then cultures them until formation of a mature tissue graft. It is proposed that such systems could carry out autologous tissue grafting on site, eliminating the need for expensive GMP facilities and minimising operator handling. However, the most important aspects for bioreactor design at present address conditions for ensuring a reproducible and controlled growth environment for constructs that are millimetres to centimetres in size.

7. The Source of Cells for 3D Models

The source of cells for 3D cultures and tissue engineering usually requires a host or a donor-derived origin. A remarkable number of possibilities exist in principle for the various sites and sources, but in general terms for tissue engineering, these can be grouped into stem cells, autologous cells, allogenic cells, and xenogenic cells. For 3D in vitro models, this list can be extended to include animal-derived primary cells, cell lines, and genetically modified variants of all the above cell types. Many investigators favour the use of autologous cells for clinical implantation, principally for the avoidance of immune rejection. A widely publicised early example was the generation of engineered cartilage (44). Here, a small healthy biopsy of cartilage was taken, the cells explanted in the laboratory, expanded, and seeded on to PGA meshes and PLA scaffolds before being implanted into the defect site. However, autologous cells are not always available and even if they are, they may not be viable or capable of proliferation in vitro. As an alternative, it is possible in some cases to use allogenic cells; however, the potential for immune incompatibility must be considered. Xenogenic cells can be used if the clinical requirement is for the supply of chemicals within a tissue. A good example of xenogenic transplantation is the use of pancreatic islets for insulin production (45). It is, however, necessary to physically contain such cells within a semi-permeable encapsulating membrane.

A commonly encountered problem with the need for primary cells is a lack of availability or an insufficient potential to generate sufficient numbers for clinical purposes. Therefore, the use of progenitor and multipotent stem cells holds great promise. Remarkable advances have been made in the isolation, expansion, characterisation, and targeted differentiation of progenitor cells towards a number of different lineages. The number of tissue sites from which haematopoietic, mesenchymal, or neural stem cells can be isolated includes the lung, liver, retina, pancreas, cardiovascular system, brain, spinal chord, adipose tissue, and bone marrow. Irrespective of this large potential source, a particular challenge for any application is the ability to direct cellular differentiation with great accuracy towards an intended phenotype. Variations are observed between the implantation of stem cells in vivo versus differentiation potential in vitro, e.g. neural stem cells have a greater range of expression markers following surgical implantation, compared to expansion of the same progenitor population in vitro (46, 47). Similarly, work on embryonic stem cells shows variations in response to stimuli such as the addition of growth factors in vitro, where not all cells differentiate equally (48). In contrast, recent work on the introduction of hECs in scaffolds followed by implantation shows extensive differentiation towards a functioning tissue, e.g. in vessel formation (49).

8. The Commercial Promise of 3D Culture

The potential to repair and restore tissue function by the clinical delivery of tissue engineered constructs sparked the creation of a number of biotechnology and healthcare companies in the 1990s.

The majority of products introduced were for the treatment of skin and epithelial injuries. However, a number of financial difficulties were encountered shortly after, not because the products were necessarily ineffective, but because of difficulties in getting the products from bench to clinic. Tissue engineered constructs can be notoriously expensive to manufacture and, in combination with needing to recover R&D costs, a very real threat is that such products are simply unaffordable to the consumer. In practice, the National Health Service in the UK predominantly determines whether a commercial product succeeds or fails. The situation in the United States is somewhat different, with the prevalence of private healthcare. Thus, cost is important for the fundamental design of tissue engineered products and arguably this starts with the basics of scaffold design, before considering if one needs to use autologous cells or stem cells. However, cost is not the only issue. A compounding factor in a number of countries is also the lack of clear regulatory guidance on facilitating the smooth transition of constructs from the laboratory to the clinic. A number of problems encountered over the past ten years have also been hampered by regulatory uncertainty, largely due to tissue-engineered products not fitting neatly into traditional forms of healthcare therapy such as devices or drugs (12).

9. A Combinatorial Approach

In summary, 3D culture models can only succeed by combining a number of key areas, in particular materials science, cell biology, bioreactor design, and aligning these to clinical applications and regulatory practice if intended for implantation. While the ultimate goal might be to create an identical tissue ex vivo, many strategies have made tremendous gains by focussing on a single aspect such as biomaterial design, an appropriate cell source, or the bioreactor environment. For tissue engineering, a more common approach has not been to make an exact copy of living tissues, but to generate a "nucleating environment" in which 3D structures have sufficient information for permitting cellular adhesion, proliferation, and differentiation into a mature and functioning construct. For example, epithelial–dermal sheets can be readily fabricated for skin reconstruction using microstructured fibre scaffolds (27, 28), questioning the need for nanostructured scaffolds. Conversely, the alignment of peripheral nerve axons for repairing traumatic injuries may require a scaffold with nanostructured features, due to the complexity of organising thousands of axons over several millimetres (26). Considerable effort is presently being invested on establishing methods for integrating cells into scaffolds, and investigating exactly how complex this environment needs to be for promoting the formation of new tissues. Much attention has surrounded the subject of 3D culture models and tissue engineering over the past decade, and immediate clinical and commercial expectations have frequently been unrealistic. However, during this time tremendous advances have been made in the basic development of 3D models. The aim of this book is therefore to provide an overview of the methods and techniques successfully devised for practising 3D cell culture.

References

1. Freshney, I. R. (2005) *Culture of animal Cells. A Manual of Basic Technique*, 5 ed. John Wiley & Sons, Hoboken, NJ.
2. Abbott, A. (2003) Cell culture: Biology's new dimension. *Nature* 424, 870–872.
3. Langer, R. and Tirrell, D. A. (2004) Designing materials for biology and medicine. *Nature* 428, 487–492.
4. Lee, J., Cuddihy, M. J., and Kotov, N. A. (2008) Three-dimensional cell culture matrices: State of the art. *Tissue Eng. Part B Rev.* 14, 61–86.
5. Lavik, E. and Langer, R. (2004) Tissue engineering: current state and perspectives. *Applied Microbiol. Biotechnol.* 65, 1–8.
6. Pampaloni, F., Reynaud, E. G., and Stelzer, E. H. K. (2007) The third dimension bridges the gap between cell culture and live tissue. *Nature Rev. Mol. Cell Biol.* 8, 839–845.
7. Toda, S., Watanabe, K., Yokoi, F., Matsumura, S., Suzuki, K., Ootani, A., Aoki, A., Koike, N., and Sugihara, H. (2002) A new organotypic culture of thyroid tissue maintains three-dimensional follicles with C cells for a long term. *Biochem. Biophys. Res. Commun.* 294, 906–911.
8. Hadjantonakis, A. K., Dickinson, M. E., Fraser, S. E., and Papaioannou, V. E. (2003) Technicolour transgenics: Imaging tools for functional genomics in the mouse. *Nature Rev. Genet.* 4, 613–625.
9. Timmins, N. E., Harding, F. J., Smart, C., Brown, M. A., and Nielsen, L. K. (2005) Method for the generation and cultivation of functional three-dimensional mammary constructs without exogenous extracellular matrix. *Cell Tissue Res.* 320, 207–210.
10. Castaneda, F. and Kinne, R. K. H. (2000) Short exposure to millimolar concentrations of ethanol induces apoptotic cell death in multicellular HepG2 spheroids. *J. Cancer Res. Clin. Oncol.* 126, 305–310.
11. Ivascu, A. and Kubbies, M. (2006) Rapid generation of single-tumor spheroids for high-throughput cell function and toxicity analysis. *J. Biomol. Screen.* 11, 922–932.
12. MacNeil, S. (2007) Progress and opportunities for tissue-engineered skin. *Nature* 445, 874–880.
13. Ghosh, M. M., Boyce, S., Layton, C., Freedlander, E., and MacNeil, S. (1997) A comparison of methodologies for the preparation of human epidermal-dermal composites. *Annals Plastic Surg.* 39, 390–404.
14. Griffith, L. G. and Swartz, M. A. (2006) Capturing complex 3D tissue physiology in vitro. *Nature Rev. Mol. Cell Biol.* 7, 211–224.
15. Canton, I., Sarwar, U., Kemp, E. H., Ryan, A. J., MacNeil, S., and Haycock, J. W. (2007) Real-time detection of stress in 3D tissue-engineered constructs using NF-kB activation in transiently transfected human dermal fibroblasts. *Tissue Eng.* 13, 1013–1024.
16. Allen, L. T., Tosetto, M., Miller, I. S., O'Connor, D. P., Penney, S. C., Lynch, I., Keenan, A. K., Pennington, S. R., Dawson, K. A., and Gallagher, W. M. (2006) Surface-induced changes in protein adsorption and implications for cellular phenotypic responses to surface interaction. *Biomaterials* 27, 3096–3108.
17. Wilson, C. J., Clegg, R. E., Leavesley, D. I., and Pearcy, M. J. (2005) Mediation of biomaterial-cell interactions by adsorbed proteins: A review. *Tissue Eng.* 11, 1–18.
18. Vroman, L. and Lukosevicius, A. (1964) Ellipsometer recordings of changes in optical thickness of adsorbed films associated with surface activation of blood clotting. *Nature* 204, 701.
19. Vroman, L. (1962) Effect of adsorbed proteins on wettability of hydrophilic and hydrophobic solids. *Nature* 196, 476.
20. Murray-Dunning, C. M., McKean, R., Forster, S., Ryan, A. J., McArthur, S. L., and Haycock, J. W. (2010) Three-dimensional alignment of Schwann cells using hydrolysable microfibre

scaffolds – strategies for peripheral nerve repair. *Methods Mol. Biol.* 695, 155–166.

21. France, R. M., Short, R. D., Dawson, R. A., and MacNeil, S. (1998) Attachment of human keratinocytes to plasma co-polymers of acrylic acid octa-1,7-diene and allyl amine octa-1,7-diene. *J. Mat. Chem.* 8, 37–42.
22. Sharma, S., Johnson, R. W., and Desai, T. A. (2004) Evaluation of the stability of nonfouling ultrathin poly(ethylene glycol) films for silicon-based microdevices. *Langmuir* 20, 348–356.
23. Massia, S. P. and Hubbell, J. A. (1991) An RGD spacing of 440 nM is sufficient for integrin alpha-v-beta-3-mediated fibroblast spreading and 140 nM for focal contact and stress fiber formation. *J. Cell Biol.* 114, 1089–1100.
24. Barry, J. J. A., Silva, M. M. C. G., Shakesheff, K. M., Howdle, S. M., and Alexander, M. R. (2005) Using plasma deposits to promote cell population of the porous interior of three-dimensional poly(d,l-lactic acid) tissue-engineering scaffolds. *Adv. Funct. Mater.* 15, 1134–1140.
25. Hollister, S. J. (2005) Porous scaffold design for tissue engineering. *Nature Mater.* 4, 518–524.
26. Sun, T., Norton, D., Vickers, N., McArthur, S. L., Mac Neil, S., Ryan, A. J., and Haycock, J. W. (2008) Development of a bioreactor for evaluating novel nerve conduits. *Biotechnol. Bioeng.* 99, 1250–1260.
27. Sun, T., Mai, S. M., Norton, D., Haycock, J. W., Ryan, A. J., and MacNeil, S. (2005) Self-organization of skin cells in three-dimensional electrospun polystyrene scaffolds. *Tissue Eng.* 11, 1023–1033.
28. Blackwood, K. A., McKean, R., Canton, I., Freeman, C. O., Franklin, K. L., Cole, D., Brook, I., Farthing, P., Rimmer, S., Haycock, J. W., Ryan, A. J., and MacNeil, S. (2008) Development of biodegradable electrospun scaffolds for dermal replacement. *Biomaterials* 29, 3091–3104.
29. Stevens, M. M. and George, J. H. (2005) Exploring and engineering the cell surface interface. *Science* 310, 1135–1138.
30. Martin, I., Wendt, D., and Heberer, M. (2004) The role of bioreactors in tissue engineering. *Trends Biotechnol.* 22, 80–86.
31. Allori, A. C., Sailon, A. M., Pan, J. H., and Warren, S. M. (2008) Biological basis of bone formation, remodeling, and repair – Part III: Biomechanical forces. *Tissue Eng. Part B Rev.* 14, 285–293.
32. Engler, A. J., Sen, S., Sweeney, H. L., and Discher, D. E. (2006) Matrix elasticity directs stem cell lineage specification. *Cell* 126, 677–689.
33. Sun, T., Norton, D., Mckean, R. J., Haycock, J. W., Ryan, A. J., and MacNeil, S. (2007) Development of a 3D cell culture system for investigating cell interactions with electrospun fibers. *Biotechnol. Bioeng.* 97, 1318–1328.
34. Geiger, B., Spatz, J. P., and Bershadsky, A. D. (2009) Environmental sensing through focal adhesions. *Nature Rev. Mol. Cell Biol.* 10, 21–33.
35. Zhang, S. (2008) Designer self-assembling peptide nanofiber scaffolds for study of 3-D cell biology and beyond. *Adv. Cancer Res.* 99, 335–340.
36. Ma, P. X. and Zhang, R. Y. (1999) Synthetic nano-scale fibrous extracellular matrix. *J. Biomed. Mater. Res.* 46, 60–72.
37. Sachlos, E., Gotora, D., and Czernuszka, J. T. (2006) Collagen scaffolds reinforced with biomimetic composite nano-sized carbonate-substituted hydroxyapatite crystals and shaped by rapid prototyping to contain internal microchannels. *Tissue Eng.* 12, 2479–2487.
38. VunjakNovakovic, G., Freed, L. E., Biron, R. J., and Langer, R. (1996) Effects of mixing on the composition and morphology of tissue-engineered cartilage. *AIChE J.* 42, 850–860.
39. Unsworth, B. R. and Lelkes, P. I. (1998) Growing tissues in microgravity. *Nature Med.* 4, 901–907.
40. Wendt, D., Marsano, A., Jakob, M., Heberer, M., and Martin, I. (2003) Oscillating perfusion of cell suspensions through three-dimensional scaffolds enhances cell seeding efficiency and uniformity. *Biotechnol. Bioeng.* 84, 205–214.
41. Jasmund, I., Simmoteit, R., and Bader, A. (2001) An improved oxygenation hollow fiber bioreactor for the cultivation of liver cells. *Animal Cell Technology: from Target to Market* 1, 545–547. Kluwer Academic Publishers, London.
42. Demarteau, O., Jakob, M., Schafer, D., Heberer, M., and Martin, I. (2003) Development and validation of a bioreactor for physical stimulation of engineered cartilage. *Biorheology* 40, 331–336.
43. Rubin, J., Rubin, C., and Jacobs, C. R. (2006) Molecular pathways mediating mechanical signaling in bone. *Gene* 367, 1–16.
44. Rodriguez, A., Cao, Y. L., Ibarra, C., Pap, S., Vacanti, M., Eavey, R. D., and Vacanti, C. A. (1999) Characteristics of cartilage engineered from human pediatric auricular cartilage. *Plast. Reconstr. Surg.* 103, 1111–1119.
45. O'Connell, P. (2002) Pancreatic islet xenotransplantation. *Xenotransplantation* 9, 367–371.

46. O'Connor, S. M., Stenger, D. A., Shaffer, K. M., Maric, D., Barker, J. L., and Ma, W. (2000) Primary neural precursor cell expansion, differentiation and cytosolic Ca2+ response in three-dimensional collagen gel. *J. Neurosci. Methods* 102, 187–195.
47. Gage, F. H. (2000) Mammalian neural stem cells. *Science* 287, 1433–1438.
48. Teng, Y. D., Lavik, E. B., Qu, X. L., Park, K. I., Ourednik, J., Zurakowski, D., Langer, R., and Snyder, E. Y. (2002) Functional recovery following traumatic spinal cord injury mediated by a unique polymer scaffold seeded with neural stem cells. *Proc Natl. Acad. Sci.* 99, 3024–3029.
49. Levenberg, S., Golub, J. S., Amit, M., Itskovitz-Eldor, J., and Langer, R. (2002) Endothelial cells derived from human embryonic stem cells. *Proc Natl. Acad. Sci.* 99, 4391–4396.

Chapter 2

Scaffolds for Tissue Engineering and 3D Cell Culture

Eleonora Carletti, Antonella Motta, and Claudio Migliaresi

Abstract

In tissue engineering applications or even in 3D cell cultures, the biological cross talk between cells and the scaffold is controlled by the material properties and scaffold characteristics. In order to induce cell adhesion, proliferation, and activation, materials used for the fabrication of scaffolds must possess requirements such as intrinsic biocompatibility and proper chemistry to induce molecular biorecognition from cells. Materials, scaffold mechanical properties and degradation kinetics should be adapted to the specific tissue engineering application to guarantee the required mechanical functions and to accomplish the rate of the new-tissue formation. For scaffolds, pore distribution, exposed surface area, and porosity play a major role, whose amount and distribution influence the penetration and the rate of penetration of cells within the scaffold volume, the architecture of the produced extracellular matrix, and for tissue engineering applications, the final effectiveness of the regenerative process. Depending on the fabrication process, scaffolds with different architecture can be obtained, with random or tailored pore distribution. In the recent years, rapid prototyping computer-controlled techniques have been applied to the fabrication of scaffolds with ordered geometry. This chapter reviews the principal polymeric materials that are used for the fabrication of scaffolds and the scaffold fabrication processes, with examples of properties and selected applications.

Key words: Scaffold, Materials, Degradable polymers, Degradation, Fabrication, Tissue engineering

1. Tissue Engineering

Following a broadly accepted definition, tissue engineering is "an interdisciplinary field that applies the principles of engineering and life sciences to the development of biological substitutes that restore, maintain, or improve tissue function or a whole organ" (1). Other definitions have been coined as well: MacArthur and Oreffo defined tissue engineering as "understanding the principles of tissue growth, and applying this to produce functional replacement tissue for clinical use" (2), while Nerem and Sambanis stated "tissue engineering is an emerging multidisciplinary and

John W. Haycock (ed.), *3D Cell Culture: Methods and Protocols*, Methods in Molecular Biology, vol. 695,
DOI 10.1007/978-1-60761-984-0_2, © Springer Science+Business Media, LLC 2011

interdisciplinary field involving the development of bioartificial implants and/or the fostering of tissue remodeling with the purpose of repairing or enhancing tissue or organ function" (3). In practice, the term identifies a procedure for which materials, cells, growth factors, and/or other bioactive molecules, combined together, make implantable parts whose role is to promote repair and regeneration of the tissue of the implant site while degrading. Materials are in most cases of polymeric origin, sometimes ceramics. They are processed in order to produce 3D structures, i.e., scaffolds, possessing proper shape, size, architecture, and physical properties, tailored to the specific functions. The term regenerative medicine is often used instead of tissue engineering, although it should relate to cell therapies and the use of stem cells to produce tissues (4).

2. Scaffolds and Materials

2.1. Scaffold Morphology

Tissues possess different structures and properties that a tissue engineering scaffold should be tailored to. Moreover, scaffold should host cell adhesion, proliferation, and extracellular matrix (ECM) production: in conclusion, the scaffold should surrogate the missing ECM. Tissue engineering products can be designed to conduct, induct, or block tissues responses and architectures. Depending on the final purpose, barriers (membranes or tubes), gels, or 3D-matrices can be developed (5).

Hydrogels can encapsulate and represent a proper environment for isolated cells. Collagen gels, for instance, can be used for the preservation and immunoprotection of xenografted and homografted cells, typically used for transplantation. Semipermeable gels can be a support for cells in systems where communication between cells should be minimized. Some gel scaffolds can be directly injected in to an implantation site, for instance for bone and cartilage tissue engineering. In this case, thermoresponsiveness typical of some polymer gels formulations can be exploited to induce in situ hardening of the materials from their sol state after injection (*6*).

Scaffolds represent the space available for the tissue to develop and the physical support for cell growth. Scaffold mechanical properties should allow shape maintenance during tissue regeneration and enable stress transfer and load bearing. Moreover, during the first stage of tissue reconstruction, wound contraction forces act against the process, and enough mechanical strength and stiffness of the scaffold is required. Scaffold degradation profile has to be chosen properly to guarantee the required support during tissue formation (7).

Scaffold architecture should permit cell intrusion, nutrient and waste product permeation, and new capillary network formation.

Scaffold porosity is a fundamental characteristic for providing available space for cells to migrate and for vascularization of the tissue. Furthermore, the larger the surface available, the more cell interactions will arise. In general, the biological activity of a scaffold is determined by ligand density. Scaffold composition and porous fraction, that is the total surface of the structure exposed to cells, determine the ligand density. Highly specific surface areas allow for cell attachment and anchorage, and a high pore volume fraction enables cell growth, migration, and effective transportation of fluids and nutrients. In particular, microporosity is important for capillary ingrowth and interactions between cells and matrix, while macroporosity is relevant to nutrient supply and waste removal of cell metabolism.

Compatible to the structure of the tissue, scaffolds should be designed to have a high porosity, high surface area, a fully interconnected geometry, structural strength, and a specific three-dimensional shape. Besides, scaffold materials should be biocompatible and degradable or resorbable, so as to allow replacement of newly formed tissue in the long term. Every tissue requires a defined structure matrix design with specific material properties.

In addition to the size of the pores, the morphology can significantly influence the performance of an implanted matrix and the rate of tissue ingrowth. The optimum porosity is strictly connected to the tissue type, and diverse tissue architectures can be associated with a different microenvironment. Cell dimensions together with cell activity, phenotypic expression, and ECM production must be taken into account when designing a scaffold for tissue regeneration.

In bone tissue regeneration, for instance, the minimum pore size required is considered to be about 100 μm due to cell size, migration conditions, and transport. However, pore sizes bigger than 300 μm are recommended to improve new bone formation and to develop a net of capillaries (8).

The investigation of two different cell types, fibroblast and endothelial cells, in respect of a range of defined pore sizes, from 5 to 90 μm, can demonstrate how fibroblasts, using a bridging mechanism, spread over the closest cells able to fill even large pores. Endothelial cells cannot use this bridging system, preferring a pore size closer to their own dimensions (9). Vascular smooth muscle cells cultured on PLLA scaffolds privileged pore sizes ranging from 63 to 150 μm (9, 10). MC3T3-E1 mouse clonal osteogenic cells and four different scaffolds, with variable specific surface areas, confirmed a linear relationship between cell attachment and specific surface area, indicating that over the range of pore sizes analyzed (95.9–150.5 μm), short-term cell viability was dependent on the specific surface area available for binding (11).

The rate of degradation is also strictly connected to the degree of porosity. For instance, in polyester scaffold materials, a high porosity can reduce the risk of accumulation of acidic degradation products, thus reducing any possible further degradation reaction. Moreover, scaffold heterogeneity has shown to induce cells to adhere in a nonuniform way, and a nonhomogeneous distribution of ECM proteins is not easily obtained (10). Tissues generated from nonuniform pore architectures also showed inferior biomechanical properties if compared with tissues derived from scaffold having a uniform pore structure. Generally, cells tend to follow the scaffold geometry: if pores are equiaxial, they distribute, forming a spherical structure, while in the case of elongated pores they also align along the pore main axis (12).

Thus, scaffolds should perform in the body as devices able to support and possibly induce a complex pattern of events whose final goals are tissue repair and tissue function recovery. Additionally, scaffolds should be produced by using reproducible, controlled, and cost-effective processes; the possibility of including biological components, such as cells and growth factors, has also to be considered. Policy restrictions and intrinsic technical difficulties associated with the fabrication procedure limit the usage of preseeded scaffolds in well-defined applications and will be likely restricted to a few authorized institutions. For these considerations, the implantation of tissue engineering scaffolds directly in the body without preseeded cells has to be alternatively considered and, in many practical cases, preferred. Most scaffolds are produced from natural or synthetic polymer. Ceramic materials are also used, especially in bone tissue engineering applications and often in combination with polymers, thus forming composite materials with improved mechanical and biological properties.

2.2. Natural Polymers

Naturally derived polymer materials (13, 14), such as collagen, fibrin, fibroin, glycosaminoglycans (GAGs), chitosan, alginates, and starch, can be directly extracted from plants, animals, or human tissues or derived thereof; they generally exhibit good biocompatibility, low toxicity, and chronic inflammatory response. Batch-to-batch variability and difficult processing does occur for some polymers, which can be considered a drawback.

2.2.1. Collagen

An example of a commonly used natural material in tissue engineering is collagen. Collagen is a fibrous protein with a long, stiff, triple-stranded helical structure, whose main roles are to provide mechanical support to the connective tissues and to be a template for the cell distribution and capillary formation. There are three main collagen types: type I (found in skin and bone), type II (cartilage), type III (blood-vessel walls). A common method for producing porous collagen scaffolds is by "freeze-drying," starting with a collagen suspension in the presence of salt crystals. Collagen scaffolds have been investigated for blood vessels

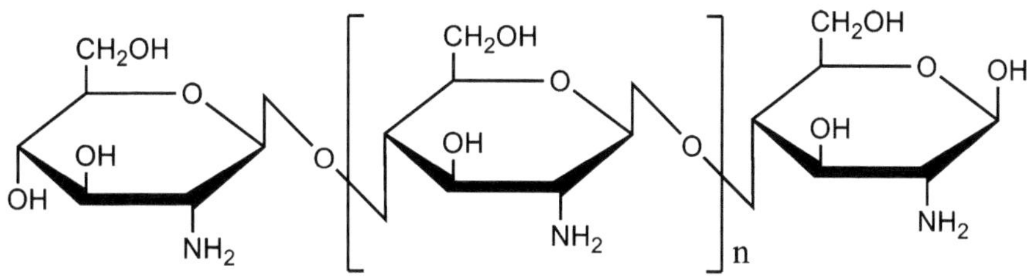

Fig. 1. Chitosan chemical formula.

(extruded collagen tubes), tendons and ligaments, dermal tissue for burn treatment, and peripheral nerve regeneration (porous collagen–GAG copolymer) (11, 13, 15–18).

2.2.2. Chitosan

Chitosan (Fig. 1) is the N-deacetylated derivative of chitin, a natural polysaccharide commonly located in the exoskeleton of crustaceans and insects (19, 20). It is a linear polysaccharide composed of β (1–4) linked d-glucosamine with randomly dispersed *N*-acetyl-d-glycosamine groups. Chitosan can undergo degradation by enzymes such as chitosanase and lysozyme, and its degradation rate depends on the amount of residual acetyl content, indicated by the deacetylation degree (21). The molecular weight and degree of deacetylation affects the physical and mechanical properties of the polymer (22, 23). Chitosan hydrogels can be produced either by ionic bonding or alternatively, to increase the strength, by covalent cross-linking, using cross-linkers such as glutaraldehyde.

Chitosan hydrogels obtained by photocrosslinking of a chitosan aqueous solution have also been used as local drug delivery carrier for agents such as FGF-2 and paclitaxel to control angiogenesis. Chitosan can be shaped into membranes or 3D structures. Biological evaluation by culturing hepatocyte cells on chitosan scaffolds have shown that cells maintain their morphology and metabolic activity, as measured by albumin secretion and urea synthesis. Chitosan materials have also been used to produce scaffolds for bone tissue regeneration and to support chondrocyte attachment and growth. Degradable porous scaffolds from naturally derived chitosan and alginate polymers have been developed for bone tissue engineering applications, and both mechanical and biological properties have been improved.

2.2.3. Glycosaminoglycans

Glycosaminoglycan (GAGs) are glycoproteins widely present inside the ECM of the body; many studies have proved the success of combining GAG and collagen to form nanofibrous scaffolds (17, 24). GAGs are long unbranched chains consisting of disaccharide units containing carboxylic and/or sulfate ester groups. These functional groups bridge and link collagens to form an interpenetrating network of ECM. Hyaluronan (HA, Fig. 2) is

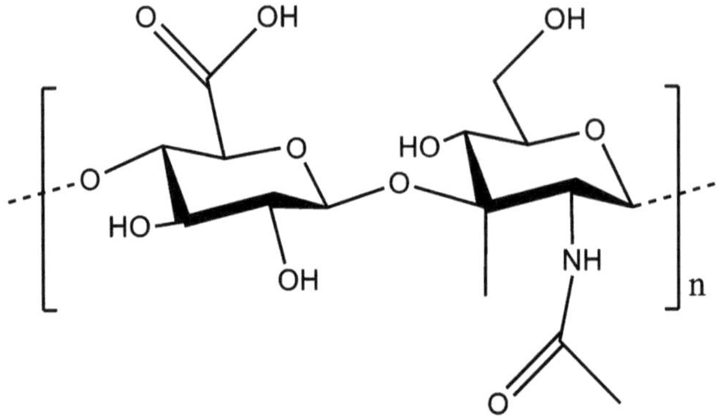

Fig. 2. Hyaluronan chemical formula.

an anionic, nonsulfated glycosaminoglycan naturally found in mammalian tissues, in particular being a fundamental compound of mammalian connective tissue. It has repeating disaccharide units of *N*-acetylglucosamine and glucoronic acid, with unbranched units ranging from 500 to several thousands. HA can bind high amounts of water, forming hydrogen bonds with the solvent. It has been used during eye surgery since 1976, and later on, hyaluronic acid injections were used to treat knee osteoarthritis (25). Furthermore, hyaluronan gels and films can be utilized to prevent postsurgical adhesion (26). Hyaluronic acid derivatives (i.e., cross-linked or esters) are used as scaffold materials for chondrocyte growth (27, 28), bone (29), and skin (30, 31) tissue regeneration. Hyaluronic acid is derived from natural sources, such as rooster combs, or can be produced by microbial fermentation and is quite easy to isolate and modify.

2.2.4. Silk Fibroin

Silks are fibrous proteins that are produced as a fiber by silkworms, other insects, and spiders. Wound sutures of silk fibers have been widely used for centuries. Due to early inflammatory process that have been attributed to sericins, i.e., the outer filament protein coating the inner fibroin brins, sericins are usually removed by washing the silk filament in water solution. The fibroin is dissolved (usually in water added with LiBr) and reprocessed. Starting from regenerated silk solutions, a variety of biomaterials such as gels, sponges, mats/nets, and films have been produced and proposed for medical applications. In particular, silks from *Bombyx mori* silkworms have been explored for use as biomaterials for many different applications (32, 33). In general, silk fibroin materials exhibit a good biocompatibility and have been demonstrated to be able to support the growth of human cells.

2.2.5. Agarose

Agarose is a polysaccharide polymer extracted from algae and is widely used in various fields of biomedical research. Its molecular structure is composed of an alternating copolymer linkage of

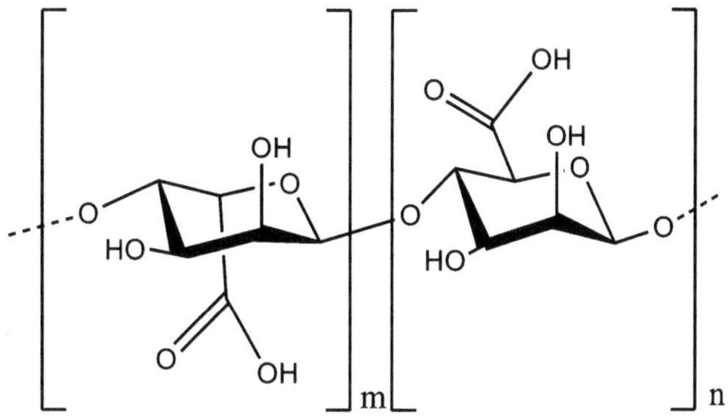

Fig. 3. Alginate chemical formula.

1,4-linked, 3-6 anhydro-α-galactose and 1,3-liked-β-d-galactose. Due to the high amount of hydroxyl groups, it is very soluble in water. A double-helix structure is formed by the interaction of two agarose chains linked together by hydrogen bonds. Agarose materials undergo enzymatic degradation by the action of agarases, and the properties of agarose gels, especially strength and permeability, depend on the concentration of agarose. Agarose is used in tissue culture systems because it permits cells to grow inside a three-dimensional suspension. Agarose gels are used in tissue engineering – especially for cartilage regeneration (34).

2.2.6. Alginate

Alginate is a naturally derived polysaccharide found in cell walls of brown algae. As for agarose, it is highly soluble in water. As shown in Fig. 3, it is a polyanion composed of two repeating monomer units: β-d-mannuronate (n) and α-l-guluronate (m). Physical and mechanical properties of alginate are highly related to chain length and properties of the guluronate blocks present inside the polymer. Alginate beads are used to encapsulate cells (35). Hepatocytes seeded on alginate porous scaffolds kept their functionality and secreted albumin (36).

2.2.7. Starch

Starches, polysaccharide carbohydrates consisting of glucose units linked together by glycosidic bonds, can be extracted from natural sources, and in particular, starches derived from corn, rice, and wheat have been extensively proposed for a wide range of biomedical applications. Starch consists of two types of molecules: the linear and helical amylose and the branched amylopectin (Fig. 4). Depending on the plant, starch generally contains 20–25% amylose and 75–80% amylopectin. Starch-based polymers have been considered a good choice as biomaterials due to their biocompatibility and degradability (37). Furthermore, they are relatively cheap and available for a large-scale supply. Starch degradation products consist of lower molecular weight chains of fructose and maltose. It has mainly been used in combination

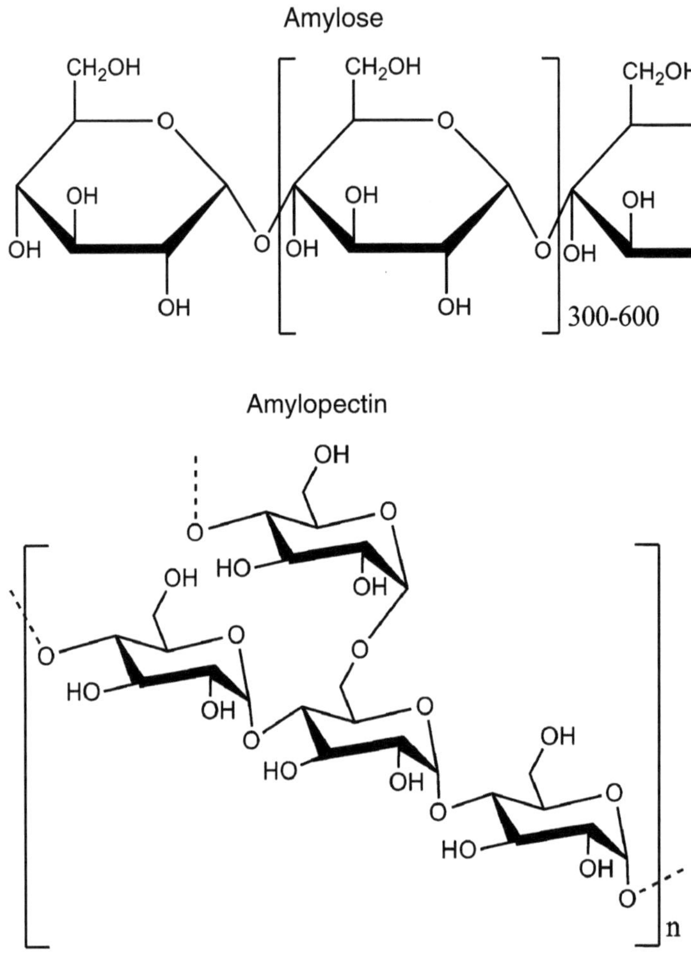

Fig. 4. The linear and helical amylose and the branched amylopectin.

with different additives, reinforcement materials, or blended with other polymers to produce composite materials with increased properties for tissue engineering scaffolds, bone cements, or hydrogels for drug delivery (38–42).

2.3. Synthetic Polymers

Synthetic degradable polymers are used as scaffold materials in tissue engineering, thanks to their high versatility, properties, reproducibility, and good workability (43). In most cases, synthetic polymers can be processed more easily than natural polymers, with more predictable results (5). In contrast, synthetic polymers are generally less biocompatible than the natural ones and are not bioactive. Degradation rate of scaffolds can be adapted to the specific application by selecting proper polymer, copolymers, or blends. Most of these polymers undergo simple hydrolytic degradation reactions. The most used degradable polymers

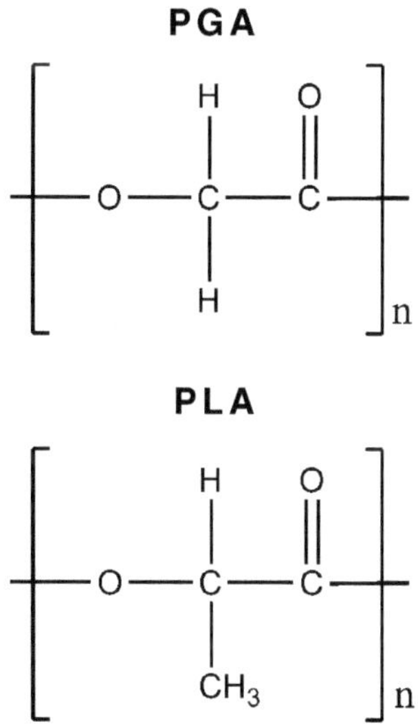

Fig. 5. Poly(glycolic acid) and poly (lactic acid) chemical formulae.

as scaffold materials are poly(glycolic acid) (PGA), poly(lactic acid) (PLA), or their copolymers or blends, as well as the aliphatic polyester polycaprolactone (PCL).

2.3.1. Poly(glycolic acid), Poly(lactic acid), and Copolymers

Poly (α-hydroxyacids), such as poly(glycolic acid) (PGA, Fig. 5a) and poly (lactic acid) (PLA, Fig. 5b) and their copolymers (PLGA), are bioabsorbable synthetic polymers widely known, studied, and successfully employed as tissue engineering scaffolds to transplant cells and regenerate different tissues (1, 5, 44–47). These linear aliphatic polyesters degrade by hydrolysis with degradation rates depending on structure, initial molecular weight, exposed surface area and size, degree of crystallinity, applied stresses, amount of residual monomer and, in the case of copolymers, the ratio of the hydroxyacid monomers. Being thermoplastic polymers, they can easily be shaped by using standard processing techniques such as molding, extrusion, solvent casting, and spin casting. Meshes having ordered or randomly distributed fibers and sponges with a specific selected porosity have been produced to meet the surface area and cellular requirements of different tissue-engineering constructs.

PGA has the simplest chemical structure; it has a high degree of crystallinity, a high melting point, and low solubility in organic solvents. It has been widely employed as a surgical suture material, and due to its hydrophilicity, it degrades fast in the presence of water (in about 4 weeks). The substitution of a hydrogen atom with a methyl group makes PLA more hydrophobic and less crystallizable. For its higher hydrophobicity, PLA degrades slower than PGA. Moreover, PLA dissolves easier in organic solvents than PGA. Due to the chiral nature of lactic acid (or lactide), two stereoisomeric forms are possible, and distinct polymers can be obtained. Poly-L-lactide (PLLA) is the product resulting from the ring-opening polymerization of L-lactide; it presents a crystallinity of about 37% with glass transition temperature between 50 and 80°C and melting between 180 and 190°C. Poly-D-lactide (PDLA) is analogous to PLLA, but it is derived from polymerization of D-lactide; it exists just at an experimental level without practical applications. Poly-D,L-lactide (PDLLA) is the racemic polymer obtained from a mixture of D- and L-lactic acids; it has an amorphous structure with glass transition temperature at about 60°C.

The degree of crystallinity strongly affects the polymer water absorption and degradation kinetics: amorphous regions absorb more water and consequently degradation occurs at a higher rate. The amorphous racemic mixture of PDLLA degrades at a higher rate than PLLA. For bone tissue regeneration, PLA, PGA, and their copolymers are often combined with ceramic materials such as bioactive glasses or hydroxyapatite. These fillers can induce bone regeneration and at the same time improve the mechanical properties of the polymer. Biocompatibility plays an important role in the long- and short-term success of all implants; for biodegradable devices, both the implant and its degradation products must be biocompatible and nontoxic. Some complications have occasionally been reported following implantation of PLA–PGA biomaterials, usually attributed to the release of acidic products during material degradation. Nevertheless, the majority of clinical studies citing complications in humans due to PLA and PGA implants report only nonspecific foreign body mild reactions (48).

2.3.2. Poly(ε-caprolactone)

Poly(ε-caprolactone) (PCL, Fig. 6) is a degradable semicrystalline aliphatic polyester that is obtained by a ring-opening polymerization of ε-caprolactone, using a catalyst such as stannous octanoate. PCL crystals melt at about 60°C, while amorphous regions have glass transition temperature at –60°C. These conditions make PCL rubbery and flexible at room and body temperature. PCL degrades by water hydrolysis in a physiological environment (49, 50). In general, the degradation rate of PCL is slow compared to PLA, due to the more hydrophobic nature of the polymer. This limits its use if fast tissue regeneration is required but on the other hand makes PCL more suitable for longer-term implantable devices. Degradation of PCL can, however, be catalyzed in the presence of enzymes, such as lipases.

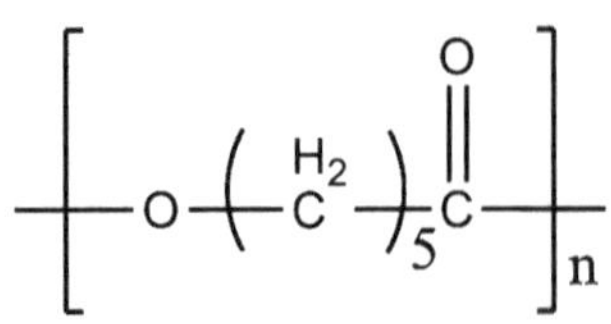

Fig. 6. Poly(ε-caprolactone) chemical formula.

2.3.3. Polyurethanes

Polyurethanes cover a wide family of polymers containing urethane bonds in the polymer chain. Polyurethanes are produced by the reaction between a diisocyanate and a polyol; a segmented block copolymer is then obtained where low-glass transition temperature, flexible segments are combined with hard segments (Fig. 7). Polyurethanes' excellent physical properties and good biocompatibility make them good candidates for many different biomedical applications. Polyurethanes were first introduced to the medical device market in the early 1950s in foam breast prostheses and cardiovascular devices. Polyurethanes are used as biomaterials in different applications such as in pacemakers, for lead insulator, catheters, total artificial heart, and heart valves. For all the above applications, degradation-stable polyurethanes are used; however, even in some cases, long-term degradation has been observed. Degradable polyurethanes have been developed for tissue engineering applications such as for myocardial repair and vascular tissues (51, 52). They are produced from diisocyanates such as lysine-diisocyanate or hexamethylene diisiocyanate. The released degradation products result in nontoxic products to the human body if compared to the conventional aromatic diisocyanates.

2.3.4. Poly(ortho ester)

Poly(ortho esters) (POE, Fig. 8) are synthetic polymers that in an acidic environment degrade by surface erosion (53). This peculiarity is used in medicinal systems to make controlled-release drug devices. Degradation occurs via hydrolysis and increases with time due to the production of acidic products that act as catalysts for the degradation reaction. The incorporation of lactide dimers in the backbone results in hydrolytic degradation of the ester bonds and the production of lactic acid that catalyzes

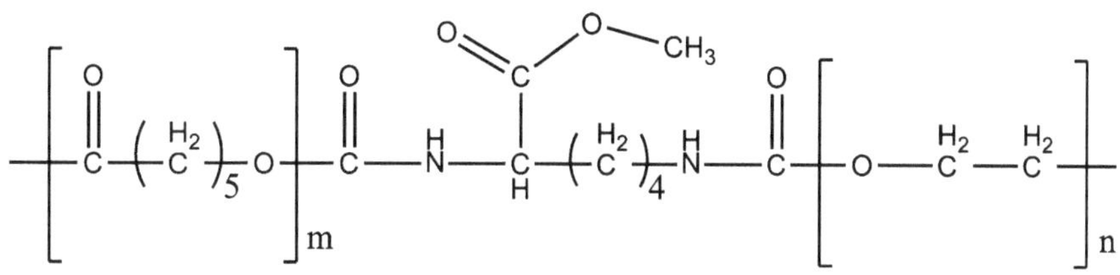

Fig. 7. Polyurethanes chemical formula.

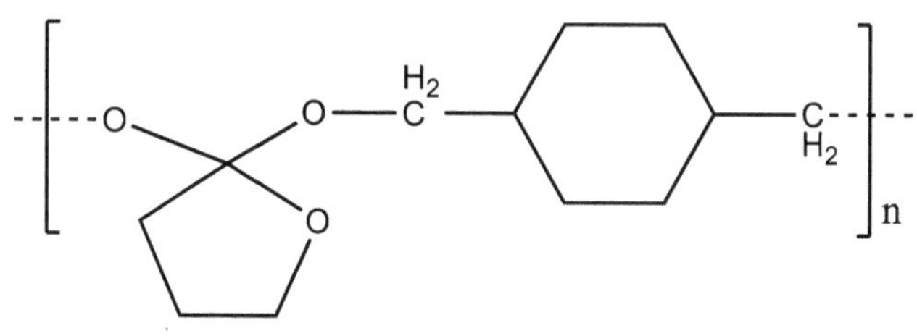

Fig. 8. Poly(ortho ester) chemical formula.

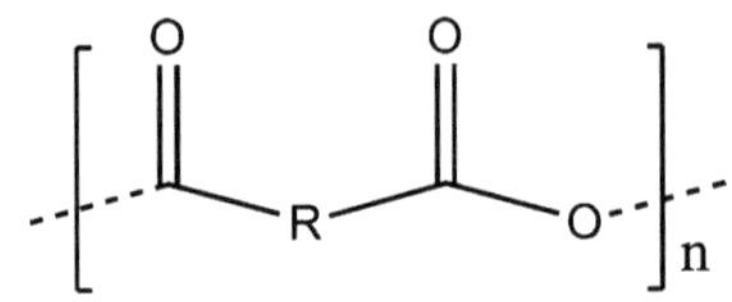

Fig. 9. Poly(anhydrides) chemical formula.

the degradation of the ortho ester bonds. Bone formation on scaffolds of poly(ortho esters) and 50:50 poly(d,l-lactide-co-glycolide) implanted into noncritical-size calvarial defects in rabbits has been compared by Andriano et al. (54), who found higher amounts of newly formed bone in the surface eroding POE with respect to the hydrolysis degradable P-G,L-LA.

2.3.5. Poly(anhydrides)

Like poly(ortho esters), poly(anhydrides) (Fig. 9) have a hydrophobic backbone that limits water diffusion and confines degradation to the surface, where degradation occurs via the hydrolysis of anhydride bonds. The mechanism of degradation makes this polymer suitable as a carrier for the zero-order release of drugs (55). For tissue engineering applications, scaffolds polymerized in situ in a tibia bone defect in rats starting from photo cross-linkable polyanhydride monomers (56) showed good adhesion of the polymer to the cortical bone and medullary cavity and minimal adverse tissue reaction to the photopolymerization reaction.

3. Degradation of Biomedical Polymers

First introduced in the early 1950s for making suture threads, degradable polymers are nowadays extensively used in medicine for numerous applications. Besides resorbable surgical sutures, these polymers are employed as carriers for the controlled release of drugs, as temporary prosthesis (e.g., bone pins, screws, and plates) and, in the last 10 years, as materials for the fabrication of scaffolds for tissue engineering procedures.

The term biodegradation refers to the "gradual breakdown of a material mediated by a specific biological activity" (57). More precisely, a material should be defined as biodegradable if that breakdown is due to cells and/or tissue activity. Following this definition, the previously listed materials cannot be defined as *biodegradable*, but solely *degradable*, their degradation being due to the reaction with water present in the body. Quite often the term biodegradation is used to define materials that degrade after implantation at a body site, less independent of the mechanism of degradation. However, what is ultimately important is the fact that the products of biodegradation (or degradation) are metabolized and eliminated from the body, i.e., resorbed.

Scaffold materials should fulfill several requirements. As already stated, a scaffold is not just a passive support for cell growth, but a device whose properties affects the regeneration cascade. Mechanical properties, surface properties, and morphology are in turn relevant to the specific application. Degradation kinetics and the rate at which scaffold properties change with degradation should always be predictable. In particular, the degradation behavior of biomaterials can follow several mechanisms and is controlled by different factors. Understanding the degradation kinetics and mechanism of biomaterials is necessary to optimize their possible usage. Particularly in drug delivery applications, a polymer is required to degrade following a well-defined mass loss profile to release the encapsulated or linked drugs and agents at specific rates.

During polymer processing and scaffold fabrication, some incipient degradation could occur, affecting also the degradation behavior. If processing techniques involve high temperature conditions or high shear stresses, this may cause degradation of the starting polymer. Alternatively, chain orientation/rearrangement induced by the process could make the material more resistant to degradation. Sterilization methods may also affect material degradation by causing cross-linking or polymer chain scission (depending on the material and sterilization method).

The physiological environment of the human body can be aggressive to many polymers, due to the presence of various biologically active species, such as enzymes, free radicals, phagocytic cells, etc. Degradation of a material implanted in a body generally occurs by hydrolysis of enzyme-catalyzed reactions (57–59). In addition, during the inflammatory reaction, in response to foreign materials, inflammatory cells generate highly reactive molecules such as hydrogen peroxide (H_2O_2), nitric oxide (NO) and hypochlorous acid (HOCl) as well as the ion superoxide (O_2^-). These products can cause polymer chain breakage and lead to degradation. Most of the synthetic polymers undergo hydrolytic degradation, while enzymes accelerate or, in many cases, are necessary for the degradation of natural polymers.

Hydrolytic degradation is due to the attack of water; oligomers and monomers are then formed and the average molecular weight is reduced. Water attack is directed to water-weak bonds as a result of direct access of water to the polymer surface or by absorption into the polymer matrix. The hydrophilic or hydrophobic nature of polymeric materials determined by their chemical structure influences their degradation rate: covalent bonds in the backbone and no hydrolyzable groups require longer times to degrade. Enzymes known as hydrolases, such as proteases, esterases, glycosidases, and phosphatases, may catalyze hydrolysis reactions.

4. Processing Techniques

A scaffold should mimic the natural ECM architecture. A big variety of matrices have been developed due to the large range of tissues to repair and the need to create cellular supports with different physical appearance, porosity, permeability, and mechanical characteristics. Different scaffold fabrication techniques have been employed to produce porous polymeric matrices as substrates for cell support, adhesion, growth, and subsequently, proliferation and differentiation on or within their structures.

A wide range of techniques commonly used in tissue engineering can generate scaffolds having a random structure with unpredictable pore sizes and reduced pore interconnections (Fig. 10). Any variation in porosity within the 3D structure cannot be controlled, and mechanical strength, structural stability, and reproducibility are generally low. Among these techniques, solvent casting, freeze drying, phase inversion, fiber bonding, melt-based technologies, and high-pressure-based methods are the most commonly used. More recently, electrospinning has also been widely investigated to produce meshes of submicrometric fibers for different tissue engineering applications (Fig. 11). In addition to the above techniques, computer-controlled microfabrication processes have been developed to generate scaffolds with complex shapes and predesigned architecture (60). A scaffold produced by these techniques is shown in Fig. 12.

Unlike the traditional techniques, which involve a constant removal of materials, the principle of solid free form (SFF)

Fig. 10. Silk fibroin/poly(ethylene glycol) sponge produced by salt leaching (by courtesy of Foss, C., University of Trento).

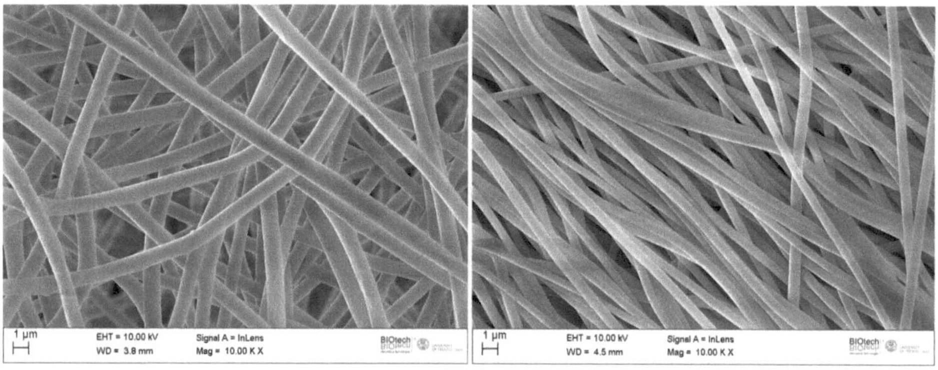

Fig. 11. Electrospun random and aligned polyamide 6 nets (by courtesy of Volpato, F. Z, University of Trento).

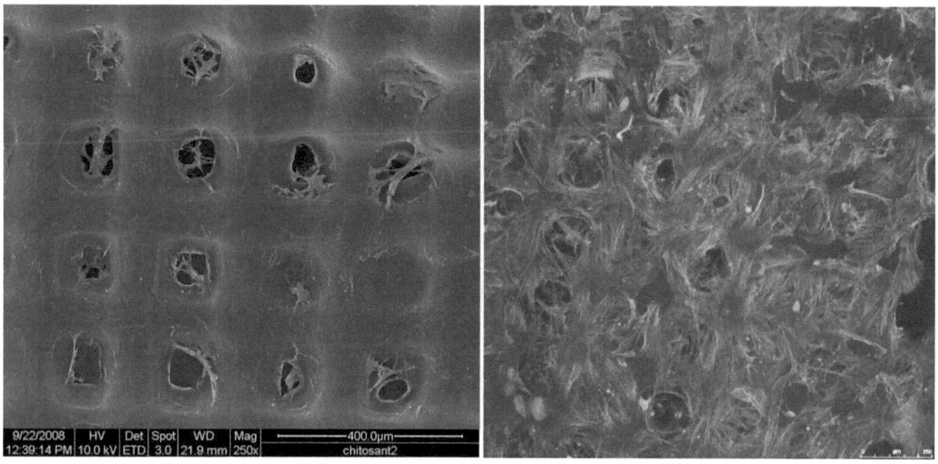

Fig. 12. A microfabricated chitosan scaffold after 16 days of human osteoblast cell culture: (*left*) SEM image, (*right*) confocal laser microscopy image (author's data).

fabrication is that each scaffold is formed by selectively adding the material layer by layer, under the control of a computer. High scaffold resolution and reproducibility is achieved by SFF. Fused deposition modeling (FDM), three-dimensional printing (3-DP), selective laser sintering (SLS), and stereolithography (SLA) are some examples of rapid prototyping techniques. Novel robotic assembly and automated 3D cell encapsulation techniques have been also implemented to introduce cells during the scaffold production (61, 62).

4.1. Solvent Casting/ Particulate Leaching

Solvent casting is a versatile method for producing thin polymeric films from solution. The method involves depositing a polymer solution onto a casting surface and subsequently evaporating the solvent leaving polymeric films of various thicknesses. Solvent casting is one of the oldest technologies in plastic material processing and is still an attractive method in the polymer field due to the high quality of obtainable films. In addition, it represents a relatively easy and controllable technique. One of the most important advantages of the process is the ability to control the thickness uniformity and distribution of the film. The final properties of the product depend on various processing parameters. One element is the polymer–solvent combination: in general, the polymer should be dissolved in a volatile solvent at a specific concentration, and the final solution must be stable with the proper viscosity. Depending on the solvent used, the chemistry of the solution can be modulated as well as the final properties of the material. Properties are also affected by the casting substrate, casting temperature, and drying conditions (i.e., indirect heating, heating by radiation, and air-stream drying) (63).

Nowadays in tissue engineering, solvent casting is often used in combination with other common methods in order to produce a 3D porous structure. Solvent casting in combination with particulate leaching has been used successfully to fabricate 3D scaffolds. Porogen agents commonly used in the field are: sodium chloride (64, 65), ammonium bicarbonate (66), and glucose (67) with different crystal sizes.

4.2. Freeze-Drying

Freeze-drying method is one of the most commonly used methods to produce porous structures (68). The process is based on solvent sublimation, usually water, under specific temperature and pressure conditions. By this method, it is possible to control pore size, porosity, interconnectivity, and a porous distribution of the sponge. In addition, by adjusting the polymer concentration the scaffold, mechanical properties can be controlled. Materials with aligned porosity in the micrometric range can be obtained by controlling quenching temperature and cooling rate (69). In particular, the process and the final structure are governed by the competition between solute interfacial concentration gradient and surface energy. More complex structures can be tailored by combining freeze-drying with particulate leaching methods (69, 70).

4.3. Phase Inversion

Phase inversion is based on the phenomenon through which, under specific conditions (solution viscosity, density, interfacial tension, and agitation speed), the phases of a liquid–liquid dispersion interchange by diffusion, and the dispersed phase becomes the continuous phase and vice versa. This behavior can be used to separate liquids, where there are two immiscible phases. In tissue engineering processing, the phase inversion method makes use of polymer solutions cast on a proper support and subsequently washed in a nonmiscible solvent to coagulate the polymer. Once in contact, the solvent and the nonsolvent induce the solution to phase-separate. In general, phase inversion is an effective method to obtain porous membranes and scaffolds by the combination of mass transfer and liquid phase separation (71–73).

A supercritical fluid, such as carbon dioxide, can be used as a nonsolvent leading to the advantage of omitting a drying step, which can produce a final porous structure free of any residual solvent. Due to the fact that carbon dioxide is nontoxic, noncorrosive, nonflammable, and cheap, it represents one of the best candidates to be used as supercritical fluid (74–77). Factors that can affect the scaffold porosity, pore size, distribution, and interconnectivity mainly depend on the solubility and diffusivity of the supercritical fluid into the polymer as well as the polymer crystallinity (78).

This technique can also be combined with a particulate porogen leaching method to increase the pore interconnectivity (79, 80). Applications as a porous matrix have been proposed for tissue engineering applications (76), in particular for bone tissue regeneration (77).

4.4. Electrospinning

Electrospinning, while not a new technique, recently has become an attractive method for tissue engineering scaffold fabrication. Electrospun products are nonwoven fibrous structures with fiber diameters in the range of 10 nm to microns (81). By this method, it is possible to obtain a high surface area-to-volume ratio. The pore size can also be modulated; however, normally the size does not exceed 10 μm, leading to a nanofibrous scaffold that hinders cell infiltration. For this reason, electrospun scaffolds can be used as a barrier to prevent cells migrating between two different tissues or as a nanostructured surface to mimic the native ECM.

The basic apparatus usually consists of three major components: a high voltage power supply, a spinneret (i.e., needle), and a collecting target (metallic plate or rotating mandrel) (82).

The resulting fibrous structure, fiber thickness, and morphology can be tailored by different parameters such as the solution properties, electric field strength, collecting target to needle distance, processing temperature, and humidity (83). When electrospinning on to a static plate, nanofibers are usually collected at random and without any evident orientation. By substituting the static plate with a rotating mandrel collector, the fiber orientation can be improved, allowing for specific fiber directions to be achieved, which can be beneficial for directional tissues like myocardium scaffolds (84). In addition, different polymeric solutions can be combined (i.e., synthetic and natural polymers) by coelectrospinning using two needles simultaneously. This technique allows for the production of multicomponent nanostructured scaffolds. In addition, core–shell nanofibres have been fabricated by using one needle and two immiscible polymer solutions to prepare fibers with a PCL shell and gelatin core (85). Lee et al. (86) reported a new procedure based on the combination of electrospinning and the salt leaching method to better modulate the final scaffold porosity in the range 50–300 μm.

4.5. Stereolithography

Maybe the precursor among all the RP techniques is SLA, which makes use of a UV laser beam to selectively polymerize a photopolymerizable liquid polymer material. The polymer solidifies from the bath once in contact with the beam, at the surface of the bath. The solidification process takes place repeatedly layer by layer. When the model is concluded, the extra resin is washed away, and the product is cured in a UV oven and polished to reduce roughness.

Nowadays, SLA is used mainly to produce anatomical models for educational reasons. The process, due to curing and shrinkage, has a low resolution and, especially in small and intricate objects, deformation phenomena can be present (87, 88). Furthermore, only a small amount of polymers can photopolymerize and are biocompatible and suitable to become tissue engineering scaffolds. Polyethylene glycol (PEG) acrylate, PEG methacrylate, polyvinyl alcohol (PVA), hyaluronic acid, and dextran methacrylate are the more common. The development of micrometric stereolithography has great potential for the fabrication of scaffolds with micron resolution.

4.6. Selective Laser Sintering

Through this technique, a CO_2 laser beam is addressed over a polymeric powder, and a sinterization process is activated. Basically, the beam can increase the local temperature of the polymer, and particles can fuse to each other. The laser scans over the powder along the cross-sectional profiles, taking by the slice data, and successive layers are formed over the previous one introducing extra powder as the preceding layer is completed. Simple and linear bulk components can be easily fabricated by SLS, while sheet-like structures undergo shrinkage. Ultra-high molecular weight polyethylene (UHMWPE) has been used to produce implants by the SLS technique, and it has been observed that degradation (that is breakage of molecular chains, oxidation, and cross-linking) occurs during sintering, especially if the starting polymer powder density is not high enough (89).

Calcium phosphate (CaP) powder has been used in combination with SLS to produce bone implants (90).

4.7. 3D Printing

During 3D printing (3DP), a first layer of polymeric powder is distributed over a deposition base while an inkjet print head dispenses the binder solution over it; following this, a new layer of powder is spread over on the previous one and the printing cycle is repeated. Layers attach when fresh binder is added. At the end of the process, the nonbonded extra powder is removed (91, 92).

The technique has been widely investigated for tissue engineering applications and drug-delivery systems, and the major advantage related to 3DP is the possibility to work inside an ambient environment. Removing the extra powder from a complicated shaped model is difficult, so the technique is suited for easier geometries without internal holes. The printer resolution depends on the nozzle dimension and on the print head movement resolution. The layer thickness is determined by powder size, and in general, surface roughness and aggregation properties of the powdered materials influence the final resolution. Natural biopolymers have been used in combination with water as a binder, eliminating the problem to use organic solvents (93). Further postprocessing step to keep the piece water-resistant is obviously necessary.

4.8. Shape Deposition Manufacturing

Shape deposition manufacturing (SDM) makes use of clinical imaging data to produce scaffolds by a computer numerical controlled cutting machine (94). Cells and growth factors can be added during the three-dimensional scaffold production process. The technique was initially designed for bone tissue regeneration and represents one of the most interesting techniques among all the assembly technology-based systems. Osteogenic scaffolds produced by SDM based on blends of polycaprolactone (PCL) and poly-d,l-lactide-co-glycolide (PGLA) together with HA grains have been reported (95).

4.9. Robotic Microassembly

The robotic microassembly technique has been studied and developed at the National University of Singapore (96). Scaffold structure is generated by building small block units that have a different design and have been previously fabricated via lithography (or another microfabrication technique). The blocks are then connected together using a specific precision robotic system having microgripping capabilities. Finally, a scaffold having the desired material, chemical, and physical properties is obtained.

4.10. Fused Deposition Modeling

FDM is a heat-based manufacturing technology that has been applied to the building of 3D scaffolds (97, 98). The system includes a head-heated liquefier fixed to a moving system; the head can extrude the material pumping the polymeric filament through a nozzle directly on a platform, following a previously predetermined trajectory. Such scaffolds are built up layer over layer in the vertical direction, where the layer thickness depends on the nozzle inner diameter. The technique is limited to thermoplastic polymers with proper viscous properties, and cells cannot be included during the process. Hutmacher and coworkers have used FDM to fabricate bioresorbable scaffolds of PCL also in combination with HA and TCP particles, thus forming composite materials scaffolding supports for bone tissue applications. Another similar technique called precision extruding deposition (PED) has also been developed: in this processing system, the polymer can be extruded directly without a previous filament preparation.

References

1. Langer, R. and Vacanti, J. P. (1993) Tissue engineering. *Science* **260**, 920–926.
2. Macarthur, B. D. and Oreffo, R. O. C. (2005) Bridging the gap. *Nature* **433**, 19.
3. Nerem, R. M. and Sambanis, A. (1995) Tissue engineering: from biology to biological substitutes. *Tissue Eng.* **1**, 3–13.
4. Ringe, J., Kaps, C., Burmester, G. and Sittinger, M. (2002) Stem cells for regenerative medicine: advances in the engineering of tissues and organs. *Naturwissenschaften* **89**, 338–351.
5. Mooney, D. J., Sano, K., Kaufmann, P. M., Majahod, K., Schloo, B., Vacanti, J. P. and Langer, R. (1997) Long-term engraftment of hepatocytes transplanted on biodegradable polymer sponges. *J. Biomed. Mater. Res.* **37**, 413–420.
6. Klouda, L. and Mikos, A. G. (2008) Thermoresponsive hydrogels in biomedical applications. *Eur. J. Pharm. Biopharm.* **68**, 34–45.
7. Hutmacher, D. W. (2000) Scaffolds in tissue engineering bone and cartilage. *Biomaterials* **21**, 2529–2543.
8. Karageorgiou, V. and Kaplan, D. (2005) Porosity of 3D biomaterial scaffolds and osteogenesis. *Biomaterials* **26**, 5474–5491.
9. Salem, A. K., Stevens, R., Pearson, R. G., Davies, M. C., Tendler, S. J. B., Roberts, C. J., et al. (2002) Interactions of 3T3 fibroblasts and endothelial cells with defined pore features. *J. Biomed. Mater. Res.* **61**, 212–217.
10. Zeltinger, J., Sherwood, J. K., Graham, D. A., Mueller, R. and Griffith, L. G. (2001) Effect of pore size and void fraction on cellular adhesion, proliferation, and matrix deposition. *Tissue Eng.* 7, 557–572.
11. O'brien, F. J., Harley, B. A., Yannas, I. V. and Gibson, L. J. (2005) The effect of pore size on cell adhesion in collagen-GAG scaffolds. *Biomaterials* **26**, 433–441.
12. Zhou, Y., Hutmacher, D. W., Varawan, S. L. and Lim, T. M. (2006) Effect of collagen-I modified composites on proliferation and differentiation of human alveolar osteoblasts. *Aust. J. Chem.* **59**, 571–578.
13. Lee, C. H., Singla, A. and Lee, Y. (2001) Biomedical applications of collagen. *Int. J. Pharm.* **221**, 1–22.
14. Bensaid, W., Triffitt, J. T. and Blanchat, C. (2003) A biodegradable fibrin scaffold for mesenchymal stem cell transplantation. *Biomaterials* **24**, 2497–2502.
15. Pitaru, S., Tal, H., Soldinger, M., Grosskopf, A. and Noff, M. (1988) Partial regeneration of periodontal tissues using collagen barriers. Initial observations in the canine. *J. Periodontol.* **59**, 380–386.
16. Negri, S., Fila, C., Farinato, S., Bellomi, A. and Pagliaro, P. P. (2007) Tissue engineering: chondrocyte culture on type 1 collagen support. Cytohistological and immunohistochemical study. *J. Tissue Eng. Regen. Med.* **1**, 158–159.
17. Zhong, S. P., Teo, W. E., Zhu, X., Beuerman, R., Ramakrishna, S. and Yung, L. Y. L. (2007) Development of a novel collagen-GAG nanofibrous scaffold via electrospinning. *Mater. Sci. Eng. C.* **27**, 262–266.
18. Flanagan, T. C., Wilkins, B., Black, A., Jockenhoevel, S., Smith, T. J. and Pandit, A. S. (2006) A collagen-glycosaminoglycan co-culture model for heart valve tissue engineering applications. *Biomaterials* **27**, 2233–2346.
19. Di Martino, A., Sittinger, M. and Risbud, M. V. (2005) Chitosan: a versatile biopolymer for orthopaedic tissue-engineering. *Biomaterials* **26**, 5983–5990.
20. Chun, H. J., Kim, G. W. and Kim, C. H. (2008) Fabrication of porous chitosan scaffold in order to improve biocompatibility. *J. Phys. Chem. Solids.* **69**, 1573–1576.
21. Duarte, M. L., Ferreira, M. C., Marvão, M. R. and Rocha, J. (2001) Determination of the degree of acetylation of chitin materials by 13C CP/MAS NMR spectroscopy. *Biol. Macromol.* **28**, 359–363.
22. Zhou, H. Y., Chen, X. G., Kong, M., Liu, C. S., Cha, D. S. and Kennedy, J. F. (2008) Effect of molecular weight and degree of chitosan deacetylation on the preparation and characteristics of chitosan thermosensitive hydrogel as a delivery system. *Carbohydr. Polym.* **73**, 265–273.
23. Hsu, S. H., Whu, S., Tsai, S. L., Wu, Y. H., Chen, H. W. and Hsieh, K. H. (2004) Chitosan as scaffold materials: effects of molecular weight and degree of deacetylation. *J. Polym. Res.* **11**, 141–147.
24. Birk, D. E., Lande, M. A. and Fernandez-Madrid, F. R. (1981) Collagen and glycosaminoglycan synthesis in aging human keratocyte cultures. *Exp. Eye Res.* **32**, 331–339.
25. Weiss, C., Balazs, E. A., St.Onge, R. and Denlinger, J. L. (1981) Clinical studies of the intraarticular injection of HealonR (sodium hyaluronate) in the treatment of osteoarthritis

of human knees. *Semin. Arthritis Rheum.* **11**, 143–144.

26. Holzman, S. and Connolly, R. J. (1994) Effect of hyaluronic acid solution on healing of bowel anastomoses. *J. Invest. Surg.* 7, 431–437.
27. Grigolo, B., Lisignoli, G., Piacentini, A., Fiorini, M., Roseti, L., Major, E. O., et al. (2001) Evidence for re-differentiation of human chondrocytes seeded on a hyaluronan derivative scaffold. *Arthritis Res.* **3**, P7.
28. Marcacci, M., Kon, E., Zaffagnini, S., Iacono, F., Filardo, G. and Delcogliano, M. (2006) Autologous chondrocytes in a hyaluronic acid scaffold. *Oper. Tech. Orthop.* **16**, 266–270.
29. Giordano, C., Sanginario, V., Ambrosio, L., Silvio, L. D. and Santin, M. (2006) Chemical-physical characterization and in vitro preliminary biological assessment of hyaluronic acid benzyl ester-hydroxyapatite composite. *J. Biomater. Appl.* **20**, 237–252.
30. Heden, P., Sellman, G., Von Wachenfeldt, M., Olenius, M. and Fagrell, D. (2009) Body shaping and volume restoration: the role of hyaluronic acid. *Aesthetic Plast. Surg.* **33**, 274–282.
31. Duranti, F., Salti, G., Bovani, B., Calandra, M. and Rosati, M. L. (1998) Injectable hyaluronic acid gel for soft tissue augmentation. A clinical and histological study. *Dermatol Surg.* **24**, 1317–1325.
32. Motta, A., Fambri, L. and Migliaresi, C. (2002) Regenerated silk fibroin films: thermal and dynamic mechanical analysis. *Macromol. Chem. Phys.* **203**, 1658–1665.
33. Motta, A., Migliaresi, C., Lloyd, A. W., Denyer, S. P. and Santin, M. (2002) Serum protein adsorption on silk fibroin fibres and membranes: surface opsonization and binding strength. *J. Bioact. Compat. Polym.* **17**, 23–35.
34. Rahfoth, B., Weisser, J., Sternkopf, F., Aigner, T., Von Der Mark, K. and Bräuer, R. (1998) Transplantation of allograft chondrocytes embedded in agarose gel into cartilage defects of rabbits. *Osteoarthr. Cartil.* **6**, 50–65.
35. Orive, G., Hernandez, R. M., Gascon, A. R., Calafiore, R., Chang, T. M. S., de Vos, P., Hortelano, G., Hunkeler, D., Lacik, I., Pedraz, J. L. (2004) History, challenges and perspectives of cell microencapsulation. *Trends Biotechnol.* **22**, 87–92.
36. Dvir-Ginzberg, M., Gamlieli-Bonshtein, I., Agbaria, R. and Cohen, S. (2003) Liver tissue engineering within alginate scaffolds: effects of cell-seeding density on hepatocyte viability, morphology, and function. *Tissue Eng.* **9**, 757–766.
37. Kaur, L., Singh, J. and Liu, Q. (2007) Starch – a potential biomaterial for biomedical applications. In *Nanomaterials and nanosystems for biomedical applications*, Reza, M. ed., Springer, Netherlands.
38. Pereira, C. S., Cunha, A. M., Reis, R. L., Vazquez, B. and San Roman, J. (1988) New starch-based thermoplastic hydrogels for use as bone cements or drug-delivery carriers. *J. Mater. Sci. Mater. Med.* **9**, 825–833.
39. Reis, R. L., Mendes, S. C., Cunha, A. M. and Bevis, M. J. (1997) Processing and in-vitro degradation of starch/EVOH thermoplastic blends. *Polym. Int.* **43**, 347–353.
40. Marques, A. P., Reis, R. L. and Hunt, J. A. (2003) Evaluation of the potential of starch-based biodegradable polymers in the activation of human inflammatory cells. *J. Mater. Sci. Mater. Med.* **14**, 167–173.
41. Gomes, M., Reis, R. L., Cunha, A. M., Blitterswijk, C. A. and De Bruijn, J. D. (2001) Cytocompatibility and response of osteoblastic-like cells to starch-based polymers: effect of several additives and processing conditions. *Biomaterials* **22**, 1911–1917.
42. Gomes, M. E., Ribeiro, A. S., Malafaya, P. B., Reis, R. L. and Cunha, A. M. (2001) A new approach based on injection moulding to produce biodegradable starch-based polymeric scaffolds: morphology, mechanical and degradation behaviour. *Biomaterials* **22**, 883–889.
43. Gunatillake, P. A. and Adhikari, R. (2003) Biodegradable synthetic polymer for tissue engineering. *Eur. Cell Mater.* **5**, 1–16.
44. Yang, F., Cui, W., Xiong, Z., Liu, L., Bei, J. and Wang, S. (2006) Poly(l,l-lactide-co-glycolide)/tricalcium phosphate composite scaffold and its various changes during degradation in vitro. *Polym. Degrad. Stab.* **91**, 3065–3073.
45. Schieker, M., Seitz, H., Drosse, I., Seitz, S. and Mutschler, W. (2006) Biomaterials as scaffold for bone tissue engineering. *Eur. J. Trauma.* **32**, 114–124.
46. Mooney, D. J., Baldwin, D. F., Suh, N. P., Vacanti, J. P. and Langer, R. (1996) Novel approach to fabricate porous sponges of poly(d,l-lactic-co-glycolic acid) without the use of organic solvents. *Biomaterials* **17**, 1417–1422.
47. Athanasiou, K. A., Niederauer, G. G. and Agrawal, C. M. (1996) Sterilization, toxicity, biocompatibility and clinical applications of polylactic acid/polyglycolic acid copolymers. *Biomaterials* **17**, 93–102.
48. Bostman, O., Partio, E., Hirvensalo, E. and Rokkanen, P. (1992) Foreign-body reactions

to polyglycolide screws. *Acta Orthop.* **63**, 173–176.

49. Jenkins, M. J. and Harrison, K. L. (2008) The effect of crystalline morphology on the degradation of polycaprolactone in a solution of phosphate buffer and lipase. *Polym. Adv. Technol.* **19**, 1901–1906.
50. Hongfan, S., Lin, M., Cunxian, S., Xiumin, C. and Pengyan, W. (2006) The in vivo degradation, absorption and excretion of PCL-based implant. *Biomaterials* **27**, 1735–1740.
51. Mcdevitt, T. C., Woodhouse, K. A., Hauschka, S. D., Murry, C. E. and Stayton, P. S. (2003) Spatially organized layers of cardiomyocytes on biodegradable polyurethane films for myocardial repair. *J. Biomed. Mater. Res. A.* **66**, 586–595.
52. Stankus, J. J., Guan, J. and Wagner, W. R. (2004) Fabrication of biodegradable elastomeric scaffolds with sub-micron morphologies. *J. Biomed. Mater. Res. A.* **70A**, 603–614.
53. Heller, J., Barr, J., Ng, S. Y., Shen, H. R., Schwach, A. K., Emmahal, S., et al. (2000) Poly(ortho esters): their development and some recent applications. *Eur. J. Pharm. Biopharm.* **50**, 121–128.
54. Andriano, K. P., Tabata, Y., Ikada, Y. and Heller, J. (1999) In vitro and in vivo comparison of bulk and surface hydrolysis in absorbable polymer scaffolds for tissue engineering. *J. Biomed. Mater. Res.* **48**, 602–612.
55. Chasin, M. and Langer, R. (1990) Biodegradable polymers as drug delivery systems. In *Drugs and the pharmaceutical sciences*, Chasin, M. and Langer, R. eds., Marcel Dekker, New York.
56. Burkoth, A. K. and Anseth, K. S. (2000) A review of photocrosslinked polyanhydrides: in situ forming degradable networks. *Biomaterials* **21**, 2395–2404.
57. Williams, D. F. and Zhong, S. P. (1994) Biodeterioration/biodegradation of polymeric medical devices in situ. *Int. Biodeter. Biodegrad.* **34**, 95–130.
58. Labow, R. S., Tang, Y., Mccloskey, C. B. and Santerre, J. P. (2002) The effect of oxidation on the enzyme-catalyzed hydrolytic biodegradation of poly(urethane)s. *J. Biomater. Sci. Polym. Ed.* **13**, 651–665.
59. Coleman, J. W. (2001) Nitric oxide in immunity inflammation. *Int. Immunopharm.* **1**, 1397–1406.
60. Hutmacher, D. W., Sittinger, M. and Risbud, M. V. (2004) Scaffold-based tissue engineering: rationale for computer-aided design and solid free-form fabrication systems. *Trends Biotechnol.* **22**, 354–362.
61. Fedorovich, N. E., Alblas, J., De Wijn, J. R., Hennink, W. E., Verbout, A. J. and Dhert, W. J. (2007) Hydrogels as extracellular matrices for skeletal tissue engineering: state-of-the-art and novel application in organ printing. *Tissue Eng.* **13**, 1905–1925.
62. Mironov, V., Boland, T., Trusk, T., Forgacs, G. and Markwald, R. R. (2003) Organ printing: computer-aided jet-based 3D tissue engineering. *Trends Biotechnol.* **21**, 157–161.
63. Siemann, U. (2005) Solvent cast technology: a versatile tool for thin film production. *Prog. Colloid Polym. Sci.* **130**, 1–14.
64. Kim, H. J., Kim, U. J., Leisk, G. G., Bayan, C., Georgakoudi, I. and Kaplan, D. L. (2007) Bone regeneration on macroporous aqueous-derived silk 3D scaffolds. *Macromol. Biosci.* **7**, 643–655.
65. Pattison, M. A., Wurster, S., Webster, T. J. and Haberstroh, K. M. (2005) Three-dimensional, nano-structured PLGA scaffolds for bladder tissue replacement applications. *Biomaterials* **26**, 2491–2500.
66. Yoon, J. J., Song, S. H., Lee, D. S. and Park, T. G. (2004) Immobilization of cell adhesive RGD peptide onto the surface of highly porous biodegradable polymer scaffolds fabricated by a gas foaming/salt leaching method. *Biomaterials* **25**, 5613–5620.
67. Shin, M., Abukawa, H., Troulis, M. J. and Vacanti, J. P. (2008) Development of a biodegradable scaffold with interconnected pores by heat fusion and its application to bone tissue engineering. *J. Biomed. Mater. Res. A.* **84A**, 702–709.
68. Lv, Q. and Feng, Q. (2006) Preparation of 3-D regenerated fibroin scaffolds with freeze drying method and freeze drying/foaming technique. *J. Mater. Sci. Mater. Med.* **17**, 1349–1356.
69. Zhang, H., Hussain, I., Brust, M., Butler, M. F., Rannard, S. P. and Cooper, A. I. (2005) Aligned two- and three-dimensional structures by directional freezing of polymers and nanoparticles. *Nat. Mater.* **4**, 787–793.
70. Xu, C. Y., Inai, R., Kotaki, M. and Ramakrishna, S. (2004) Aligned biodegradable nanofibrous structure: a potential scaffold for blood vessel engineering. *Biomaterials* **25**, 877–886.
71. Faizal, C. K. M., Kikuchi, Y. and Kobayashi, T. (2009) Molecular imprinting targeted for [alpha]-tocopherol by calix[4]resorcarenes derivative in membrane scaffold prepared by phase inversion. *J. Memb. Sci.* **334**, 110–116.
72. Ma, D. and Mchugh, A. J. (2007) The interplay of phase inversion and membrane formation in the drug release characteristics of a membrane-based delivery system. *J. Memb. Sci.* **298**, 156–168.

73. Kim, S. Y., Kanamori, T., Noumi, Y., Wang, P. C. and Shinbo, T. (2004) Preparation of porous poly(d,l-lactide) and poly(d,l-lactide-co-glycolide) membranes by a phase inversion process and investigation of their morphological changes as cell culture scaffolds. *J. Appl. Polym. Sci.* **92**, 2082–2092.
74. Duarte, A. R. C., Mano, J. F. and Reis, R. L. (2009) Dexamethasone-loaded scaffolds prepared by supercritical-assisted phase inversion. *Acta. Biomater.* **5**, 2054–2062.
75. Tsivintzelis, I., Pavlidou, E. and Panayiotou, C. (2007) Porous scaffolds prepared by phase inversion using supercritical CO2 as antisolvent: I. Poly(l-lactic acid). *J. Supercrit. Fluids.* **40**, 317–322.
76. Duarte, A. R. C., Mano, J. F. and Reis, R. L. (2009) Preparation of starch-based scaffolds for tissue engineering by supercritical immersion precipitation. *J Supercrit. Fluids.* **49**, 279–285.
77. Duarte, A. R. C., Caridade, S. G., Mano, J. F. and Reis, R. L. (2009) Processing of novel bioactive polymeric matrixes for tissue engineering using supercritical fluid technology. *Mater. Sci. Eng. C.* **29**, 2110–2115.
78. Tsivintzelis, I., Pavlidou, E. and Panayiotou, C. (2007) Biodegradable polymer foams prepared with supercritical CO_2-ethanol mixtures as blowing agents. *J. Supercrit. Fluids.* **42**, 265–272.
79. Jang, J. H. and Shea, L. D. (2003) Controllable delivery of non-viral DNA from porous scaffolds. *J. Control. Release.* **86**, 157–168.
80. Sheridan, M. H., Shea, L. D., Peters, M. C. and Mooney, D. J. (2000) Bioabsorbable polymer scaffolds for tissue engineering capable of sustained growth factor delivery. *J. Control. Release.* **64**, 91–102.
81. Li, D. and Xia, Y. (2004) Electrospinning of nanofibers: reinventing the wheel? *Adv. Mater.* **16**, 1151–1170.
82. Liang, D., Hsiao, B. S. and Chu, B. (2007) Functional electrospun nanofibrous scaffolds for biomedical applications. *Adv. Drug Deliv. Rev.* **59**, 1392–1412.
83. Huang, Z. M., Zhang, Y. Z., Kotaki, M. and Ramakrishna, S. (2003) A review on polymer nanofibers by electrospinning and their applications in nanocomposites. *Comp. Sci. Technol.* **63**, 2223–2253.
84. Zong, X., Bien, H., Chung, C. Y., Yin, L., Fang, D., Hsiao, B. S., et al. (2005) Electrospun fine-textured scaffolds for heart tissue constructs. *Biomaterials* **26**, 5330–5338.
85. Zhang, Y. Z., Huang, Z. M., Xu, X., Lim, C. T. and Ramakrishna, S. (2004) Preparation of core-shell structured PCL-r-gelatin bi-component nanofibers by coaxial electrospinning. *Chem. Mater.* **16**, 3406–3409.
86. Lee, Y. H., Lee, J. H., An, I. G., Kim, C., Lee, D. S., Lee, Y. K., et al. (2005) Electrospun dual-porosity structure and biodegradation morphology of Montmorillonite reinforced PLLA nanocomposite scaffolds. *Biomaterials* **26**, 3165–3172.
87. Wang, W. L., Cheah, C. M., Fuh, J. Y. H. and Lu, L. (1996) Influence of process parameters on stereolithography part shrinkage. *Mater. Des.* **17**, 205–213.
88. Harris, R. A., Hague, R. J. M. and Dickens, P. M. (2003) Crystallinity control in parts produced from stereolithography injection mould tooling. *Proc. Inst Mech Eng L: J. Mater. Des. Appl.* **217**, 269–276.
89. Rimell, J. T. and Marquis, P. M. (2000) Selective laser sintering of ultra high molecular weight polyethylene for clinical applications. *J. Biomed. Mater. Res.* **53**, 414–420.
90. Vail, N. K., Swain, L. D., Fox, W. C., Aufdemorte, T. B., Lee, G. and Barlow, J. W. (1999) Solid freeform fabrication proceedings. *Mater. Des.* **20**, 123–132.
91. Curodeau, A., Sachs, E. and Caldarise, S. (2000) Design and fabrication of cast orthopedic implants with freeform surface textures from 3-D printed ceramic shell. *J. Biomed. Mater. Res.* **53**, 525–535.
92. Giordano, C., Russell, A., Wu, B. M., Borlanda, S. W., Cima, L. G., Sachs, E. M., et al. (1997) Mechanical properties of dense polylactic acid structures fabricated by three dimensional printing. *Biomater. Sci.* **8**, 63–75.
93. Lam, C. X. F., Mox, M., Teoh, S. H. and Hutmacher, D. W. (2002) Scaffold development using 3D printing with a starch-based polymer. *Mater. Sci. Eng. C.* **20**, 49–56.
94. Sachs, E. M., Haggerty, J. S., Cima, M. J. and Williams, P. A. (1993) Three-dimensional printing techniques. US Patent 5,204,055.
95. Marra, K. G., Szem, J. W., Kumta, P. N., Dimilla, P. A. and Weiss, L. W. (1999) In vitro analysis of biodegradable polymer blend/hydroxyapatite composites for bone tissue engineering. *J. Biomed. Mater. Res. A.* **47**, 324–335.
96. Hutmacher, D. W. (2000) Polymeric scaffolds in tissue engineering bone and cartilage. *Biomaterials* **21**, 2529–2543.
97. Comb, J. W., Priedeman, W. R. and Turley, P. W. (1994) FDM technology process improvements. In: Proceedings of Solid Freeform Fabrication Symposium, Austin, Tex, 42–49.
98. Zein, I., Hutmacher, D. W., Tan, K. C. and Teoh, S. H. (2002) Fused deposition modeling of novel scaffold architectures for tissue engineering applications. *Biomaterials* **23**, 1169–1185.

Chapter 3

Tracking Nanoparticles in Three-Dimensional Tissue-Engineered Models Using Confocal Laser Scanning Microscopy

Vanessa Hearnden, Sheila MacNeil, and Giuseppe Battaglia

Abstract

Here we describe a method for imaging the position of nanoparticles within a 3D tissue-engineered model using confocal laser scanning microscopy (CLSM). The ability to track diffusion of nanoparticles in vitro is an important part of trans-dermal and trans-mucosal drug delivery development as well as for intra-epithelial drug delivery. Using 3D tissue-engineered models enables us to image diffusion in vitro in a physiologically relevant way; not possible in two-dimensional monolayer cultures (MacNeil, Nature 445:874–880, 2007; Hearnden et al., Pharmaceutical Res. 26(7):1718–1728, 2009). CLSM enables imaging of viable in vitro models in three dimensions with good spatial and axial resolution (Georgakoudi et al., Tissue Eng 14:1–20, 2008; Schenke-Layland et al., Adv. Drug Del. Rev. 58:878–896, 2006). Here we show that fluorescently labelled nanoparticles can be visualised, quantified, and their position within the cell can be determined using CLSM.

Key words: Confocal laser scanning microscopy, Three-dimensional tissue-engineered epithelial models, Nanoparticles, Diffusion, Polymersomes, Auto-fluorescence, Co-localisation

1. Introduction

Tissue engineering has many applications ranging from tissue replacement for patients (1) to studies of cell behaviours in three dimensions and toxicity testing (2). It is hoped that physiologically accurate and reproducible tissue-engineered models will reduce the amount of in vivo testing needed by becoming an in vitro tool which is better able to predict substance effects in vivo and enable high throughput screening of drugs and materials as they are developed. Tracking nanoparticle diffusion is clearly important when developing a nanoparticle delivery system

John W. Haycock (ed.), *3D Cell Culture: Methods and Protocols*, Methods in Molecular Biology, vol. 695,
DOI 10.1007/978-1-60761-984-0_3, © Springer Science+Business Media, LLC 2011

in order to understand the speed and method of diffusion of the particles. Imaging where a therapeutic agent could potentially be delivered is possible with fluorescently labelled materials. CLSM is a microscopy technique which enables good three-dimensional resolution and can penetrate into the tissue to see nanoparticles within the model.

Here we will describe the production of the nanoparticles studied in our laboratory: polymersomes. Polymersomes are self-assembling polymer structures capable of encapsulating (and delivering intracellularly) both hydrophobic and hydrophilic molecules (3, 4). We explain how polymersomes are applied to a tissue-engineered model and cover details of how these models can be imaged using CLSM (5).

CLSM gives good spatial and axial resolution enabling tracking of nanoparticles in three dimensions (6, 7). By fluorescently labelling the nanoparticles their distribution within tissue-engineered constructs can be monitored. Three-dimensional images (*z*-stacks) comprised of several *x*-*y* slices enable a 3D reconstruction of the model to be generated.

The epithelial models used in our laboratory, described in detail in Chapter 9 (S. MacNeil et al.) display high levels of auto-fluorescence when excited at 488 nm (8). This auto-fluorescence is useful when focusing the image and as a reference point for the position of the nanoparticles; however, a fluorophore with excitation outside the auto-fluorescent range is needed to get clear images of the nanoparticle's position. We describe a technique to determine which wavelengths the sample is excited by and the fluorescence emission profiles. Using a counterstain the co-localisation of fluorophores can give further information about the location of the nanoparticle of interest.

2. Materials

2.1. Production of Rhodamine-Labelled Polymersomes

1. Polymer film preparation: 1.9 mg/mL $PMPC_{25}$-$PDPA_{70}$ (2-(methacryloyloxy)ethyl phosphorylcholine)-poly(2-(diisopropylamino)ethyl methacrylate) block copolymer (Biocompatibles UK Ltd., Farnham, Surrey, UK (9)), 0.1 mg/mL Rho-$PMPC_{30}$-$PDPA_{60}$ (synthesized by Department of Chemistry, University of Sheffield), made up in a 2:1 chloroform:methanol solution in a glass vial.
2. Film rehydration: Rehydrate 5 mg of polymer film into 1 mL of 100 mM phosphate-buffered saline (PBS), brought to pH 2.0 using 1 M HCl, filter sterilised and store at room temperature.
3. NaOH made up at 1 M, filter sterilised, store at room temperature.
4. Sonicator.

5. LipsoFast-Basic and gas tight syringe (Avestin Inc. Ottawa, Canada).
6. Polycarbonate membrane (Avestin Inc. Ottawa, Canada).
7. Sepharose, stored at 5°C (Sigma-Aldrich, Dorset, UK).
8. Liquid chromatography column, Luer lock, non-jacketed, 1.5 × 30 cm. (Sigma-Aldrich, Dorset, UK).
9. 70% Industrial methylated spirits.

2.2. Applying the Nanoparticles (Polymersomes) to a 3D Tissue Construct

1. Tissue culture insert, Thincert® 12-well 8.0 μm pore size (Greiner Bio-One, Gloucestershire, UK).

2.3. Preparation of the Sample for Imaging of Three-Dimensional Tissue-Engineered Models Using Confocal Laser Scanning Microscopy

1. Zeiss Laser Scanning Confocal Microscope LSM 510 META (Carl Zeiss Inc., Germany).
2. Water-dipping lenses (Carl Zeiss Inc., Germany).
3. Six-well tissue culture plate.
4. 60 μm Dish for live cell imaging (Ibidi Integrated Biodiagnostics, Glasgow, UK).
5. Glass bottomed 24-well plates No. 0, Uncoated (MatTek Corporation, Ashland, MA, USA).

2.4. Determining the Auto-fluorescence of a Sample

1. Zeiss Laser Scanning Confocal Microscope LSM 510 META (Carl Zeiss Inc., Germany).
2. LSM 510 META Software (Carl Zeiss Inc., Germany).

2.5. Co-localisation of Fluorophore and Other Component

1. DAPI (4′,6-diamidino-2-phenylindole) stored in single use aliquots at 1 mg/mL at −20°C. DAPI is light sensitive. Used at a 2 μg/mL in cell culture medium.

3. Methods

3.1. Production of Rhodamine-Labelled Polymersomes

1. $PMPC_{25}$-$PDPA_{70}$ block copolymer is dissolved in a glass vial in a 2:1 chloroform:methanol solution at a 1.9 mg/mL. Rho-$PMPC_{30}$-$PDPA_{60}$ polymer (block copolymer with covalently attached rhodamine fluorescent dye) is dissolved at 0.1 mg/mL (5% of non-labelled polymer).
2. A copolymer film is formed by evaporating the solvent in a vacuum oven at 50°C. The time this takes depends on the volume of sample and efficiency of vacuum pump (see Note 1).
3. The film is then rehydrated using pH 2 PBS (100 mM) giving a copolymer suspension of 5 mg/mL. The pH is checked and extra drops of 1 M HCl added until the solution reaches pH 2.

4. Once the film has fully dissolved the pH is increased gradually to pH 7.4 using 1 M NaOH (see Note 2).
5. The copolymer solution is sonicated for 10 min.
6. The size of polymersomes can be adjusted by manually extruding the solution 31 times with an extruder through specific size membranes.
7. The solution is then purified by gel permeation chromatography using a sepharose 4B size exclusion column. The column is washed with 70% IMS to sterilise and then washed three times with PBS to remove all IMS.
8. The final PBS wash should pass through the column until the top surface of the sepharose is dry. The polymersome solution is then added to the top of the column and allowed to enter into the sepharose. Once the polymer solution has entered the sepharose and the top surface of the sepharose is dry again PBS is added to allow the polymer solution to pass through the column. The fraction with the largest particles passes through the column first. The polymersomes are the largest molecules and can be detected by eye as a cloudy, turbid and in this case pinkish solution. If 2 mL is added to the top of the column the first 2 mL fraction of polymersomes should be collected. The remaining fractions contain micelles and unimer polymer chains; these should be discarded (see Note 3).

3.2. Applying the Nanoparticles (Polymersomes) to a 3D Tissue Construct

1. The chapter by S MacNeil et al. gives an in depth description of how to produce tissue-engineered models of skin and the oral mucosa. For this protocol the tissue-engineered construct of choice should be grown in a tissue culture insert.
2. Once the model has matured the nanoparticles should be added to the top surface of the construct within the tissue culture insert.
3. For 12-well tissue culture inserts 100 μl of the pure nanoparticle solution should be added – this can be scaled up for larger inserts.
4. Leave the nanoparticles on the constructs for the desired amount of time.

3.3. Preparation of the Sample for Imaging with Confocal Laser Scanning Microscopy

1. Tissue-engineered models are removed from tissue culture insert and washed in PBS two to three times to remove any cell culture media and excess nanoparticles which have not entered the tissue.
2. Models are then left submerged in PBS in a six-well plate.
3. Here we describe a method to image the models using an upright confocal microscopy and a water submersion dipping lens. Using a dipping lens to image the model allows it to be kept submerged in PBS and keeps the model viable for some time (see Notes 4 and 5).

3.4. Determining the Auto-fluorescence of a Sample

1. To determine the characteristics of any auto-fluorescence in a sample the CLSM can be used in "lambda mode."
2. Focus the sample and excite the sample at the first wavelength of interest. The detector should be set to detect emissions of the excitation wavelength and above, e.g. excitation at 488 nm, detection at 488 nm and above.
3. Once the image has been captured the emission profile at each wavelength can be obtained using the "mean" function on the Zeiss LSM Image Browser software. It is possible to see the emission at each wavelength using the "gallery" view (Fig. 1).
4. The fluorophore used to label the nanoparticle should be excited at a wavelength which does not excite auto-fluorescent components of the model (see Note 6).

3.5. Obtaining a Three-Dimensional Image of the Model

1. Focus on a plane of the model which has the highest fluorescence and set the detector gain and pinhole using the software (see Note 7).
2. Use the *z*-axis focus adjustment to find the top layer of your sample.
3. On the "*z* settings" tab select "Mark first/last" and mark this as the first slice of your image.
4. Next scan down using the focus adjustment to find the bottom of your model (see Note 8). Mark this as the last slice in the *z* settings.

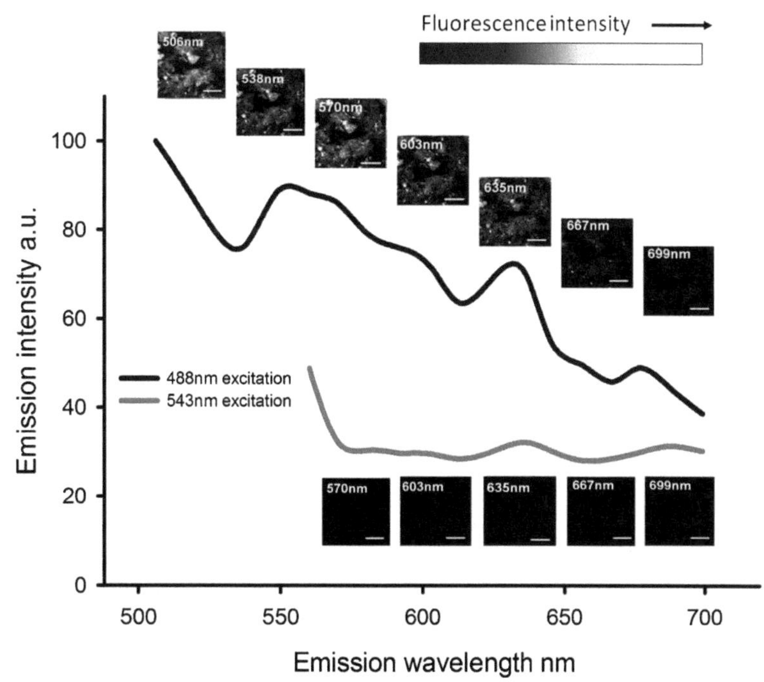

Fig. 1. Emission spectra of tissue-engineered model excited with 488 and 543 nm lasers.

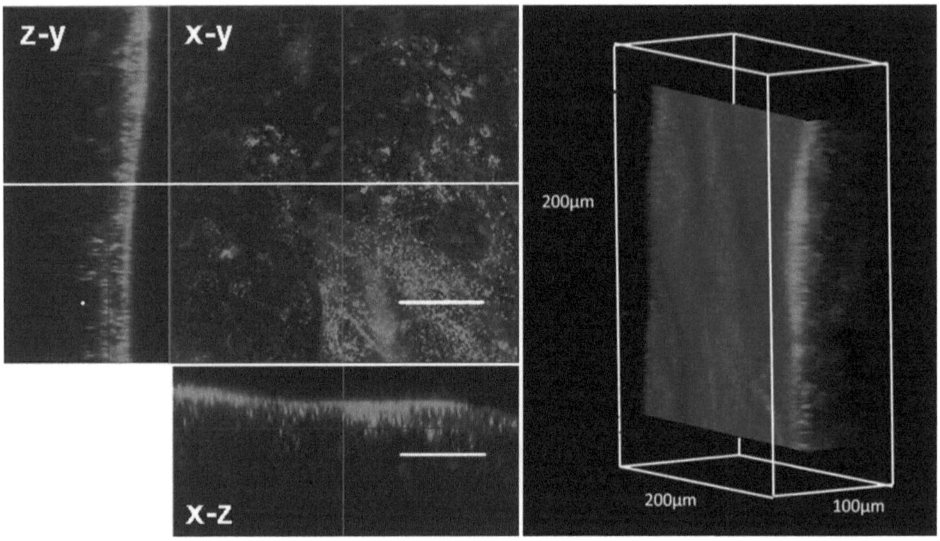

Fig. 2. Example of a 3D CLSM image. Rhodamine-labelled polymersomes diffusing into a tissue-engineered model of oral mucosa (1). *Left*: shows the image in the *x-y* plane and the cross sections through the *z* stack of images. Scale bar 50 μm. *Right*: 3D reconstruction of all *z* slices.

5. Here you can adjust the amount of slices to be imaged or the thickness of each slice. Using thinner optical slices will give a better 3D reconstruction (see Note 9).
6. If you are imaging two different fluorophores simultaneously it is important that the optical slice (determined by the pinhole) is the same for the two channels. A smaller pinhole will give higher resolution image.
7. Adjusting the speed and number of scans per slice will also improve the quality of the image (this needs to be weighed up against the length of time taken to obtain the image and viability of your sample) (Fig. 2).

3.6. Quantification of Depth of Diffusion

1. Once a three-dimensional image (*z* stack) has been obtained viewing it in the "ortho" setting allows you to see the *z-y* and *z-x* projecti.ons of the area imaged (Fig. 2) (see Note 10).
2. To estimate the depth of penetration find the first slice which is fluorescent and the last slice which is fluorescent. The distance between these two points (Z_{max}) can be measured using the "dist" function (see Note 11).

3.7. Quantification of Volume of Diffusion

1. Using the "histo" function on the Zeiss LSM Meta software it is possible to determine the area of each slice which is fluorescent over a particular emission intensity threshold.
2. Using a control image (with no nanoparticles) to determine the threshold measure the area in each slice which is fluorescent.

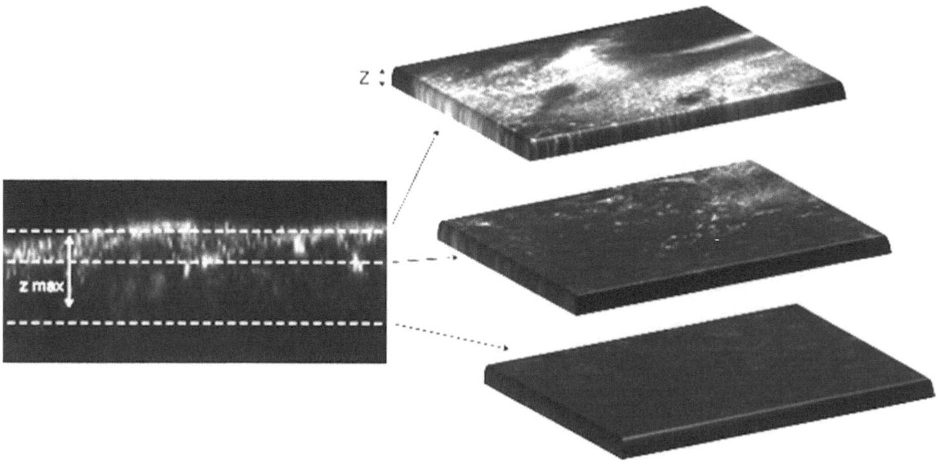

Fig. 3. Diagram to demonstrate the method used to measure Z_{max} and quantify the volume of diffusion using 3D confocal images. Area of fluorescence emission in each *x-y* slice was measured. This area is multiplied by the *z* dimension of the slice and used to calculate the volume of the nanoparticles diffused into in the whole 3D image.

3. The area of fluorescence in each slice can be used to estimate the total fluorescent volume using the following equation.

$$V = \sum_{i=1}^{N} A_{si} \times z,$$

where V = total volume of fluorescence, N = number of slices, A_s = area in *x-y* slice, z = optical slice thickness (see Fig. 3).

3.8. Co-localisation of Fluorophore and Other Counterstain

1. When two fluorophores have been imaged on different channels at the same time their degree of co-localization can be measured. In the example shown in Fig. 4a, b the tissue-engineered oral mucosa exposed to rhodamine-labelled polymersomes has been counterstained with DAPI.
2. Add DAPI to cell culture medium at a concentration of 2 μg/mL.
3. Incubate for 1 h at 37°C with 5% CO_2.
4. Models should be prepared for imaging as described in Subheading 3.3.
5. Image using the "multi-channel" mode set up one channel to excite and detect DAPI and another channel to excite and detect rhodamine fluorescence.
6. When analysing the image select the "histo" function in the image window.
7. Selecting "co-localisation" shows a scatter plot of pixel intensity for channel 1 vs. pixel intensity for channel 2.
8. A scatter plot of channel 1 intensity vs. channel 2 intensity is shown (Fig. 4a, c).

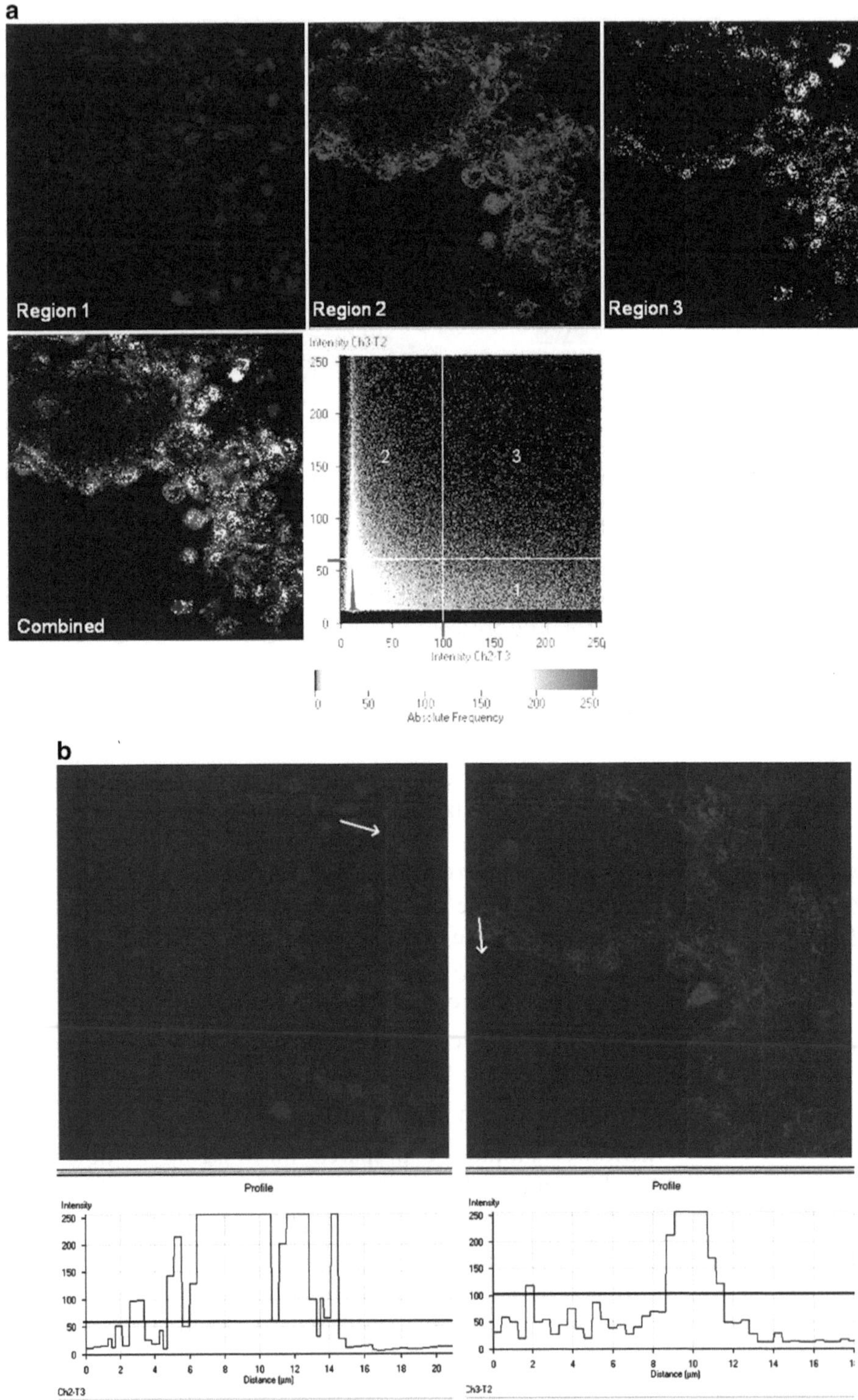

Fig. 4. Tissue-engineered oral mucosa exposed to rhodamine-labelled polymersomes, washed and counter stained with DAPI – small amount of co-localisation (region 3). (**a**) Diagram to show fluorescence profile used to determine thresholds. (**b**) Image of tissue-engineered oral mucosa submerged in rhodamine-labelled polymersomes. Region 1 = auto-fluorescence, region 2 = rhodamine-labelled polymersomes. High level of co-localisation (region 3).

c

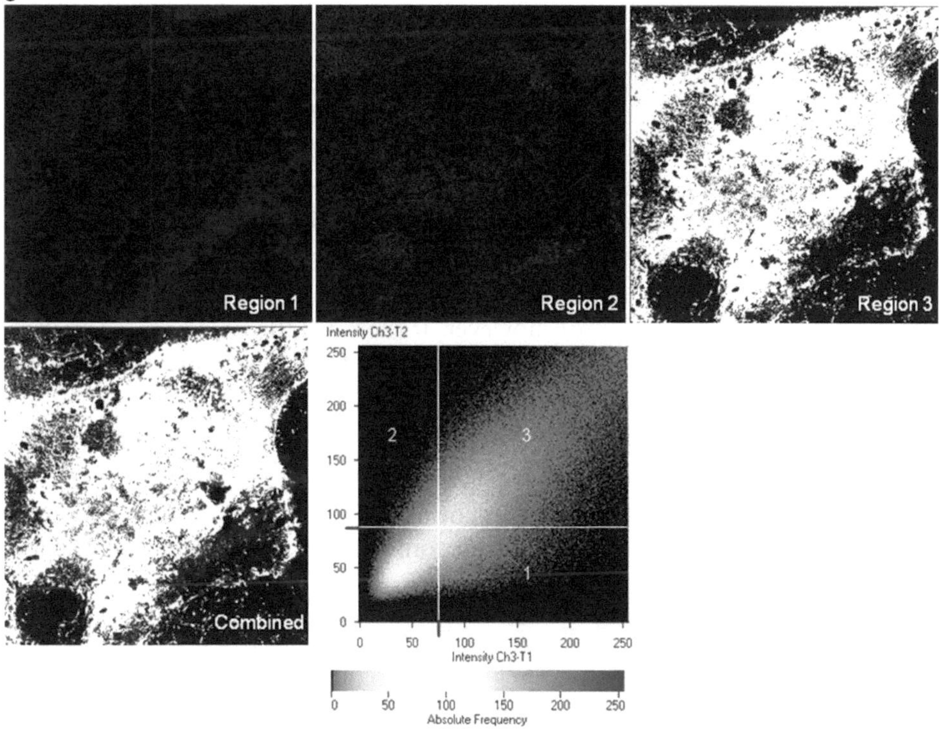

d

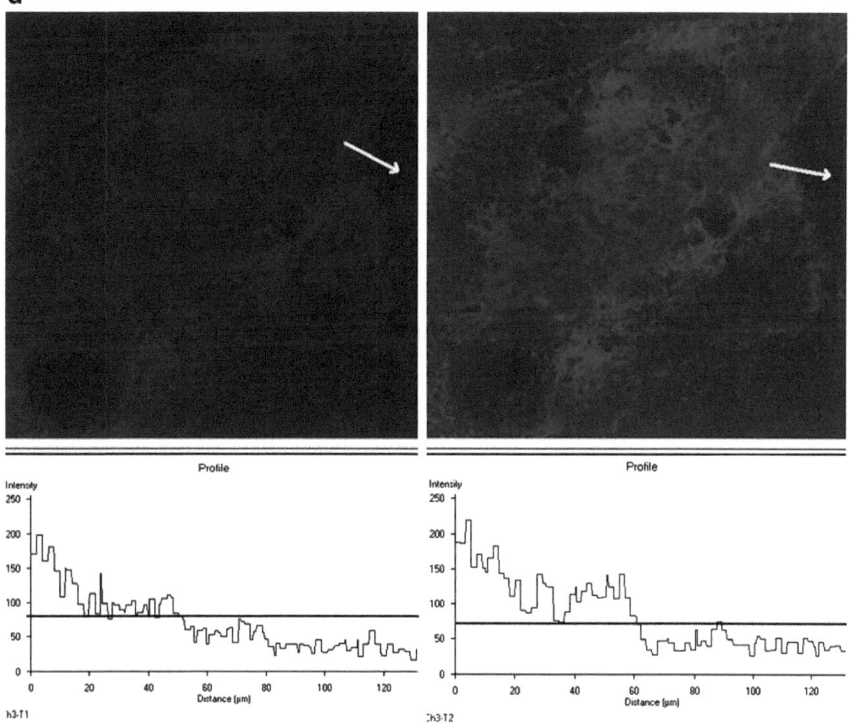

Fig. 4. (**c**) and (**d**) (continued). Diagram to show fluorescence profile used to determine thresholds.

9. It is important to remove the background fluorescence. This can be done using the profile function.
10. Figure 4b, d show the profile of fluorescence emission intensity along the arrow. From this decide which intensities are florescent and which are background.
11. Use this intensity value to move the thresholds on the scatter plot (black lines on profile = white lines on scatter plot).
12. If the two fluorophores are co-localised the scatter will cluster in a linear trend and the majority of pixels will lie in region 3 (Fig. 4c); however, if there is little or no co-localisation the scatter will be spread with pixels in all three regions and little correlation between the two channels (Fig. 4a).
13. Region 3 on the scatter plot shows the co-localising pixels (fluorescent in both rhodamine and DAPI above the threshold).
14. The correlation coefficient is calculated by the software using the following equation, Pearson's correlation coefficient (10).

$$R_{\mathrm{p}} = \frac{\sum_i (\mathrm{Ch1}_i - \mathrm{Ch1}_{\mathrm{average}}) \times (\mathrm{Ch2}_i - \mathrm{Ch2}_{\mathrm{average}})}{\sqrt{\sum_i (\mathrm{Ch1}_i - \mathrm{Ch1}_{\mathrm{average}})^2 \times (\mathrm{Ch2}_i - \mathrm{Ch2}_{\mathrm{average}})^2}}$$

15. "Show table" shows details of (1) the area of fluorescence and (2) intensity of fluorescence for each channel. The correlation coefficient shows the degree of co-localisation of the two flurophores. This can give you information about where in the cell your nanoparticle is located by counterstaining the cell structure of interest, e.g. DAPI for cell nuclei.

4. Notes

1. Ensure all solvent is evaporated before moving onto the next step.
2. The polymer may precipitate here if the pH is increased too rapidly. Vortexing and sonication helps to remove the polymer precipitates.
3. Care should be taken not to let the column run dry as this causes bubbles to form. If this happens shaking the sepharose vigorously with PBS in the column will enable it to be repacked. The volume passed through the column at any one time depends on the size of the column but there should be sufficient sepharose to allow the solution to separate as it passes through the column.
4. For models and nanoparticles which can be fixed without losing fluorescence the models can be mounted between a glass slide and coverslip and imaged with oil immersion or non-immersion lenses.

5. The models stay viable for some time submerged in PBS; however, for repeated or sterile imaging phenol red free media can be used and the model can be imaged in a sealed Petri (Ibidi) or glass bottomed well plate (MatTek) using a non-immersion lens. However we have found some background and reflection when using this method.
6. There will always be emission at wavelengths just above the excitation – this is the reflection of the excitation laser.
7. The range indicator is useful to show whether the image is overexposed (overexposed areas shown in red). If overexposed adjust the detector gain or the pinhole until there is no red.
8. This may not be the bottom of the model but instead the limit of the penetration of that laser, depending on the thickness of the sample.
9. The minimum optical slice is dependant on the pinhole (adjustable) and the lens used.
10. It is important to set up the software with the correct objective when obtaining the image to ensure the 3D reconstruction is accurate.
11. If the model is not flat you may need to measure Z_{max} on several z-x and z-y projections and chose the longest Z_{max}.

References

1. Bhargava, S., Chapple, C. R., Bullock, A. J., Layton, C., and MacNeil, S. (2004) Tissue-engineered buccal mucosa for substitution urethroplasty. *BJU Int.* 93, 807–811.
2. Moharamzadeh, K., Brook, I. M., Van Noort, R., Scutt, A. M., and Thornhill, M. H. (2007) Tissue-engineered oral mucosa: a review of the scientific literature. *J. Dent. Res.* 86, 115–124.
3. Lomas, H., Massignani, M., Abdullah, K. A., Canton, I., Lo Presti, C., MacNeil, S., Du, J., Blanazs, A., Madsen, J., Armes, S. P., Lewis, A. L., and Battaglia, G. (2008) Non-cytotoxic polymer vesicles for rapid and efficient intracellular delivery. *Faraday Discuss.* 139, 143.
4. Lomas, H., Canton, I., MacNeil, S., Du, J., Armes, S. P. A., Ryan, A. J., Lewis, A. L., and Battaglia, G. (2007) Biomimetic pH sensitive polymersomes for efficient DNA encapsulation and delivery. *Adv. Mater.* 19, 4238–4243.
5. Hearnden, V., Lomas, H., MacNeil, S., Thornhill, M., Murdoch, C., Lewis, A. L., Madsen, J., Blanazs, A., Armes, S. P., and Battaglia, G. (2009) Diffusion studies of nanometer polymersomes across tissue engineered human oral mucosa. *Pharm. Res.* 26(7), 1718–1728.
6. Georgakoudi, I., Rice, W. L., Hronik-Tupaj, M., and Kaplan, D. L. (2008) Optical spectroscopy and imaging for the noninvasive evaluation of engineered tissues. *Tissue Eng.* 14, 1–20.
7. Muller, M. (2006) *Introduction to confocal microscopy.* 2nd ed., SPIE – The international society for optical engineering, USA.
8. Schenke-Layland, K., Riemann, I., Damour, O., Stock, U. A., and König, K. (2006) Two-photon microscopes and in vivo multiphoton tomographs – powerful diagnostic tools for tissue engineering and drug delivery. *Adv. Drug Deliv. Rev.* 58, 878–896.
9. Du, J., Tang, Y., Lewis, A. L., and Armes, S. P. (2005) pH-sensitive vesicles based on a biocompatible zwitterionic diblock copolymer. *J. Am. Chem. Soc.* 127, 17982–17983.
10. Zeiss, C. Chapter 4 Operating in Expert Mode. In *Carl Zeiss LSM 510 Operating Manual*, pp 280–285.

Chapter 4

Using Immuno-Scanning Electron Microscopy for the Observation of Focal Adhesion-substratum interactions at the Nano- and Microscale in S-Phase Cells

Manus J.P. Biggs, R. Geoff Richards, and Matthew J. Dalby

Abstract

It is becoming clear that the nano/microtopography of a biomaterial in vivo is of first importance in influencing focal adhesion formation and subsequent cellular behaviour. When considering next-generation biomaterials, where the material's ability to elicit a regulated cell response will be key to device success, focal adhesion analysis is an useful indicator of cytocompatibility and can be used to determine functionality. Here, a methodology is described to allow simultaneous high-resolution imaging of focal adhesion sites and the material topography using field emission scanning electron microscopy. Furthermore, through the use of BrdU pulse labelling and immunogold detection, S-phase cells can be selected from a near-synchronised population of cells to remove artefacts due to cell cycle phase. This is a key factor in adhesion quantification as there is natural variation in focal adhesion density as cells progress through the cell cycle, which can skew the quantitative analysis of focal adhesion formation on fabricated biomaterials.

Key words: Nanotopography, Microtopography, Electron microscopy, Osteoblast

1. Introduction

Quantification of focal adhesion frequency and area is accepted as one method for measuring total cell adhesion and for determining cytocompatibility. Typically, this has been accomplished via light microscopy, utilising immunofluorescent detection of focal adhesion proteins in adherent cells, but the practical resolution limitations of this technique restrict the simultaneous visualisation of adhesion sites and topographical structures to less than approximately 0.35 μm. The technique described below utilises a 1.4 nm gold colloid–fluorescein conjugate, which allows focal adhesion analysis via scanning electron microscopy (SEM) and subsequent

John W. Haycock (ed.), *3D Cell Culture: Methods and Protocols*, Methods in Molecular Biology, vol. 695, DOI 10.1007/978-1-60761-984-0_4, © Springer Science+Business Media, LLC 2011

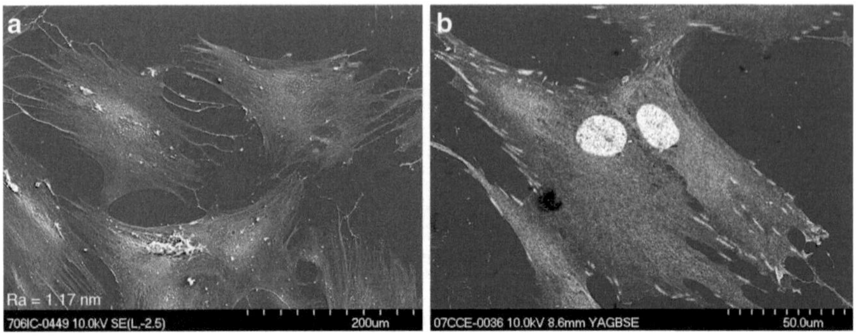

Fig. 1. HOBs cultured on planar pMMA substrates. (**a**) Secondary-electron micrograph of non-immunolabelled osteoblasts shows them as well-spread flattened cells. Adhesion plaques and nuclei are not readily discernable. (**b**) Backscattered electron image of two dual immunolabelled HOBs. The central nuclei and elongated peripheral adhesion sites are visible as areas of high contrast, following deposition of silver-enhanced gold colloids.

verification by standard immunofluorescent techniques in a single sample. SEM studies utilising dual vinculin and S-phase nuclei labelling improve on immunofluorescent techniques by allowing regulation of depth of signal formation, increasing the spatial resolution available, and facilitating the simultaneous visualisation of topography, whole cells, and adhesion sites. Here, the technique has been optimised for the analysis of focal adhesion quantification in S-phase primary human osteoblasts (HOBs), example of which can be seen in Fig. 1. This approach reduces intrasample deviation in focal adhesion formation, allowing both qualitative and quantitative data to be collected from a synchronised cellular population.

S-phase identification relies on the successful incorporation of BrdU into the nuclear DNA of dividing cells. The cell fixation and permeabilisation regimes used in the described protocol are required to facilitate BrdU penetration of the nucleus and are optimised to ensure preservation of cellular protein structure coupled with adequate BrdU detection. Of particular importance is the activity of the DNase solution used for DNA scission, decreased activity of which results in decreased nuclear signal (Fig. 2a).

Past EM studies have relied on tritiated thymidine detection coupled with a two-level vinculin detection system and subsequent resin embedding for duel S-phase, vinculin labelling (1, 2). This protocol was not only time consuming, but also involved nuclear emulsion development, resin curing and handling of radioisotopes, factors which have all been eliminated from the technique described below. Many investigators have found BrdU to be as reliable in S-phase detection as tritiated thymidine, and when incorporated into the cell nucleus is a very stable antigen, giving a strong and reliable signal regardless of the fixation method or further tissue processing (3).

Examples of the development of this protocol and its application to biomaterial science can be seen in (4–6).

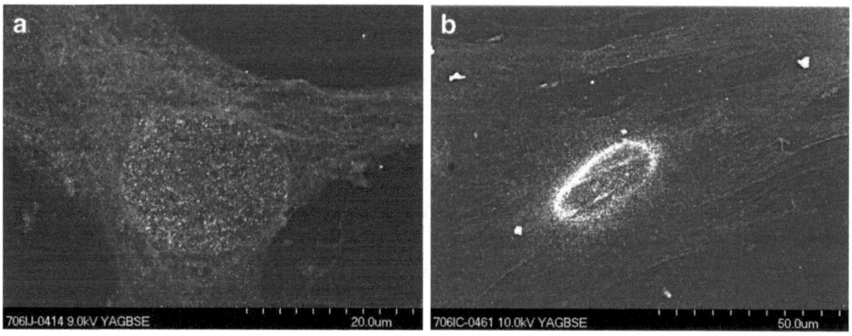

Fig. 2. Immuno-SEM labelling of S-phase nuclei in HOBs. (**a**) DNase with reduced activity results in incomplete scission of the DNA molecule, a reduced availability of the BrdU antigen, and subsequent reduced colloid deposition. (**b**) nuclear permeation is reduced with colloids approaching 5 nm in diameter. Here nuclear labelling appears as a peripheral nuclear ring.

2. Materials

2.1. Cell Culture

1. Trypsin solution: 0.04% trypsin, 0.03% EDTA dissolved in HEPES-buffered saline, pH 7.4.
2. Osteoblast growth medium (PromoCell®, Heidelberg, Germany): Supplemented with 10% foetal calf serum, herein described as "OGM."
3. BrdU solution: 10 μM 5-bromo-2-deoxyuridine (BrdU) (Sigma, St. Louis, MD) dissolved in OGM.
4. Dimethyl-sulfoxide (DMSO; Sigma, St. Louis, MD).
5. HEPES-buffered saline: 2 M 4-(2-hydroxyethyl)-1-piperazineethanesulfonic acid, 0.3 M sucrose, 5 M NaCl, 30 mM MgC_{12}, and 0.1% phenol red, dissolved in ROH_2O, pH 7.4.

2.2. Immunocyto-chemistry for SEM

1. PIPES buffer: 0.1 M 1,4-Piperazine bis (2-ethanosulfonic acid) (PIPES), pH 7.4.
2. Triton X-100 buffer: 0.5% Triton X-100, dissolved in 0.1 M PIPES buffer, pH 7.4.
3. Paraformaldehyde solution: 4% Paraformaldehyde, 1% sucrose dissolved in 0.1 M PIPES buffer, pH 7.4.
4. BSA solution: 1% bovine serum albumin (BSA), 0.1% Tween 20 dissolved in 0.1 M PIPES buffer, pH 7.4.
5. Primary antibody solution: 1 M Tris (hydroxymethyl) aminomethane, 50 U/mL of DNase I, 1:100 concentration of monoclonal mouse anti-BrdU IgG2 (Amersham Biosciences, Uppsala, Sweden), and 1:200 concentration of monoclonal mouse anti-human vinculin IgG1 (Sigma, St. Louis, MD) dissolved in ROH_2O.

6. Blocking solution: 5% Goat serum, 1% BSA, 0.1% Tween 20 dissolved in PIPES buffer, pH 7.4.
7. Avidin/Biotin Blocking Kit (SP-2001) (Vector Laboratories, Burlingame, CA).
8. Secondary antibody solution: 1% BSA, 0.1% Tween 20, and 1:100 concentration of biotin conjugated monoclonal horse anti-mouse IgG, (Vector Laboratories, Burlingame, CA) dissolved in 0.1 M PIPES buffer, pH 7.4.
9. Tertiary antibody solution: 1% BSA, 0.1% Tween 20, 1:300 concentration of streptavidin conjugated goat anti-mouse 1.4 nm gold-fluorescein Fluoronanogold™ (Nanoprobes, Yaphank, NY) dissolved in 0.1 M PIPES buffer, pH 7.4.
10. Glutaraldehyde solution: 2.5% Glutaraldehyde, dissolved in PIPES buffer, pH 7.4.
11. Silver developing solution (SE100) (Sigma-Aldrich, UK).
12. Osmium tetroxide solution: 1% Osmium tetroxide dissolved in PIPES buffer, pH 6.2.

2.3. Sample Preparation for Scanning Electron Microscopy

1. Ethanol/water series: 50, 60, 70, 80, 90, 96, and 100%.
2. Fluorisol/ethanol series: 25, 50, 75, and 100% (fluorisol – 1,1,2-trichloro, 1,2,2-trifluoroethane).
3. Critical point drier (Polaron E3100) (Quorum Technologies, Sussex, UK).
4. Aluminium stubs and adhesive carbon disks (Baltec, FL, Liechtenstein).
5. Baltec CED030 carbon thread evaporator (Baltec, FL, Liechtenstein).
6. Hitachi S-4700 field emission scanning electron microscope (FESEM) fitted with an Autrata yttrium aluminium garnet (YAG) backscattered electron (BSE) scintillator type detector.

3. Methods

Previously, problems with signal intensity have been noted when using colloids <2 nm in diameter. Poor protein labelling with these colloids is probably due to impaired silver enhancement, as the colloid becomes enveloped by the conjugated immunoglobulin, effectively adsorbing a layer of protein that resists silver reduction at the colloid surface (7). Here, this problem of small colloid probe envelopment is overcome by the use of a three-level detection system with a tertiary gold colloid conjugated to a streptavidin molecule. Streptavidin is smaller than a whole immunoglobulin molecule and does not aggregate (8), allowing penetration of the nucleus to reach antigens inaccessible to larger gold probes, an example of which can be seen in Fig. 2b.

3.1. Cell Culture and Sample Preparation for Focal Adhesion Immunodetection by SEM

1. Primary HOBs derived from a femoral head biopsy of an 84-year-old Caucasian female and a knee biopsy of a 74-year-old Caucasian female (PromoCell®, Heidelberg, Germany) are cultured in 75 cm^2 culture flasks until 80% confluent.
2. 0.015 g of BrdU is dissolved first in 1 mL of DMSO and then added to 49 mL of HEPES saline to create a 1 mM BrdU solution. This stock solution is diluted as needed in OGM to give a final concentration of 10 μM BrdU.
3. OGM is removed from the culture, and HOBs are rinsed for 30 s in 10 mL HEPES-buffered saline solution. This is removed, and HOBs are subsequently trypsinised in 3 mL of trypsin solution for approx. 5 min. The cell suspension is diluted in 10 mL of OGM and centrifuged at $400 \times g$ for 5 min. The supernatant is removed and the cell pellet resuspended in OGM.
4. HOBs are seeded onto untreated experimental substrates at a density of 1×10^5 cells per mL (see Note 1). Cells are maintained in OGM, which is replaced every 2 days. Following 7 days of culture, HOBs are maintained for 4 days without replacing the OGM, inducing a brief period of serum starvation (see Note 2). OGM is subsequently introduced to the culture following 4 days of serum starvation and HOBs allowed to metabolise for 17 h, giving rise to a population of HOBs with a synchronised nuclear cycle (see Note 3).
5. Following 17 h in culture, the OGM is removed and HOBs are cultured in fresh OGM encompassing 10 μM BrdU for 3 h.

3.2. Immunocytochemistry for Focal Adhesion Immunodetection by SEM

1. HOBs are rinsed for 2 min in 0.1 M PIPES buffer, pH 7.4 before being rinsed with Triton X-100 buffer (4 × 1 min) to permeabilise the cell membrane (see Note 4). Cells are then stabilised in paraformaldehyde buffer for 5 min and rinsed three times for 2 min in 0.1 M PIPES buffer to remove unreacted aldehyde. Non-specific binding sites are blocked with blocking solution for 15 min. Cells are then incubated overnight at 22°C with primary antibody solution (see Note 5).
2. Cells are rinsed six times for 2 min in BSA solution. Non-specific binding sites are blocked with blocking solution for 15 min, followed by 10 min avidin blocking and then 10 min biotin blocking at 22°C. HOBs are incubated in secondary antibody solution for 1.5 h. HOBs are rinsed six times for 2 min in BSA solution. The cells are incubated in a further tertiary antibody solution overnight for 12 h at 22°C (see Note 6). Samples are then rinsed six times for 2 min in PIPES buffer.

3. The samples are fixed permanently in 2.5% glutaraldehyde solution for 5 mins (see Note 7) and rinsed three times for 2 min in PIPES buffer. Gold probes are silver-enhanced with a silver developing solution for 5 min (see Note 8). The samples are then immediately rinsed twice in ultra-pure water to remove any unreacted enhancer. Additional contrasting of the cell is accomplished by staining the cells with osmium tetroxide solution for 1 h at 22°C. The cells are rinsed in ultra-pure water for 2 min.

3.3. Sample Dehydration and Mounting for SEM

HOBs on experimental substrates are dehydrated through an ethanol/water series (50, 60, 70, 80, 90, 96, and 100%), followed by a fluorisol/ethanol series (25, 50, 75, and 100%). The samples are critical-point-dried for 1 h (see Note 9), mounted on adhesive carbon pads and aluminium stubs, and coated with a 12 nm layer of carbon. The samples are imaged using an Hitachi S-4700 field emission scanning electron microscope fitted with an Autrata yttrium aluminium garnet backscattered electron scintillator type detector (see Note 10). Images are taken in both the BSE and secondary electron (SE) modes, with accelerating voltages between 2 and 10 kV (Fig. 3); BSE and SE images are taken with an emission current of 50 μA, an aperture of 100 μm (apt1), and a working distance in the range of 10–12 mm (9) (see Note 11).

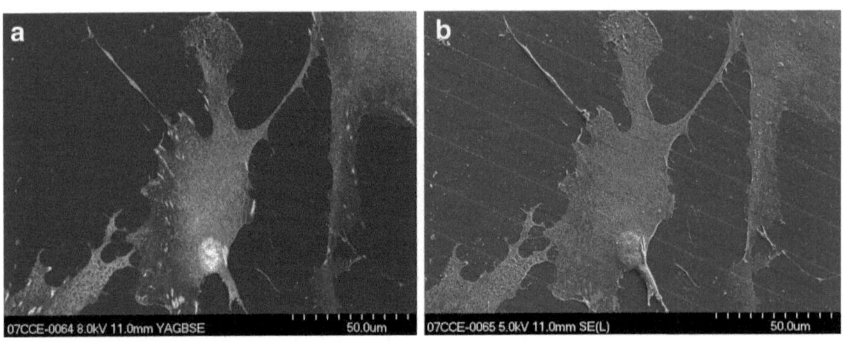

Fig. 3. Backscattered and secondary-electron imaging of S-phase HOBs on a microgrooved topography. (**a**) Backscattered electron detection allows simultaneous visualisation of ventral adhesion formation as well as colloid accumulation in S-phase nuclei. (**b**) Secondary-electron detection coupled with low-voltage imaging (5 kV) allows resolution of low-aspect ratio topographical features.

4. Notes

1. When seeding HOBs onto experimental substrates, release the cell suspension at a slow steady rate and ensure that the pipette and substrate are in close approximation.
2. Following 2 days of serum starvation, check cells daily to ensure that toxicity is not a concern.
3. Cell cycle length varies with cell type. Post serum starvation culture and BrdU incubation times must be optimised for each experiment to ensure high levels of S-phase cells.
4. All PIPES buffer used is 0.1/mol concentration unless otherwise stated.
5. Ensure activity of DNase I. Insufficient DNA scission greatly diminishes BrdU detection.
6. Larger colloid diameter reduces background noise and unspecific colloid deposition. Nucleoskeletal elements, however, greatly hamper nuclear colloid distribution with colloids larger than 5 nm in diameter.
7. Glutaraldehyde fixation is necessary for permanent protein cross-linking.
8. Silver enhancing time must be optimised for colloid species. Silver deposition varies with compound activity. Dated reagents may require longer enhancing times.
9. The chamber should be flushed with CO_2 at least four times during this period.
10. Signal-to-noise ratio of adhesion sites is greatly enhanced with BSE imaging, allowing the simultaneous visualisation of nuclear and vinculin-associated colloid deposition.
11. Low-voltage imaging facilitates the use of short working distances and high resolution imaging of low-aspect topographical features.

References

1. Meredith, D. O., Owen, G. R., ap Gwynn, I., and Richards, R. G. (2004) Variation in cell-substratum adhesion in relation to cell cycle phases. *Exp. Cell Res.* **293,** 58–67.
2. Owen, G. R., Meredith, D. O., ap Gwynn, I., and Richards, R. G. (2002) Simultaneously identifying S-phase labelled cells and immunogold-labelling of vinculin in focal adhesions. *J. Microsc.* **207,** 27–36.
3. Mokry, J. and Nemecek, S. (1995) Immunohistochemical detection of proliferative cells. *Sb Ved Pr Lek Fak Karlovy Univerzity Hradci Kralove* **38,** 107–113.
4. Biggs, M. J., Richards, R. G., McFarlane, S., Wilkinson, C. D., Oreffo, R. O., and Dalby, M. J. (2008) Adhesion formation of primary human osteoblasts and the functional response of mesenchymal stem cells to 330 nm deep microgrooves. *J. R. Soc. Interface* **5,** 1231–1242.
5. Biggs, M. J., Richards, R. G., Wilkinson, C. D., and Dalby, M. J. (2008) Focal adhesion

interactions with topographical structures: a novel method for immuno-SEM labelling of focal adhesions in S-phase cells. *J. Microsc.* **231,** 28–37.

6. Biggs, M. J., Richards, R. G., Gadegaard, N., Wilkinson, C. D., and Dalby, M. J. (2006) Regulation of implant surface cell adhesion: characterization and quantification of S-phase primary osteoblast adhesions on biomimetic nanoscale substrates. *J. Orthop. Res.* **25,** 273–282.
7. Owen, G. R., Meredith, D. O., ap Gwynn, I., and Richards, R. G. (2001) Enhancement of immunogold-labelled focal adhesion sites in fibroblasts cultured on metal substrates: problems and solutions. *Cell Biol. Int.* **25,** 1251–1259.
8. Laitinen, O. H., Hytonen, V. P., Nordlund, H. R., and Kulomaa, M. S. (2006) Genetically engineered avidins and streptavidins. *Cell Mol. Life Sci.* **63,** 2992–3017.
9. Richards, R. G. and ap Gwynn, I. (1995) Microjet impingement followed by scanning electron microscopy as a qualitative technique to compare cellular adhesion to various biomaterials. *Cell Biol. Int.* **19(12),** 1015–1024.

Chapter 5

3D Sample Preparation for Orthopaedic Tissue Engineering Bioreactors

Sarah H. Cartmell, Sarah Rathbone, Gemma Jones, and L. Araida Hidalgo-Bastida

Abstract

There are several types of bioreactors currently available for the culture of orthopaedic tissue engineered constructs. These vary from the simple to the complex in design and culture. Preparation of samples for bioreactors varies depending on the system being used. This chapter presents data and describes tried and tested methodologies for the preparation of 3D samples for a Rotatory Synthecon Bioreactor (Cellon), a plate shaker, a perfusion system, and a Bose Electroforce Systems Biodynamic Instrument for the in vitro culture of bone and ligament tissue.

Key words: Synthecon, Biodynamic Instrument, Perfusion, Bioreactor, 3D sample preparation, In vitro

1. Introduction

Preparing 3D samples for perfusion culture involves challenges at different levels. Firstly, widely used sterilisation methods such as steam (autoclaving), gamma rays, and ethylene oxide are not suitable for some materials involved in scaffold and bioreactor fabrication. Alternative sterilisation methods suitable for biomaterials have been designed for this application (1).

Establishing the best protocol to seed efficiently is the next step. Although static seeding has been used as a standard in 3D constructs, dynamic regimes such as shaking, centrifuge, and perfusion seeding have demonstrated better results regarding efficiency and homogeneity (2). Jones and Cartmell demonstrated that effective cell adherence and viability could be obtained using plate shaker and rotatory bioreactor regimes (2). Detailed information regarding these findings is described in this chapter.

John W. Haycock (ed.), *3D Cell Culture: Methods and Protocols*, Methods in Molecular Biology, vol. 695,
DOI 10.1007/978-1-60761-984-0_5, © Springer Science+Business Media, LLC 2011

Bioreactors have been designed for improving the growth of cells using perfused conditions (3). Configurations of flow-through perfusion bioreactors have been predicted (4, 5). Optimising cell seeding in these perfusion bioreactors includes Wendt et al. work on oscillating bi-directional perfused seeding, which indicates that perfusion bioreactors can be used to seed homogeneously through 3D scaffolds (6). Another advantage of perfusion seeding, its high reproducibility, has also been proved by Timmins et al. (7). Alvarez-Barreto et al. also reported increased seeding efficiencies and homogeneous distribution of cells when seeding by perfusion (8). Moreover, oscillatory perfusion demonstrated to promote stronger cell adhesion (8).

More complex bioreactors that have the capability to apply mechanical stimulus at the same time as providing nutrients to the developing construct also require specific seeding regimes. We describe in the following chapter our tried and tested methodology for seeding cells onto constructs prior to loading in a Bose Biodynamic Instrument.

In order to assess the best seeding regime, appropriate assays to quantify the efficacy of these are needed. These tests will equally evaluate the outcome of the perfusion culture of 3D constructs.

There are several methods to assess cell viability, proliferation, and concentration. Picogreen DNA quantification allows the determination of cell number and concentration at different time points. The calcein or live/dead assay is a two-colour fluorescence method that measures cell viability by simultaneously determining the number of live and dead cells. Live cells are coloured as intracellular esterases convert to fluorescent green the calcein acetoxymethyl (calcein AM) permeated into the cell through intact membranes. Dead cells are recognised as damaged membranes allow the reaction of ethidium homodimer-1 and nucleic acids resulting in an intense fluorescent red colour. This method can be used to quantify live and dead cells or to visualise them under a fluorescent microscope to evaluate morphology (1, 9).

Scanning electron microscopy (SEM) has been used not only to characterise material surfaces but also to analyse cell morphology, distribution, and coverage. In the case of scaffolds, it is a concern that the biological performance may be altered by the material characteristics and so this information allows to modify the material to adapt for the needed response. As this type of microscopy requires a dried and metal-coated sample, it must be considered that this technique may alter the morphology, especially on materials that undergo changes in hydrated conditions (10).

2. Materials

2.1. Cell Culture Reagents: Perfusion Bioreactor/Plate Shaker/Synthecon/ Bose Biodynamic Instrument

1. MG63 human osteoblast – like cells (perfusion bioreactor/ plate shaker/Synthecon).
2. L929 murine fibroblast cell line (Bose Biodynamic Instrument).
3. Culture medium: low-glucose Dulbecco's Modified Eagle's Medium (DMEM) (perfusion bioreactor/plate shaker/ Synthecon), high-glucose DMEM w/o l-glutamine, with sodium pyruvate (Bose Biodynamic Instrument), 10% foetal bovine serum, 2 mM glutamine, 100 U/mL penicillin, 100 μg/mL streptomycin, and 0.25 μg/mL amphotericin B.
4. Trypsin–EDTA solution.
5. Phosphate-buffered saline (PBS).
6. Tissue culture flasks.
7. Rectangular non-tissue culture treated four-well tissue culture dishes (Bose Biodynamic Instrument).

2.2. Scaffold Fabrication, Sterilisation, and Conditioning: Perfusion Bioreactor

1. The porous hydroxyapatite scaffold used in the perfusion bioreactor in these studies was manufactured by Dr. Irene Turner and Dr. Jon Gittings (from the Department of Engineering and Applied Science, University of Bath, Bath, BA27AY, UK, e-mail I.G.Turner@bath.ac.uk) using hydroxyapatite powder (TCP130, Thermophos UK Ltd).
2. Autoclave.
3. Culture medium: low-glucose DMEM, 100 U/mL penicillin, 100 μg/mL streptomycin, and 0.25 μg/mL amphotericin B.

2.3. Scaffold: Plate Shaker/Synthecon

1. Spongostan Standard (Johnson and Johnson) – an absorbable gelatin sponge with a pore size of 100–300 μm, cut under sterile conditions to 1 × 1 × 0.5 cm rectangular dimensions scaffolds using a sharp sterile blade.

2.4. Scaffold: Bose Biodynamic Instrument

1. Phosphate-based degradable glass fibre/polycaprolactone film composites (manufactured by Giltech Ltd) cut to 1 × 6 cm strips.

2.5. Scaffold Seeding: Perfusion Bioreactor

1. Supplemented low-glucose DMEM with FCS, l-glutamine, and antibiotics/antimycotics solution as described in Subheading 2.1.
2. Trypsin–EDTA solution.
3. PBS.

2.6. Picogreen DNA Assay: Perfusion Bioreactor/Plate Shaker/Synthecon

1. Picogreen Quant-iT dsDNA assay kit.
2. Lysis buffer prepared using 1× TE buffer from kit with 1% Triton X-100.
3. 96-Well plates.
4. Fluorescence plate reader.

2.7. Confocal with Live/Dead Stain: Perfusion Bioreactor/Plate Shaker/Synthecon/Bose Biodynamic Instrument

1. Live/Dead® viability/cytotoxicity kit for mammalian cells (Molecular Probes, Paisley, UK): ethidium homodimer and calcein AM diluted to 4 and 2 μM, respectively, using PBS.
2. Olympus IX 71 confocal scanning microscope, Fluoview 300 software – 10× objective.
3. PBS.

2.8. SEM: Bose Biodynamic Instrument

1. Glutaraldehyde solution for electron microscopy, 70%.
2. Ethanol solutions of 70, 90, 95, and 100% (v/v).
3. Critical point dryer.
4. Gold coater.
5. SEM.

2.9. Bioreactors/Instruments Used for 3D Seeding (**See **Figs. 1 and 2)

1. Perfusion bioreactor.
2. Synthecon.
3. Plate shaker.
4. Roller.
5. Bose Biodynamic Instrument.

3. Methods

3.1. Cell Culture Prior to Cell Seeding: Perfusion Bioreactor/Plate Shaker/Synthecon/Bose Biodynamic Instrument

1. MG63 (for perfusion bioreactor/Synthecon/plate shaker) or L929 cells (for Bose Biodynamic Instrument), both commercially obtained from ECACC, were cultured in 75-cm^2 cell culture flasks at 37°C in a humidified atmosphere 5% CO_2 in air (to 80% confluency) prior to passaging to increase cell number.
2. Culture medium: low-glucose (for MG63 cells) or high-glucose (for L929 cells) DMEM, 10% (v/v) foetal calf serum, 2 mM glutamine, and 100 U/mL penicillin, 100 μg/mL streptomycin, and 0.25 μg/mL amphotericin B – changed every 2–3 days.
3. Serial passages were made by trypsination, centrifugation, and re-suspension of pellets of sub-confluent monolayers and then re-plated at a dilution of 1:3.

3.2 Scaffold Fabrication and Conditioning: Perfusion Bioreactor

1. 15-mm diameter × 10-mm height hydroxyapatite (HA) scaffolds were manufactured by Dr. Irene Turner and Dr. Jon Gittings (University of Bath).
2. A commercially available hydroxyapatite powder was used to fabricate the HA scaffolds.
3. The composition of the powder was precipitated to calcium phosphate with a CaO/P_2O_5 weight ratio of 1.30 in order to achieve the molar ratio of ~1.67 Ca:P for the formation of HA.
4. The porous nature of the scaffold was created using a reticulated foam technique.
5. The dense cap at the top of the scaffold was created by "capping" the foam with tissue paper prior to sintering with the calcium phosphate powder.
6. The scaffolds were sterilised by autoclaving. Sterile HA scaffolds were stored at room temperature until use.

3.3. Scaffold Fabrication and Conditioning: Plate Shaker/Synthecon

1. Spongostan Standard, an absorbable gelatin sponge (already sterile from supplier) with a pore size of 100–300 μm, was cut under sterile conditions to 1 × 1 × 0.5 cm rectangular dimensions scaffolds using a sharp sterile blade.
2. These scaffolds were pre-wetted* by rolling overnight in culture medium (plus 10% FCS and antibiotics) before use.

3.4. Scaffold Fabrication and Conditioning: Bose Biodynamic Instrument

1. Water soluble glass fibre/PCL scaffolds were produced by solvent casting at Giltech Ltd., cut into rectangles of 1 cm × 6 cm.
2. Samples were then plasma treated and gamma sterilised prior to cell seeding.

3.5. Scaffold Seeding: Perfusion Bioreactor

1. MG63 cells concentrated cell suspensions of 1×10^5 cells per sample were prepared.
2. Cells were added to the HA scaffolds either statically or dynamically using the bioreactor using the following regimes:
   (a) 100,000 cells were suspended in 60 μL medium (containing FCS and antibiotics) and were placed onto the HA construct with the "cap top" facing downwards in a static tissue culture plastic bijou. 5 mL of medium (containing FCS and antibiotics) were then added directly to the samples and they were then incubated at 37°C, 5% CO_2.
   (b) 100,000 cells were suspended in 60 μL medium (containing FCS and antibiotics) and were placed onto the HA construct with the "cap top" facing downwards in a static tissue culture plastic bijou. The samples were then incubated at 37°C, 5% CO_2 for 2 h to allow cells to adhere in static

culture. After 2 h, 5 mL of medium (containing FCS and antibiotics) were then added directly to the samples. Samples were then returned to incubator for culture.

(c) HA constructs with the "cap top" facing downwards were placed into a stainless steel bioreactor (see Figs. 1 and 3). 100,000 cells were suspended in 5 mL of medium (containing FCS and antibiotics) and were placed into the reservoir feeding the scaffold in the bioreactor. Cells were perfused through the scaffold/bioreactor chamber at a flow rate of 0.01 mL/min and oscillated every 10 min for a 2-h period at 37°C, 5% CO_2. After 2 h, the samples were removed from the bioreactor and cultured in static bijous in 5 mL medium (containing FCS and antibiotics). Samples were then returned to incubator for culture.

(d) HA constructs with the "cap top" facing downwards were placed into stainless steel bioreactor (see Figs. 1 and 3). 100,000 cells were suspended in 60 μL of media (containing FCS and antibiotics) and were placed directly onto the scaffold in the bioreactor. The samples were incubated for 2 h at 37°C, 5% CO_2. After 2 h, 5 mL medium (containing FCS and antibiotics) were then perfused through the scaffold/bioreactor chamber at a flow rate of 0.01 mL/min and oscillated every 10 min for a further 2-h period at 37°C, 5% CO_2. After this 4-h period, the samples were removed from the bioreactor and either cultured in static bijous in 5 mL medium (containing FCS and antibiotics) for the 4-day analysis or analysed immediately for the 4-h analysis.

3. At 4 h or 4 days, cell-seeded constructs were removed and analysed for picogreen DNA quantification and live dead confocal imaging.

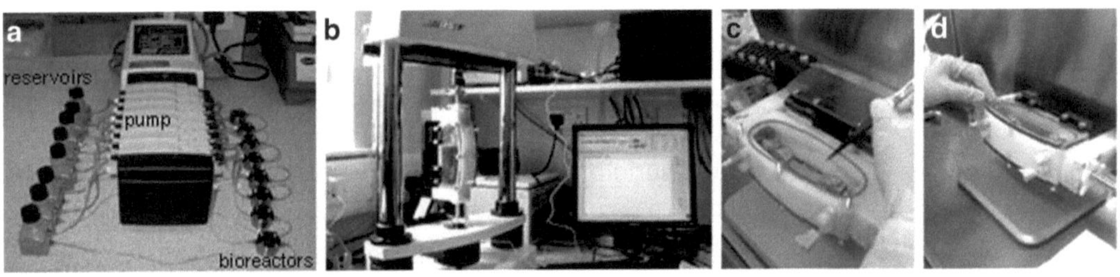

Fig. 1. Bioreactors used in the described methodology: (**a**) Eight stainless steel perfusion bioreactors set up in parallel with separate culture medium reservoirs for each bioreactor. (**b**) Bose Biodynamic Instrument set up in ELF 3200 mechanical tester during loading regime of PCL/glass fibre composite. (**c**) Removing the PCL/glass fibre composite film from the Bose biodynamic chamber in a sterile laminar flow hood. (**d**) Setting up the Bose Biodynamic Instrument in the sterile laminar flow hood.

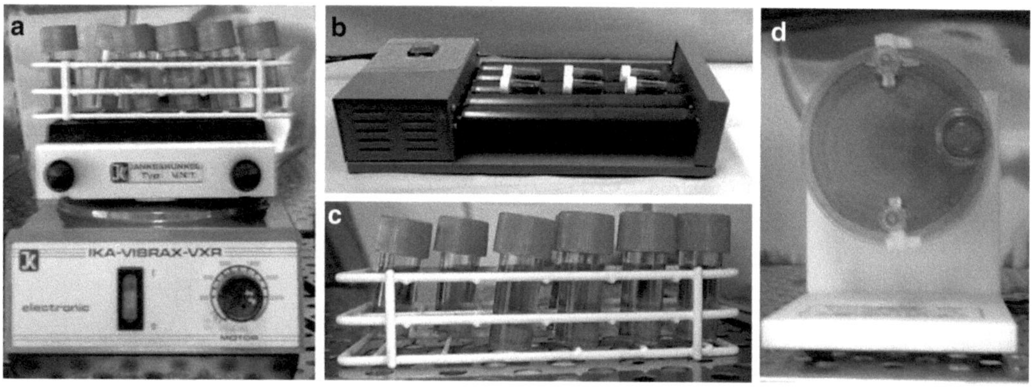

Fig. 2. Bioreactors used in the described methodology: (**a**) Plate shaker with bijou vessels containing culture medium and cell-seeded spongostan scaffolds. (**b**) Roller with bijou vessels containing culture medium and cell-seeded spongostan scaffolds. (**c**) Static culture with bijou vessels containing culture medium and cell-seeded spongostan scaffolds. (**d**) Synthecon (Cellon) rotatory system.

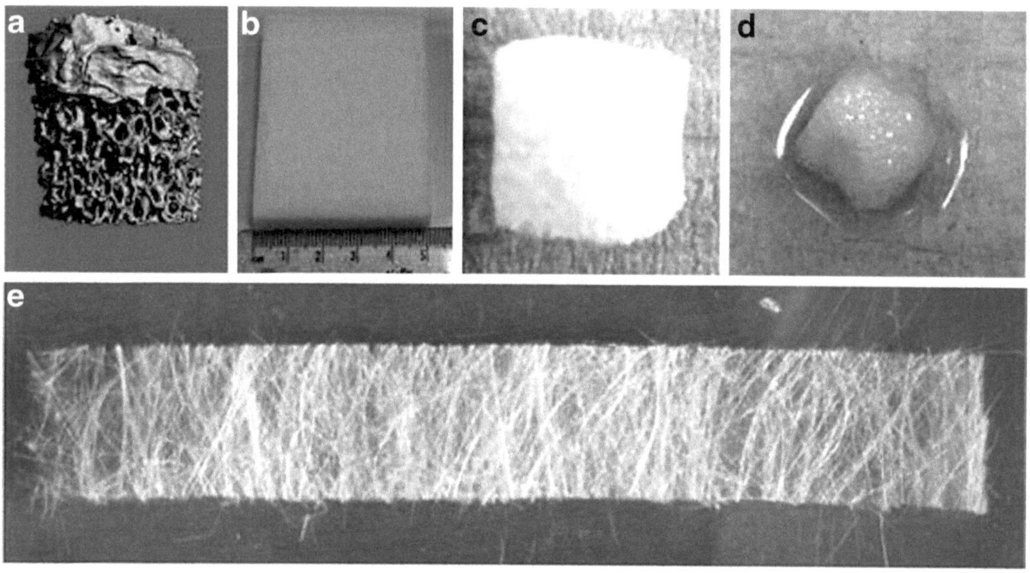

Fig. 3. Scaffolds used in the described methodology: (**a**) Hydroyapatite scaffold manufactured at University of Bath by Dr. Irene Turner and Dr. Jon Gittings. (**b**) Spongostan Standard (Johnson and Johnson) prior to being cut to size. (**c**) Spongostan Standard (Johnson and Johnson) cut to size. (**d**) Spongostan Standard scaffold seeded with 50 μL volume cell suspension. (**e**) PCL/glass fibre composite (manufactured at Giltech Ltd.) film used in Bose biodynamic chamber.

3.6. Scaffold Seeding: Synthecon/Plate Shaker

1. 400,000 MG63 cells were added to each Spongostan Standard scaffold in bijou containers (a sterile vessel that can hold up to 8 mL total volume with filter cap to allow sterile gas exchange) in the following two ways:
   (a) 50 μL of cell suspension in medium was pipetted directly onto the top of the scaffold, and left for 1 h before 5 mL medium applied.

(b) 5 mL of cell suspension in medium was placed in a bijou container. The scaffold was then placed into this cell suspension.

2. These samples were then cultured for 3 days in five different conditions:

   (a) Static

   (b) Plate shaker 60 rpm*

   (c) Plate shaker 120 rpm*

   (d) Roller 33 rpm*

   (e) Rotating bioreactor 9 rpm*

   *rpm speeds are approximate.

3. In the case of rotating bioreactor, in group A, the ten-scaffold samples were taken after 1 h of cell adherence, from their bijou containers and placed into a Synthecon vessel (the "reactor") with 50 mL of medium (plus 10% FCS and antibiotics as described above).

4. In group B, the 4×10^6 ($10 \times 400{,}000$) MG63 cells were suspended in 50 mL medium (plus FCS and antibiotics) and placed into the Synthecon vessel. Ten pre-wetted spongostan scaffolds were also then placed into this vessel (giving an equivalent of 5 mL medium per scaffold). The vessels in each group A and B were then rotated for 3 days at approximately 9 rpm.

5. The plate shaker moved in a horizontal circular motion with a 4-mm diameter agitation stroke. The plate shaker had an electronic feedback speed control to ensure constant delivery of 60 or 120 rpm. The roller rolled the samples at 33 rpm around the centre of the bijou axis at a 16-mm amplitude.*

6. Each of the experimental groups above contained six-sample repeats in each group with the exception of the rotating bioreactor group where ten-sample repeats were included (increased sample repeat number due to the method of bioreactor use). For example, the plate shaker 120 rpm group (Shaker 120) contained 12 spongostan scaffolds, each seeded with 400,000 MG63 cells. Six of these scaffolds had the cells applied in a 50-μL volume and the other six scaffolds had the cells applied in a 5-mL volume. Five of the medium, bijou, and scaffold portions of each of these individual samples were analysed separately for DNA content. The remaining one scaffold was analysed using laser scanning confocal microscopy for cell viability.

3.7. Scaffold Seeding and Culture: Bose Biodynamic Instrument

1. L929 fibroblasts were suspended in culture medium (containing FCS and antibiotics) and seeded at a density of 100,000 per 1×1 cm section of the 1×6 cm composite.In order to seed these cells, the PCL/glass fibre composite films

were laid flat in non-tissue culture treated rectangular four-well tissue culture dishes. The cell suspension was added via a pipette dropwise over the composite. The cells were allowed to adhere for 45 min prior to the addition of 10 mL of culture medium.

2. All cell/scaffold composites were cultured for 12 days. Half of the samples was cultured statically for this period. The other half was maintained in static culture until day 7, when they was transferred to the Bose biodynamic chamber (see Fig. 1). Half of the culture medium (total culture medium was 200 mL) was changed every 3 days throughout the incubation period.
3. Cell proliferation was analysed using a DNA assay at 4 h and 7 days. Cell viability (confocal) and morphology (SEM) were analysed at 7 days and 12 days, respectively.
4. Samples that were placed into the Bose biodynamic chamber at day 7 were loaded by placing in an ELF3200 Bose tensile testing machine and cyclically strained at 1% tensile strain for 1 h each day for 5 days (days 7–12) at 0.5 Hz using a sinusoidal waveform.

3.8. Picogreen DNA Assay: Perfusion Bioreactor

1. A picogreen DNA quantitation assay was used to quantitate the number of cells located either: (a) on the construct or (b) still suspended in the medium.
2. For group (a), scaffolds were removed from the bijou container/bioreactor and placed directly into 1 mL of 1% Triton X-100 solution.
3. In group (b), the medium was removed, centrifuged, and the cells resuspended in 1 mL 1% Triton X-100 solution. Each sample group was vortexed to ensure complete cell lysis and therefore full DNA detection.
4. The picogreen protocol provided by the manufacturer was followed. Standards were made using TE buffer (200 mM Tris–HCL, 20 mM EDTA, pH 7.5) and Triton X-100. Standards and experimental samples in 100 μL volumes were added to a 96-well plate, along with 100 μL picogreen solution. After mixing, the plate was read on a fluorescence plate reader at 435–529 nm.

3.9. Picogreen DNA Assay: Synthecon/Plate Shaker/Bose Biodynamic Instrument

1. A picogreen DNA quantitation assay was used to quantitate the number of cells located either (a) on the construct, (b) on the inside surface of the bijou container, and (c) still suspended in the medium.
2. For group (a), scaffolds were removed from the bijou container and placed directly into 1 mL of 1% Triton X-100 solution.

3. In group (c), the medium was removed, centrifuged, and the cells resuspended in 1 mL 1% Triton X-100 solution, also another 1 mL 1% Triton X-100 solution placed in the bijou container for group (b). Each sample group was vortexed to ensure complete cell lysis and therefore full DNA detection.
4. The picogreen protocol provided by the manufacturer was followed. Standards were made using TE buffer (200 mM Tris–HCL, 20 mM EDTA, pH 7.5) and Triton X-100. Standards and experimental samples in 100 μL volumes were added to a 96-well plate, along with 100 μL picogreen solution. After mixing, the plate was read on a fluorescence plate reader at 435–529 nm.
5. The picogreen data was then calculated by adding up the sum of the fluorescence data from each of the locations (bijou, scaffold, and medium) for each individual sample. This sum was then given to equal 100% of the DNA for that particular sample.
6. The location of cells in each sample was then calculated in percentiles for each of the positions analysed – either in the bijou container, in the scaffold, or in the medium. A number of five cell/scaffold samples from each group was analysed for picogreen (see Figs. 4–6).

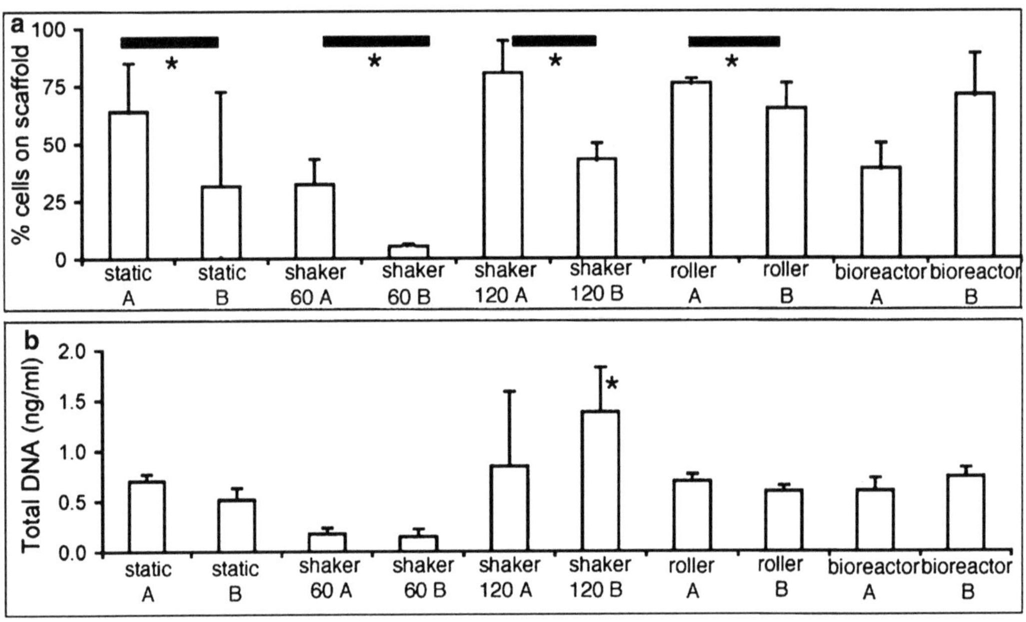

Fig. 4. The average amount of cells remaining on the scaffold after 3 days of culture as measured by a picogreen DNA assay. *Static*=static culture for 3 days, *shaker 60*=samples shaken on a plate shaker at 60 rpm for 3 days, *shaker 120*=samples shaken on a plate shaker at 120 rpm for 3 days, *roller*=samples rolled at 33 rpm for 3 days, *bioreactor*=samples cultured in a Synthecon rotatory bioreactor at 9 rpm for 3 days. (**a**) Scaffolds seeded with 50 μL cell suspension. (**b**) Samples seeded with cells suspended in 5 mL, except for Synthecon samples (bioreactor) as the cells were suspended in 50 mL volumes. Groups indicated by *arrows* and *asterisks* are significantly different, $p<0.01$. Figure reproduced from (2) (© Società Italiana Biomateriali – with permission from the publisher).

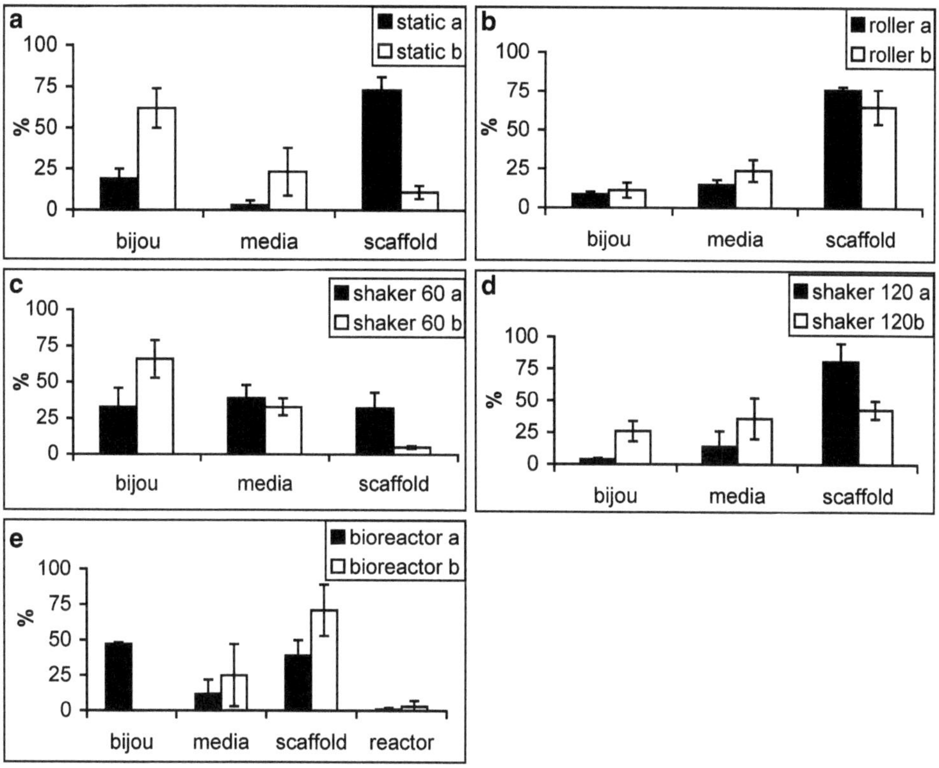

Fig. 5. Graphs (**a**–**e**) demonstrate the average percentage of cells in each of the following three locations after a 3-day culture period as measured by a picogreen DNA assay: *bijou* = cells adhered to the inside of the bijou container housing the scaffold and cells for 3 days, *medium* = cells suspended in the medium after a 3-day culture period, and *scaffold* = percentage of cells adhered to spongostan scaffold after 3-day culture period. In graph (**e**) – "*bijou*" refers to the initial seeding container for the small volume cell suspension, "*reactor*" refers to the Synthecon vessel that housed the cells plus ten scaffolds for 3 days. (**a**) *Static A* = 50 μL initial cell seeding volume in static culture for 3 days, *static B* = 5 mL initial cell seeding volume in static culture for 3 days. (**b**) *Roller A* = 50 μL initial cell seeding volume rolled at 33 rpm for 3 days, *roller B* = 5 mL initial cell seeding volume rolled at 33 rpm for 3 days. (**c**) *Shaker 60 A* = 50 μL initial cell seeding volume shaken at 60 rpm for 3 days, *shaker 60 B* = 5 mL initial cell seeding volume shaken at 60 rpm for 3 days. (**d**) *Shaker 120 A* = 50 μL initial cell seeding volume shaken at 120 rpm for 3 days, *shaker 120 B* = 5 mL initial cell seeding volume shaken at 120 rpm for 3 days. (**e**) *Bioreactor A* = 50 μL initial cell seeding volume then transferred to 50 mL medium volume and rotated at 9 rpm for 3 days, *bioreactor B* = cells seeded into 50 mL medium volume with all six scaffolds at once and rotated at 9 rpm for 3 days. Figure reproduced from (2) (© Società Italiana Biomateriali – with permission from the publisher).

3.10. Confocal Microscopy for Live/Dead Stained Cells: Perfusion Bioreactor/Plate Shaker/Synthecon/Bose Biodynamic Instrument

1. Constructs were incubated for 30 min with 2 μmol/L calcein AM and 4.0 μM ethidium homodimer.
2. Following incubation, samples are visualised under a fluorescent confocal microscope. Figure 7 shows a typical live/dead image of a bone construct after 4 days of static culture. Figures 8 and 9 show typical live/dead images of cells on spongostan scaffolds and PCL/glass composites, respectively.

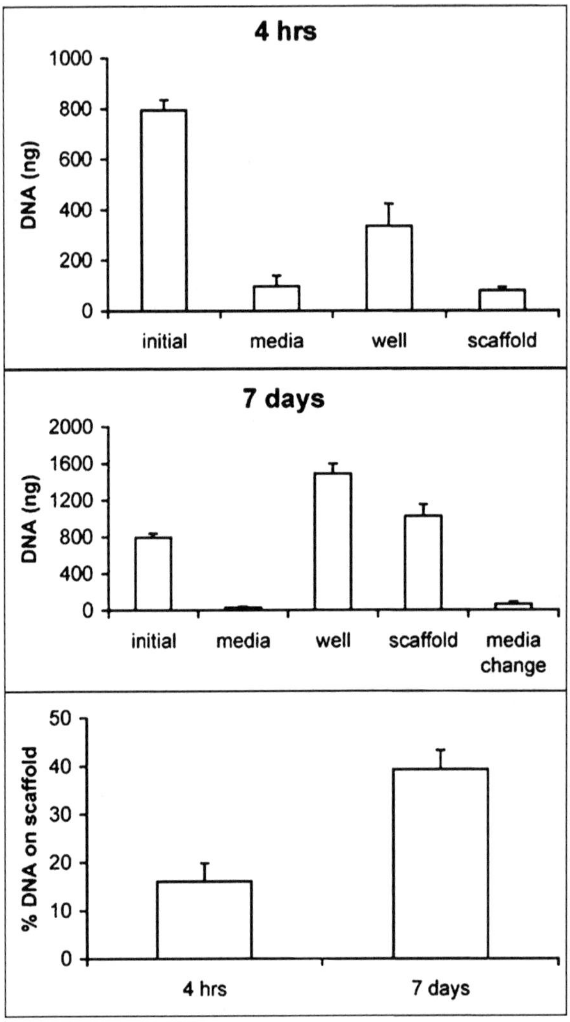

Fig. 6. Graphs demonstrate DNA quantity from cell population on PCL/glass fibre composites. *Initial*: starting cell population prior to placing on scaffold, *medium*: cells in the culture medium, *well*: cells adhered to well, *scaffold:* cells adhered to scaffold. *Top*: Data from 4 h seeded/cultured scaffolds. *Middle*: Data from 7-day seeded/cultured scaffolds. *Bottom*: 4-h and 7-day data demonstrated as percentage of total cells seeded on scaffold after static culture prior to incubation in Bose Biodynamic Instrument.

3.11. SEM: Bose Biodynamic Chamber

1. PCL/glass fibre constructs containing cells are processed prior to imaging under SEM. Constructs were fixed in 3 mL of 2.5% glutaraldehyde and stored at 4°C in a sealed container for 1 h.
2. Glutaraldehyde was then removed and samples are washed in PBS.
3. Samples were then serially dehydrated through ethanol solutions of 70, 90, 95, and 100% (v/v) before incubating with critical point drying.

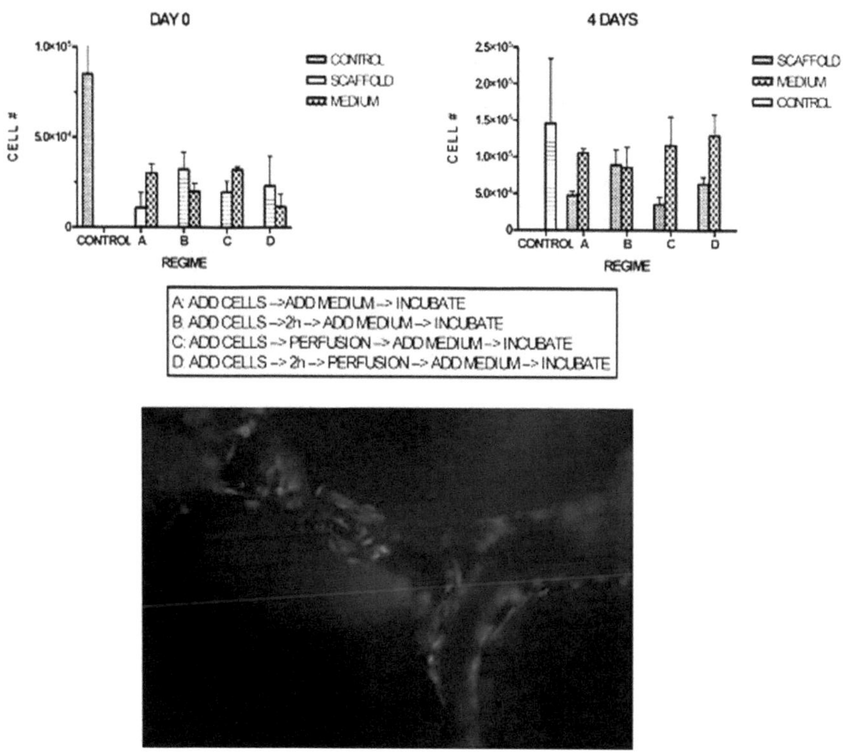

Fig. 7. Seeding efficiency and cell proliferation of four seeding regimes in a perfusion bioreactor: (A) static seeding in 60 μL followed by the addition of 5-mL culture medium, (B) static seeding in 60 μL incubated for 2 h before adding 5-mL culture medium, (C) oscillatory perfused seeding (0.01 mL/min) in 5-mL culture medium followed by static culture, and (D) static seeding in 60 μL incubated for 2 h before oscillatory perfusion and then static culture, at (A) day 0 and (B) day 4. BOTTOM IMAGE shows live cell distribution of a statically cell seeded and cultured HA construct after 4 days (all cells were live).

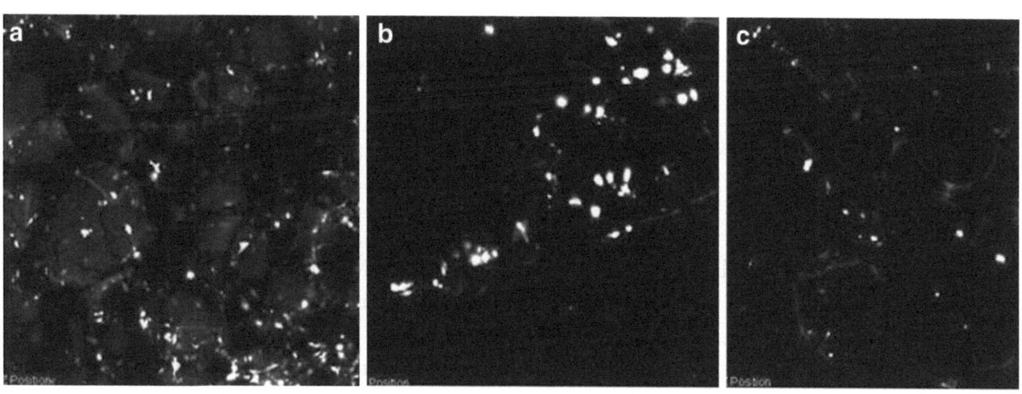

Fig. 8. Representative images of fluorescent cell viability. Cells are adhered to spongostan scaffold after 3 days of culture. All cells visualised were viable. (**a**) Scaffolds cultured in Synthecon bioreactor for 3 days with large initial cell seeding density (4×10^6 cells in 50 mL plus 10 spongostan scaffolds). (**b**) Scaffolds rolled at 33 rpm for 3 days with 50 μL initial cell seeding volume. (**c**) Scaffolds cultured in static conditions and cells seeded in 5 mL initial volume. Figure reproduced from (2) (Società Italiana Biomateriali – with permission from the publisher).

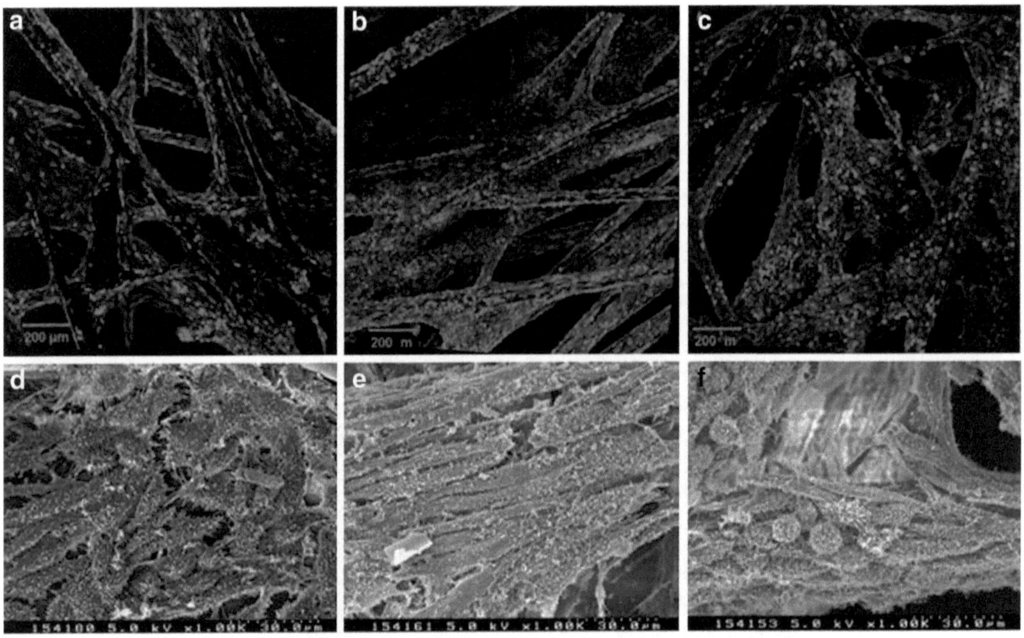

Fig. 9. Live/dead and SEM images of L929 fibroblasts seeded onto PCL/glass fibre composite films and cultured in static or dynamic conditions. All cells visualised were live. (**a**) Confocal image of 7-day static culture. (**b**) Confocal image of 12-day static culture. (**c**) Confocal image of 12-day dynamic culture (static culture for 7 days and then 1 h/day 1% strain 0.5% Hz loading using Bose Biodynamic Instrument). (**d**) SEM image of 7-day static culture. (**e**) SEM image of 12-day static culture. (**f**) SEM image of 12-day dynamic culture (static culture for 7 days and then 1 h/day 1% strain 0.5% Hz loading using Bose Biodynamic Instrument).

4. Samples were sputter-coated with gold at an argon pressure of 13 Pa for 3 min at a current rate of 14 mA, prior to SEM examination.

5. The constructs were then imaged on an SEM with an accelerating voltage of 10 and at a spot size of 40. Random areas were selected to evaluate the samples.

4. Notes

1. Prior to cell seeding, the scaffold may either be dry or it may be "pre-wetted" with culture medium first. Pre-wetting is performed to remove air bubbles and to allow proteins from the foetal calf serum in the medium to adhere to the surface of the scaffolds. This general protein coating will then give the cells an anchor to attach to, which is especially beneficial in the case of non-surface-treated hydrophobic polymer scaffolds. Pre-wetting the scaffold usually involves removing excess medium by dabbing the scaffold with a sterile absorbent wipe prior to cell seeding. The cells may then be placed directly onto the scaffold in a small (typically 50 μL) or large (typically 5 mL)

suspension in the medium. After placing the cells onto the scaffold, the next step can either be static (where the samples are left in tissue culture plastic container in the incubator) or dynamic (where the samples are gently agitated to further improve cell migration into the scaffold).

2. The Stuart roller (SRT1) from SLS (cat. No. 403/0238/02) had been modified by shortening the length of the roller in order for its size to be accommodated in the incubator. A cooling system of tubing wrapped around the motor enabled cold water from a tap supply to be administered to the roller during operation to prevent a change from the constant 37°C in the incubator.

Acknowledgements

The authors wish to acknowledge a private local charity, the Royal Society, and the BBSRC (BBS/S/M/2006/13131 and BB/F013892/1) for financial assistance. The authors also wish to acknowledge the contribution of Mr. Michael Vipin and Dr. Nikki Kuiper (Keele University, UK) for providing the perfusion bioreactor seeding efficiency data. We wish to acknowledge Swann Morton for assistance in gamma sterilisation of the PCL/glass fibre composites and Dr. Alex Fotheringham at Herriot Watt University for plasma sterilisation. We also wish to acknowledge the contribution of the hydroxyapatite scaffold fabrication by Dr. Irene Turner and Dr. Jon Gittings at the University of Bath. Finally, we wish to acknowledge the financial and technical assistance of Mr. David Healy at Giltech Ltd. for the fabrication of the PCL/glass fibre scaffolds.

References

1. Stephens, J. S., Cooper, J. A., Phelan, F. R., Jr., and Dunkers, J. P. (2007) Perfusion flow bioreactor for 3D in situ imaging: investigating cell/biomaterials interactions. *Biotechnol. Bioeng.* 97, 952–961.
2. Jones, G. and Cartmell, S. H. (2006) Optimization of cell seeding efficiencies on a three-dimsional gelatin scaffold for bone tissue engineering. *J. Appl. Biomater. Biomech.* 4, 172–180.
3. Cartmell, S. H., Porter, B. D., Garcia, A. J., and Guldberg, R. E. (2003) Effects of medium perfusion rate on cell-seeded three-dimensional bone constructs *in vitro*. *Tissue Eng.* 9, 1197–1203.
4. Porter, B., Zauel, R., Stockman, H., Guldberg, R., and Fyhrie, D. (2005) 3-D computational modeling of media flow through scaffolds in a perfusion bioreactor. *J. Biomech.* 38, 543–549.
5. Whittaker, R. J., Booth, R., Dyson, R., Bailey, C., Parsons Chini, L., Naire, S., Payvandi, S., Rong, Z., Woollard, H., Cummings, L. J., Waters, S. L., Mawasse, L., Chaudhuri, J. B., Ellis, M. J., Michael, V., Kuiper, N. J., and Cartmell, S. (2009) Mathematical modelling of fibre-enhanced perfusion inside a tissue-engineering bioreactor. *J. Theor. Biol.* 256, 533–546.
6. Wendt, D., Stroebel, S., Jakob, M., John, G. T., and Martin, I. (2006) Uniform tissues engineered by seeding and culturing cells in 3D scaffolds under perfusion at defined oxygen tensions. *Biorheology* 43, 481–488.

7. Timmins, N. E., Scherberich, A., Fruh, J. A., Heberer, M., Martin, I., and Jakob, M. (2007) Three-dimensional cell culture and tissue engineering in a T-CUP (tissue culture under perfusion). *Tissue Eng.* 13, 2021–2028.
8. Alvarez-Barreto, J. F., Linehan, S. M., Shambaugh, R. L., and Sikavitsas, V. I. (2007) Flow perfusion improves seeding of tissue engineering scaffolds with different architectures. *Ann. Biomed. Eng.* 35, 429–442.
9. Johnson, I. (2002) Fuorescent staining of living cells, in *Microscopy and Histology for Molecular Biologists* (Kiernan, J. A., and Mason, I., Eds.), pp. 92–101, Portland Press, London.
10. Vlierberghe, S. V., Cnudde, V., Dubruel, P., Masschaele, B., Cosijns, A., Paepe, I. D., Jacobs, P. J., Hoorebeke, L. V., Remon, J. P., and Schacht, E. (2007) Porous gelatin hydrogels: 1. Cryogenic formation and structure analysis. *Biomacromolecules* 8, 331–337.

Chapter 6

Quantification of mRNA Using Real-Time PCR and Western Blot Analysis of MAPK Events in Chondrocyte/Agarose Constructs

David A. Lee, June Brand, Donald Salter, Oto-Ola Akanji, and Tina T. Chowdhury

Abstract

In vitro models of chondrocyte mechanobiology have been used to compare the intracellular signalling pathways altered in normal and osteoarthritis-affected cartilage. However, differences in the model system and type of loading configuration have led to complicated pathways. This chapter is a follow-on of previous studies from our group utilising 3D agarose as a physiological model to study mechanotransduction pathways. Experimental methods are described to assess targets at the protein and gene expression level by Western blot analysis and real-time PCR, respectively. This chapter provides a quantitative gene expression approach to explore the intracellular pathways activated by both mechanical loading and inflammatory mediators and examine upstream phosphorylation events. Ultimately, development of methods used to analyse mechano-sensitive pathways will provide important information for the identification of appropriate pharmacological and physiotherapeutic agents for the treatment of osteoarthritis.

Key words: Mechanotransduction, Chondrocyte, Agarose, MAPK, mRNA, cDNA, RT-qPCR, Western blotting

1. Introduction

Numerous studies demonstrate that chondrocyte function and anabolic/catabolic transcriptional activities are tightly regulated by distinct cell deformations (1–3). These studies utilised a number of in vitro models and bioreactor systems as tools to explore mechanotransduction pathways. In general, static compression inhibits the expression profiles and biosynthesis of matrix proteins, whereas dynamic compression stimulates expression of extracellular matrix components (4–9). Several groups including our own have shown

John W. Haycock (ed.), *3D Cell Culture: Methods and Protocols*, Methods in Molecular Biology, vol. 695, DOI 10.1007/978-1-60761-984-0_6, strong interplay between the signalling pathways activated by both mechanical loading and interleukin-1β (IL-1β) in 3D agarose gels and chondrocyte monolayers (10–15). There is now increasing evidence demonstrating involvement of the mitogen-activated protein kinases (MAPKs) in mediating the transcriptional response of chondrocytes to mechanical loading and/or IL-1β (16–20). The molecular mechanisms underlying the specific mechanotransduction pathways are complex and will vary depending on the type of mechanical stimuli and pathological environment of the tissue. Furthermore, the nature of the mechanical loading regime and type of model system will additionally determine whether mechanical signals will prevent or induce gene transcription of anabolic/catabolic mediators (21, 22). Ultimately, elucidation of the intracellular pathways in an appropriate model and bioreactor system will facilitate the successful identification of pharmacological and physiotherapeutic agents to treat osteoarthritis.

Previous chapters in this series have described methods for the preparation of 3D chondrocyte-seeded constructs and the assessment of strain transfer to the cells and biological response using biochemical methods (23–27). However, elucidation of complex intracellular signalling cascades induced by mechanical loading typically requires analysis at the molecular level, to assess changes in mRNA expression levels of target genes in agarose constructs (28–30). In addition, assessment of cellular protein is required to identify phosphorylation events involving MAPK signalling (28). However, the transfer of existing methods for these analyses to the agarose system is complicated by the low cell density and the presence of high levels of agarose within the 3D system. This chapter describes the experimental techniques employed for monitoring alterations of gene expression profiles by real-time quantitative polymerase chain reaction (qPCR) and phosphorylation events by Western blot analysis.

2. Materials

2.1. Isolation of Bovine Chondrocytes

1. Front feet from approximately 18-month-old steers (obtain from local slaughterhouse).
2. 70% (v/v) industrial methylated spirits diluted in water.
3. Dissection kit: Metal dissection tray, sterile scalpels fitted with no. 20, no. 11, and no. 15 blades.
4. Earle's balanced salt solution (EBSS).
5. Culture medium: Dulbecco's minimal essential medium supplemented with 20% (v/v) foetal calf serum, 2 mM L-glutamine, 20 mM HEPES, 50 U/mL penicillin/50 μg/mL streptomycin, and 150 μg/mL L-ascorbic acid. Sterilise by filtration through a 0.2-μm pore size filter. Aliquot and store at –20°C.

6. Pronase solution: 700 U/mL pronase in culture medium.
7. Collagenase solution: 100 U/mL collagenase type XI in culture medium.
8. Trypan Blue solution.
9. 35-mm diameter sterile petri dishes.
10. 50-mL capacity sterile polypropylene centrifuges tubes.
11. 70-μm pore size cell sieve.
12. Class II safety cabinet.
13. Roller mixer within an incubator or in a warm room at 37°C.
14. Centrifuge: Up to 2,000×*g*, compatible with 50-mL centrifuge tubes.
15. Haemocytometer.
16. Inverted microscope with phase-contrast facility.

2.2. Preparation and Culture of Chondrocyte/Agarose Constructs

1. Culture medium without 20% FCS supplemented with 1× ITS (Sigma-Aldrich, Poole, UK).
2. Isolated chondrocytes.
3. Agarose suspension: 6% (w/v) Agarose type VII in EBSS, sterilised by autoclaving.
4. Mould for preparing constructs (see Note 1).
5. Water bath, maintained at 37°C.
6. Roller mixer within an incubator or in a warm room at 37°C.
7. Compressive cell-strain bioreactor (see Note 2).

2.3. Protein Extraction

1. Lysis buffer: 1% Igepal, 1 mM activated Na_3VO_4 (see Note 3) and one protease inhibitor cocktail tablet (Roche Diagnostics, UK) dissolved in 10 mL of 0.01 M PBS. Store at 4°C.
2. Wash buffer: 1 mM activated Na_3VO_4 (see Note 3) in PBS. Store at 4°C.
3. Sterile needles (21 G, 0.8 mm).
4. Modified Protein Lowry assay kit with Folin's phenol reagent.
5. Microplate reader, e.g. MR5000 Dynatech, BMG FLUOstar Galaxy.
6. Prepare sample buffer in dH_2O with 0.5 M Tris–HCl, pH 6.8, 10 % (w/v) SDS, 1 M DTT, 0.1 M glycerol, 0.05 % (w/v) bromophenol blue. Store aliquots at –20°C.

2.4. SDS–Polyacrylamide Gel Electrophoresis

1. Separating buffer: 1.5 M Tris–HCl, 0.5% (w/v) SDS, pH 8.8. Store at room temperature.
2. Stacking buffer: 0.5 M Tris–HCl, 0.5% (w/v) SDS, pH 6.8. Store at room temperature.

3. Running buffer (10×): 125 mM Tris–HCl, 960 mM glycine, 0.5% (w/v) SDS. Store at room temperature.
4. Acrylamide/bis-acrylamide 40% solution for electrophoresis, 19:1. Store at 4°C.
5. TEMED. Store at room temperature.
6. Molecular weight markers (Amersham Biosciences, Buckinghamshire, UK). Store at -20°C.
7. Bio-Rad Protein Blotting cells (Bio-Rad, Mini Trans Blot Cell or equivalent).

2.5. Western Blotting for Active MAPK

1. Setup buffer: 25 mM Tris–HCl, 190 mM glycine, and 20% (v/v) methanol. Store at room temperature.
2. Transfer buffer: 25 mM Tris–HCl, 190 mM glycine, 20% (v/v) methanol, and 0.05% (w/v) SDS. The pH should be 8.3, but do not adjust. Refrigerate in advance and store at 4°C.
3. Nitrocellulose/PVDF transfer membranes (Millipore, Immobilon-P).
4. Blocking buffer: 1% (w/v) Non-fat dry milk (Marvel), 1× TBS, and 0.1% Tween-20 in TBS-T. Store at room temperature.
5. Wash buffer: 1× TBS, 0.1% Tween-20 in TBS-T. Store at room temperature.
6. Primary antibody: Polyclonal rabbit antibody for phospho-p38 MAP Kinase (Thr180/Tyr182) (New England Biolabs, Hertfordshire, UK). Store at -20°C. Use at 1:1,000 dilution in primary antibody dilution buffer.
7. Primary antibody dilution buffer: 1× TBS, 0.1% Tween-20 with 5% BSA.
8. Secondary antibody: Anti-rabbit IgG conjugated to horseradish peroxidase (HRP) and anti-biotin, antibody conjugated to HRP (New England Biolabs, Hertfordshire, UK). Store at -20°C.
9. Enhanced Chemiluminescence (ECL) Plus Western blotting detection system (Amersham Biosciences UK) and Bio-Max ML Film (Kodak).

2.6. Stripping and Reprobing Blots for Total MAPK

1. Stripping buffer: 62.5 mM Tris–HCl, pH 6.8, 2% (w/v) SDS. Store at room temperature. Warm to 70°C and add 100 mM β-mercaptoethanol.
2. Wash buffer: 0.1% (w/v) BSA in TBS-T. Store at room temperature.
3. Primary antibody: Control p38 MAP Kinase (Polyclonal rabbit IgG) (New England Biolabs, Hertfordshire, UK). Store at -20°C.

4. Anti-α tubulin (New England Biolabs, Hertfordshire, UK). Use as a loading control.

2.7. RNA Extraction

1. Buffer QG (Qiagen, Sussex, UK). Store at room temperature.
2. RNeasy® kit (Qiagen, Sussex, UK) Contains buffer RW1 and buffer RPE. Store at room temperature.
3. Isopropanol. Store at room temperature.
4. DNA-*free* DNase kit (Ambion, Applied Biosystems): Contains DNase buffer, DNase I enzyme, and DNase inactivating agent. Store at −20°C.
5. DNase/RNase-free water (Qiagen, Sussex, UK): Store at room temperature.
6. RNaseZAP (Sigma-Aldrich, Poole, UK): Store at room temperature.
7. DNase/RNase-free tips with filter, ranging from 2 to 1,000 μL and 1.5-mL microtubes (Sarstedt, UK).
8. Conventional table-top microcentrifuge, e.g. Sigma 1-15.
9. Nanodrop ND-1000 spectrophotometer (LabTech, UK).

2.8. Synthesis of First-Strand cDNA Using Reverse Transcriptase

1. Stratascript® First-Strand cDNA synthesis kit (Stratagene): Store all components at −20°C. After thawing, store the 2× Brilliant® II QPCR master mix at 4°C and use within 6 months.
2. For a fast protocol, use the Brilliant® II QRT-PCR, Affinity Script™ Two-Step Master Mix (Stratagene). The master mix contains a buffer with $MgCl_2$ and dNTPs that is specifically optimised for RT-qPCR performance.
3. Reference dye (see Note 4). Store at −20°C.
4. Spectrofluorometric thermal cycler (Techne, TC-312).
5. Nuclease-free PCR-grade water (Qiagen, Sussex, UK).

2.9. Quantitative PCR Amplification of cDNA

1. Brilliant SYBR Green RT-qPCR Master Mix Kit or the Brilliant RT-qPCR kit depending on the detection assay (Stratagene). It is also possible to purchase a single kit containing all the components for RT-qPCR for either SYBR Green or probe-based assays, e.g. Taqman, Molecular Beacon chemistries.
2. Mx3000P qPCR instrument (Stratagene).
3. Striptube Picofuge (Stratagene).
4. 96-well optical plate (Stratagene).
5. Mx3000P optical strip tubes (Stratagene).
6. Mx3000P optical strip caps (Stratagene).

3. Methods

3.1. Isolation of Bovine Chondrocytes

1. Clean bovine front feet with water and immerse in 70% IMS for 5 min to sterilise skin.
2. Within the Class II safety cabinet, expose the metacarpalphalangeal joint and remove full-depth cartilage slices from the proximal articular surfaces of the joint using a no. 15 blade (see Note 5). Transfer the slices to a 60-mm Petri dish containing 8 mL of EBSS.
3. Aspirate EBSS and dice the tissue into pieces approximately 1 mm^3 and transfer to a 50-mL centrifuge tube containing 10 mL of pronase solution (see Note 3). Incubate on rollers for 1 h at 37°C.
4. Allow the tissue pieces to settle and remove pronase solution. Add 30 mL of collagenase solution and incubate on rollers for 16 h at 37°C (see Notes 6 and 7).
5. Allow any remaining tissue pieces to settle and carefully remove the collagenase solution containing isolated cells. Filter the solution through a 70-μm pore size sieve into a fresh 50-mL centrifuge tube.
6. Centrifuge at 2,000 × *g* for 5 min to pellet cells. Aspirate the supernatant and resuspend the cells in 10 mL of culture medium using a 10-mL sterile graduated pipette. Repeat this process twice.
7. Add 50 μL of cell suspension to 50-μL Trypan Blue solution, mix, and add resultant suspension to a haemocytometer. Determine total cell number and cell viability, and adjust cell concentration to 8×10^6 cells/mL.

3.2. Preparation and Culture of Chondrocyte/Agarose Constructs

1. Part-assemble the mould used for preparation of chondrocyte/agarose constructs (see Note 1).
2. Add the cell suspension, adjusted to 8×10^6 cells/mL, to an equal volume of agarose suspension. Mix cell/gel solution gently on rollers at 37°C.
3. Aliquot the chondrocyte/agarose suspension into individual wells of the mould and complete the assembly by adding the second glass slide and clamping the three components of the mould to ensure that it is sealed.
4. Incubate for 10 min at 4°C.
5. Carefully dismantle the mould to allow removal of the gelled chondrocyte/agarose constructs.
6. Transfer chondrocyte/agarose constructs into 24-well plates and equilibrate in culture for 24 h at 37°C.

7. Cytokines, e.g. IL-1β, or chemical inhibitors, e.g. SB203580, and materials for the compressive cell-strain bioreactor system are made ready (see Note 2).
8. Culture constructs either under free-swelling conditions or in the compressive cell-strain bioreactor system.
9. At the end of the culture experiment, transfer constructs into 1.5-mL microfuge tube and snap-freeze in liquid nitrogen. Store constructs at –70°C prior to protein or RNA extraction.

3.3. Protein Extraction

1. Add 350 μL of Wash buffer to three constructs in a 1.5-mL microcentrifuge. Incubate for 30 s on ice and immediately remove wash buffer.
2. Add 350 μL Lysis buffer and pulverise constructs with a sterile needle. Vortex and leave on ice for 30 min.
3. Centrifuge cell/agarose lysis solution at 10,000 × *g* for 15 min.
4. Collect approximately 350 μL of supernatant and determine protein concentration by Folin–Lowry method on the microplate reader (see Note 8).
5. 100 μL of protein sample is boiled with 85 μL of sample buffer and 5 μL Na_3VO_4 for 5 min.
6. Samples are removed from heat and left in water bath for 10 min.
7. After cooling to room temperature, samples are ready for separation by SDS–polyacrylamide gel electrophoresis (SDS–PAGE).

3.4. SDS–Polyacrylamide Gel Electrophoresis

1. The following methods for SDS–PAGE assume the use of a Bio-Rad mini gel apparatus. Prior to assembly, ensure that glass plates, comb, spacers, and casting stands are scrubbed and cleaned with 70% IMS and rinsed extensively with dH_2O. The glass slides and comb should be air-dried. Assemble glass plate sandwich into the clamp assembly and transfer the unit to the casting stand. Place a comb in the assembled gel sandwich and mark on the glass plate, 1 cm below the teeth of the comb. This is the maximum level for the separating gel.
2. Prepare the separating gel by mixing 7.5 mL of 4× separating buffer with 10-mL acrylamide/bis-acrylamide solution, 12.5 mL of dH_2O, 100 μL of 10% (w/v) ammonium persulfate solution (make up fresh), and 20-μL TEMED. Pour the gel up to the mark leaving space for the stacking gel and overlay solution with water-saturated isobutanol. The polymerisation process should take 30 min at room temperature.
3. Pour off unpolymerised overlay solution and rinse the top of the gel twice with dH_2O. Place on side to drain dH_2O completely.

4. Prepare the stacking gel by mixing 2.5 mL of 4× stacking buffer with 1.3 mL acrylamide/bis solution, 6.1 mL dH_2O, 50-μL ammonium persulfate solution, and 10-μL TEMED. Gently mix the solution and pour over the separating gel to the top of the glass plate. Align the comb in the sandwich and allow the acrylamide to polymerise for 30 min at room temperature.
5. Prepare 1 L of running buffer by diluting 100 mL of the 4× running buffer with 400 mL of dH_2O. Make sure that running buffer is mixed.
6. Once the stacking gel has set, release the clamp assemblies from the casting stand and attach them to the running stand. Fill the tank with half the running buffer and carefully remove the comb. Use a syringe fitted with a 22-gauge needle to wash the wells with running buffer.
7. Add the remaining running buffer to all chambers of the gel unit and load 50 μL of each sample to all wells. Include one well for pre-stained molecular weight markers. Use lanes on outer edges last.
8. Complete the assembly of the gel unit and connect to a power supply. Run the gel unit at 50 V until the samples have passed through the stacking gel and have collected at the top of the separating gel (takes approximately 5 h). Ensure that the gel unit is kept cold throughout this process (use an ice cooling unit or freezer packs). To speed up this process, run the gel at 80 V for the first 2 h and 100–150 V for a further 60–90 min.

3.5. Western Blotting for Active MAPK

1. The samples separated by SDS–PAGE are transferred to nitrocellulose membranes electrophoretically. A tray of setup buffer is prepared that is large enough to lay out a transfer cassette with its fibre pad and two sheets of filter paper submerged on one side. A sheet of nitrocellulose membrane and two sheets of filter paper of appropriate size are submerged in a tray containing dH_2O (Avoid not to touch the centre of the membrane). Immerse nitrocellulose membrane on top of the filter paper in a separate tray containing transfer buffer for 15 min.
2. Remove the gels from the clamp assemblies and discard the stacking gel using a razor blade. A cut is made in one corner of the separating gel for easy identification. Lay the separating gel on top of the nitrocellulose membrane.
3. Two sheets of filter paper are soaked in transfer buffer for 10 min in a separate tray. Lay one filter paper on top of the separating gel and ensure that no air bubbles are trapped in the sandwich. Complete the sandwich by laying the second pre-soaked fibre pad on top of the filter paper and close the transfer cassette.
4. The cassette is placed in the transfer tank so that the nitrocellulose membrane faces the anode. Insert the ice cooling unit

and a magnetic-stir bar and fill the tank with transfer buffer. The ice cooling unit maintains a temperature of 10–15°C.

5. Transfer at 70 V for 2 h or at 45 V overnight at 4°C.
6. Once the transfer is complete, the cassette is removed from the tank and disassembled with the top fibre pad and filter paper removed. The gel is left in place on top of the nitrocellulose membrane, and these are laid out on a glass plate so that the gel can be cut into the membrane using a razor blade (note the orientation of the gel and the visibility of the molecular weight markers). Excess gel and nitrocellulose membrane are discarded.
7. Incubate membranes in 50 mL of blocking buffer for 1 h at room temperature on a rocking platform.
8. Discard blocking buffer and wash the membrane quickly prior to the addition of primary antibody (1:1,000) in 10-mL primary antibody dilution buffer with gentle agitation for 2 h at room temperature.
9. Remove primary antibody and wash six times for 5 min each with 50 mL of TBS-T.
10. Incubate membrane with freshly prepared HRP-linked anti-rabbit IgG (1:2,000) and HRP-linked anti-biotin antibody (1:1,000) in 10 mL of blocking buffer with gentle agitation for 1 h at room temperature.
11. Discard secondary antibody and wash six times for 5 min each with 50 mL of TBS-T.
12. During the final wash, ECL reagents are equilibrated to room temperature, and the following steps are carried out in the dark room. The ECL detection reagents are mixed together (40:1) and immediately added to the blot for 1 min at room temperature. The blot is gently swirled to ensure maximal coverage.
13. Drain the blot from any excess ECL reagents and place between an acetate protector sheet which has been cut to the size of an X-ray film cassette. Gently smooth out any air bubbles.
14. The acetate containing the membrane is placed in an X-ray film cassette with film for suitable exposure for up to 1 min. Make sure that exposure time is the same for all films. An example of the image is shown in Fig. 1.

3.6. Stripping and Reprobing Blots for Total MAPK

1. Provided the signal for active p38 MAPK has been obtained, the membrane is stripped and re-probed with a phosphorylation state-independent antibody.
2. Warm 50 mL of the stripping buffer to 70°C, add the β-mercaptoethanol, and incubate mixture with the blot for 30 min with occasional agitation.
3. Wash blot three times for 10 min each with washing buffer and then block again in blocking buffer.

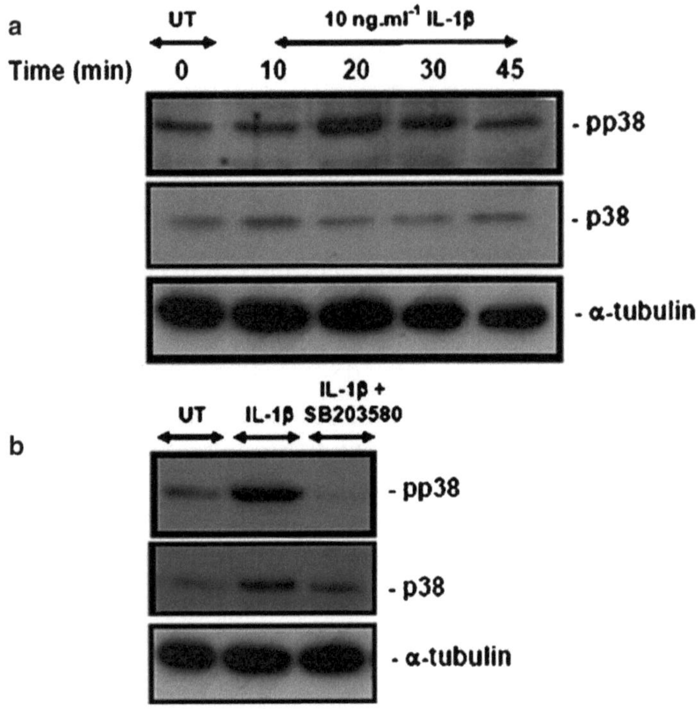

Fig. 1. Distribution of cycle threshold (Ct) values for GAPDH by chondrocyte/agarose constructs cultured under free-swelling conditions (**a–c**) or in unstrained and strained constructs subjected to dynamic compression (**d–f**). All constructs were cultured with 0 or 10 ng/mL IL-1β ng/mL and/or 10 μM SB203580 for up to 48 h. The median values are shown as lines, the 25th to 75th percentile as boxes, and range as whiskers for up to 18 replicates.

4. Reprobe membrane with anti-total p38 MAPK. Repeat steps for washing, secondary antibody and ECL detection as above. An example of the image is shown in Fig. 1.

3.7. RNA Extraction

1. Defrost a single construct on ice and incubate at 42°C for 10 min in 750-μL buffer QG. Vortex solution at least twice during incubation period.
2. Add 125 μL of isopropanol, vortex, and apply up to 750 μL of the sample onto an RNeasy® spin column.
3. Close the lid gently and centrifuge at 8,000 × *g* for 15 s at room temperature.
4. Discard flow-through and load remaining mixture onto the column and centrifuge as above.
5. Add 500 μL of buffer QG to the spin column, centrifuge at 8,000 × *g* for 15 s, and discard flow-through.
6. Using a fresh 2-mL collection tube, add 700 μL of buffer RW1, centrifuge at 8,000 × *g* for 15 s, and discard flow-through.

7. Add 500 μL of buffer RPE, centrifuge at 8,000 × *g* for 15 s, discard flow-through, and repeat this step.
8. Discard flow-through and centrifuge at full speed for 1 min.
9. To elute, transfer spin column into a new 1.5-mL collection tube. Pipette 40 μL of RNase-free water directly onto the RNeasy silica gel membrane. Close the tube gently and centrifuge at 8,000 × *g* for 1 min.
10. Immediately store eluted RNA sample on ice.
11. For DNase treatment, add 4 μL of 10× DNase buffer and 1 μL of DNase I enzyme and incubate at 37°C for 20 min.
12. Add 4 μL of DNase inactivating reagent and incubate at room temperature for 2 min.
13. Centrifuge at 14,000 × *g* for 1 min at 4°C to pellet the slurry produced by the DNase inactivating agent.
14. Transfer 40 μL of the RNA containing solution to a new 1.5-mL microfuge tube and store at −70°C prior to quantification and/or reverse transcription.
15. Quantify RNA using an appropriate method (see Note 9). The NanoDrop ND-1000 spectrophotometer is the preferred instrument when measuring up to ten samples at any one time. RiboGreen is the most accurate and convenient means of measuring absorbance for large numbers at 260 nm.
16. After blanking and setting the NanoDrop system to zero with 1 μL dH_2O, place 1 μL of RNA onto the sensor and measure the RNA absorbance. The instrument automatically calculates the RNA concentration at 260/280 nm and can measure 1 μL samples with concentrations between 2 and 3,000 ng/μL. For the chondrocyte/agarose model, expect a yield of up to 2 μg of total RNA from a single construct. Alternatively, RNA concentrations and total yield can be determined using Eqs. 1 and 2, respectively.

$$\text{RNA concentration} = 40 \times A_{260} \times \text{dilution factor} \quad (1)$$

$$\text{Total yield} = \text{concentration} \times \text{volume of sample} \quad (2)$$

3.8. Synthesis of First-Strand cDNA Using Reverse Transcriptase

1. Keep reagents and microfuge tubes on ice when setting up a 20-μL reaction mixture of the following components *in order*: 3.5 μL of RNase-free water, 10 μL of cDNA synthesis master mix (2×), 2 μL of oligo-dT primers, 1 μL of Stratascript™ RT/RNase Block enzyme mixture, and 3.5 μL of RNA (see Note 10). The optimal primer type varies for different targets and should be determined in advance (see Note 11).
2. Set up no-reverse transcriptase (NoRT) control reactions in duplicate by omitting the Stratascript™ RT. These control

reaction mixtures are important to screen for contamination of reagents or false amplification.

3. Incubate the reactions at 25°C for 5 min in a thermal cycler to allow primer annealing.
4. Incubate the reactions at 42°C for 15 min to allow cDNA synthesis (see Note 12).
5. Incubate the reaction at 95°C for 5 min on the thermal cycler to terminate cDNA synthesis.
6. First-strand cDNA synthesis reactions must be snap-cooled on ice for 5 min prior to qPCR. For long-term storage, reactions can be stored at –20°C for up to 6 months.

3.9. Quantitative PCR Amplification of cDNA

1. Prepare the experimental reaction by adding the following components in order: nuclease-free PCR-grade H_2O with adjusted final volume to 25 μL including cDNA, 12.5 μL of 2× master mix, optimised concentration of experimental probe (*x*μL), optimised concentration of upstream primer (*x*μL), optimised concentration of downstream primer (*x*μL), and 0.375 μL of the diluted reference dye (optional) (see Notes 4 and 13). A total reaction volume of 50 μL may also be used in case of low target copy numbers.
2. Gently mix the reaction without creating any bubbles using the Picofuge (do not vortex).
3. Add 2 μL of the cDNA synthesis reaction.
4. Gently mix the reaction without creating any bubbles (do not vortex). Note that bubbles interfere with fluorescence detection.
5. Centrifuge the reaction mixture briefly.
6. Place the reaction in the PCR instrument and run the appropriate PCR programme. In general, TaqMan reactions utilise the two-step cycling programme, and molecular beacons and SYBR green chemistries utilise the three-step cycling programme (see Table 1 and Note 14). Ensure that amplifications have been optimised and efficiencies validated for primer/template systems (see Note 15).
7. Note that PCRs for each sample should be run in duplicate using the 96-well thermal system of the Mx3000P qPCR instrument (Stratagene, Amsterdam, The Netherlands).
8. In order to screen for contamination of reagents or false amplification, PCR controls should be prepared for each sample by setting up identical reaction mixtures except for the addition of the template (NTC). An endogenous control reaction will therefore distinguish true negative results from PCR inhibition or failure (31). NoRT controls should additionally be included in each PCR assay.

Table 1
Three-step thermal cycling protocol for molecular beacon and SYBR green chemistries

Cycles	Duration of cycle	Temperature	Stage
1	10 min	95°C	Activation[a]
40	30 s 1 min 30 s	95°C 55–60°C[b] 72°C	Denaturation Annealing[c] Extension
1	Collect data at each temperature	Between 55 and 95°C	Melt curve

[a]Initial 10-min incubation is required to fully activate the DNA polymerase
[b]Choose an appropriate annealing temperature for the primer set used
[c]Set the temperature cycler to detect and report fluorescence during the annealing step of each cycle

3.10. Data Analysis

1. Fluorescence data should be collected during the annealing stage of amplification and data analysed using the instrument software, e.g. MxPro™ qPCR software (version 3.0, Stratagene). Check baselines and thresholds set by the software and use after manual inspection.
2. The cycle threshold (Ct) value for each duplicate reaction should be expressed as the mean value, and the results can be exported as tab-delimited text files into Microsoft Excel for further analysis. If the samples do not record a Ct value, check the characteristics of the amplification plot. Repeat reaction if necessary.
3. Remove outliers by increasing replicate number of samples and display Ct values as Box and Whisker plots. Use a Grubbs' test to examine the distribution under all treatment conditions. For example, see Fig. 2. Note that the Ct values for GAPDH remained stable with no changes detected under all culture conditions, suggesting its suitability as a reference gene.
4. Calculate the relative quantification of the target gene, by normalising each target to the reference gene and the calibrator sample by a comparative Ct approach (32), as shown in Eq. 3.

Ratio of the relative expression level

$$= \frac{\left(1 + E_{\mathrm{Target}}\right)_{\mathrm{Target}}^{\Delta \mathrm{Ct}} (\text{Mean calibrator} - -\text{sample})}{\left(1 + E_{\mathrm{Reference}}\right)_{\mathrm{Reference}}^{\Delta \mathrm{Ct}} (\text{Mean calibrator} - -\text{sample})} \quad (3)$$

Where:
E represents the efficiencies obtained for the target and reference gene (see Note 14).

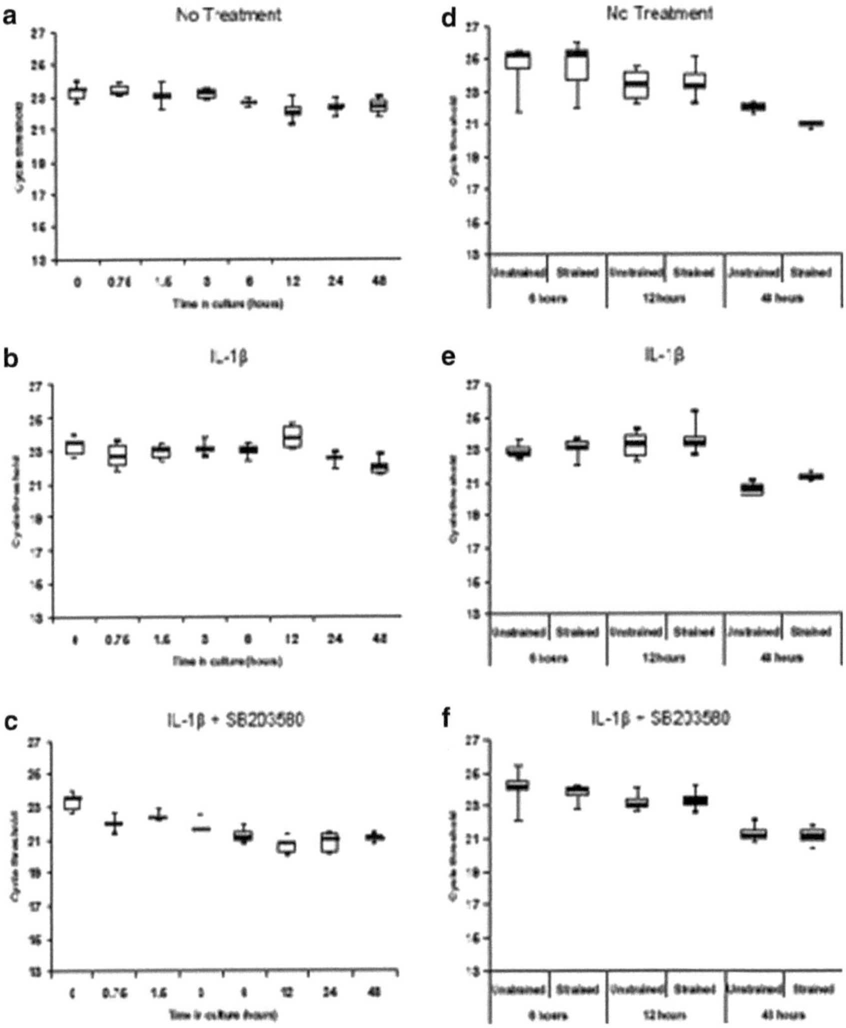

Fig. 2. p38 phosphorylation by IL-1β in chondrocyte/agarose constructs cultured under free-swelling conditions in the presence or absence of 10 ng/mL IL-1β for up to 45 min (Fig. 1a) or with IL-1β and 10 μM SB203580 for 20 min (Fig. 1b). Phospho-p38 MAPK was analysed for each test condition using a phosphorylation-state-specific anti-p38 (Thr180/Tyr182) antibody (*upper panel*) against total p38 MAPK (middle panel) and α-tubulin as a loading control (*lower panel*). All cell extracts were subjected to Western blot analysis. Each band corresponds to three constructs pooled from two separate experiments. *UT* untreated (reproduced from ref. (28)).

ΔCt_{Target} represents the difference in Ct values for the mean calibrator or sample for the target.

$\Delta Ct_{Reference}$ represents the difference in Ct values for the mean calibrator or sample for the reference gene.

5. Express the relative expression levels as ratios on a logarithmic scale (arbitrary units) following log transformation of data sets (see Notes 16 and 17). A logarithm base 2 or base 10 transformation will compress the variance so that the relevant parametric tests may be ensued (see Figs. 3 and 4 for examples of results).

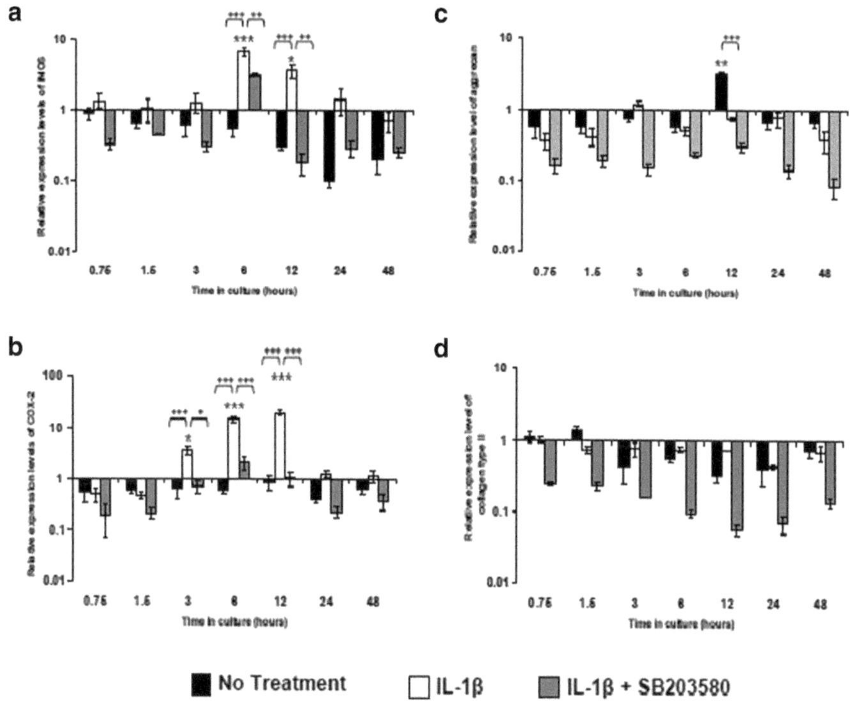

Fig. 3. Temporal profile of IL-1β on iNOS (**a**), COX-2 (**b**), aggrecan (**c**), and collagen type II (**d**) expression by chondrocyte/agarose constructs cultured under free-swelling conditions with 0 or 10 ng/mLIL-1β and/or 10 μM SB203580. *Bars* represent the mean and SEM of 6 replicates from three separate experiments. The ratio of the relative expression level for the target gene was calibrated to the mean value at time = 0 and normalized to the reference gene GAPDH. Ratios were expressed on a logarithmic scale (arbitrary units). Two-way ANOVA with post hoc Bonferroni-corrected *t*-tests was used to compare data under the different treatments, where $*p < 0.05$ for comparisons between time zero with IL-1β; *plus* indicates $p < 0.05$ for comparisons between untreated with IL-1β or IL-1β with IL-1β + SB203580.

4. Notes

1. The design for the moulds have been described in ref. (27).
2. The equipment for the compressive cell-strain bioreactor has been described in refs. (26) and (27).
3. To prepare 200 mM activated Na_3VO_4, dissolve 367.8 mg in 10-mL dH_2O and adjust pH to 10 using 1 M NaOH. At pH 10, the colour of the Na_3VO_4 should be yellow. Boil the solution for approximately 10 min until it becomes colourless. Allow to cool to room temperature and readjust pH to 10. Repeat these steps two to three times until the solution remains colourless and stabilises at pH 10. Aliquot and store at −20°C prior to use.
4. Prepare fresh dilutions of the reference dye prior to setting up the reactions, and keep all tubes containing the reference dye protected from light as much as possible. Make initial dilutions

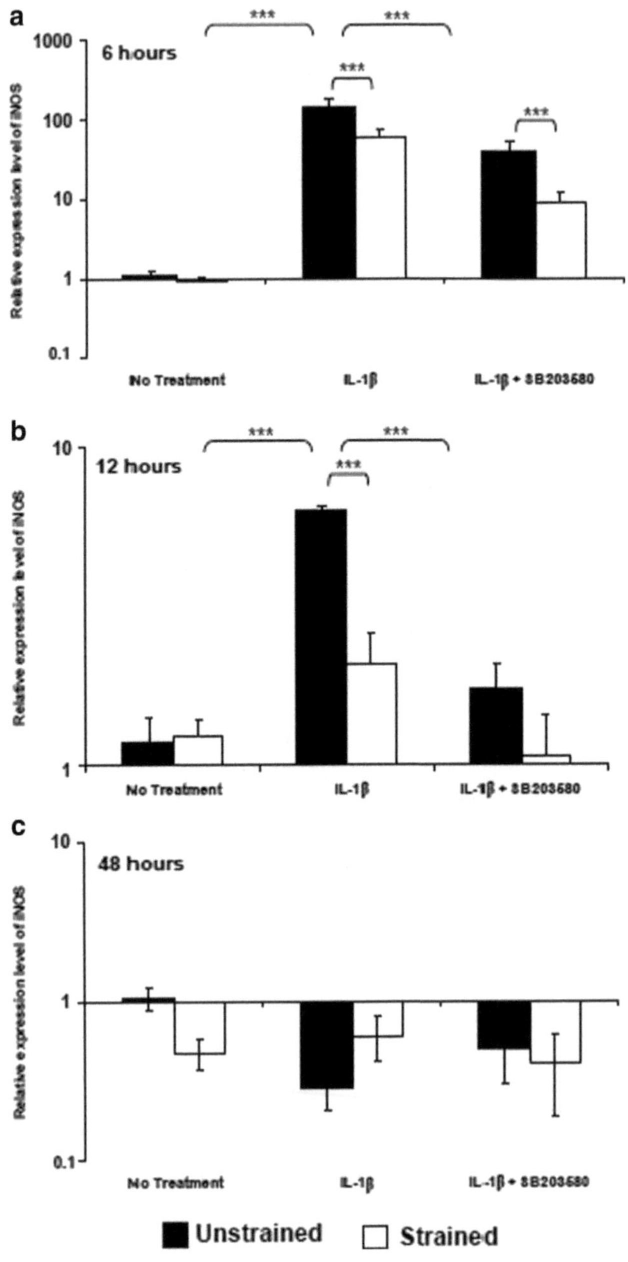

Fig. 4. Effects of 15% dynamic compressive strain on iNOS expression in unstrained and strained constructs cultured under no treatment conditions or with 10 ng/mL IL-1β and/or 10 μM SB203580 for 6 h (**a**), 12 h (**b**) and 48 h (**c**). The ratio of the relative expression level of iNOS was calibrated to the mean value for the unstrained (untreated) control and normalized to GAPDH. Ratios were expressed on a logarithmic scale (arbitrary units). Bars represent the mean and SEM of 16–18 replicates from four separate experiments. Two-way ANOVA with post hoc Bonferroni-corrected *t*-tests was used to compare data. Significant differences are indicated as follows: ***$p << 0.001$.

of the reference dye using nuclease-free PCR-grade H_2O. When using the Stratagene's Mx3000P qPCR instrument, use the reference dye at 1:50 dilution, resulting in a final concentration of 30 nM.

5. Opening the joint is best performed by making a mid-line incision using a no. 20 blade, taking care not to enter the joint space. The skin and digital extensor tendons may then be dissected away from the joint capsule. The joint may then be opened with a no. 11 blade, using the distal surface of the metacarpal bone as a guide.
6. The volumes of pronase and collagenase solution are based on the digestion of tissue from one joint. If digesting tissue from more than one joint, it is recommended that the tissue be digested in further centrifuge tubes instead of altering the volume of the solutions in each tube. Approximately 1–3 g of tissue can be obtained from a single joint, and this should yield up to 30×10^6 cells.
7. Collagenase batches vary, even when they are purchased from the same supplier. It is highly advisable to test batches prior to use to ensure that all the tissue is digested and that cell viability is maintained. It may be necessary to alter the concentration or incubation times from those stated to achieve optimal cell isolation.
8. We have identified that a significant amount of the signal, which is produced when protein extracts from agarose constructs are measured in the Folin–Lowry assay, arises from the agarose and associated medium rather than being cell-derived. This may represent protein from the defined medium as it is also present in cell-free agarose constructs. In fact, approximately 97% of the construct volume is medium, with its associated protein, while the cells represent about 0.2% of the construct volume. Thus, although a total of "400 μg of protein" is loaded, much of this will be agarose "contaminant" rather than cell-derived protein that is a smaller component of the total measurement. Westerns of "protein" extracts from agarose constructs that are cell-free and have not been incubated with serum do not show bands for phosphorylated p38.
9. The quantity of mRNA should be determined using a single method: for instance, quality of RNA by analysis of the A_{260}/A_{280} ratio and confirmation of integrity of the rRNA bands by electrophoresis in formaldehyde-agarose gels. An alternative approach is the Agilent Bioanalyser/Bio-Rad Experion microfluidic capillary electrophoresis system. However, recent reports suggest that the Agilent Bioanalyser can detect moderately degraded RNA (31). A far more reliable measure of mRNA integrity utilises the 3′:5′ assay of primer and target

probe amplicons at different positions on the mRNA transcript. Perfectly intact mRNA should result in a 3′:5′ ratio of 1–5 with denatured mRNA resulting in ratios greater than 10 (32).

10. For all procedures involving RT-qPCR, use DNase/RNase-free consumables. Maintain a dedicated set of micropipettes, which must be calibrated regularly, especially those dispensing less than 10 μL. It is essential to use the pipettor appropriate for the volume being dispensed, e.g. 1 or 2 μL. Also, use filter barrier tips for all qPCRs.
11. Use total RNA as the template for reverse transcription and include the same amount of RNA for each RT reaction. The RT reaction can be carried out using random priming, oligo-dT priming, or gene-specific priming. For random or oligo-dT priming, incubate at 20°C for 10 min followed by 50°C for 60 min. For gene-specific priming, incubate at 60°C for 15 min. Terminate the reaction at 85°C for 5 min and then place on ice for 5 min. Collect by brief centrifugation.
12. Normally, a 15-min incubation at 42°C is sufficient for most targets. Increasing the incubation time to 45 min at 42°C is optional and may increase cDNA yield for more challenging RNA targets.
13. For PCR optimization, prepare dilutions of forward and reverse primers ranging from 50 to 300 nM in Brilliant SYBR Green Master Mix and run reactions in duplicate in qPCR instrument using a three-step protocol (Table 1). Look at melt curves and amplification plots and select the primer combination that fulfils the following criteria: no primer dimer, lowest Ct, highest end point fluorescence (ΔRn), and absence of fluorescence signal in the NTC. For probe optimization, prepare dilutions ranging from 50 to 300 nm in Brilliant RT-qPCR Master Mix containing optimized forward and reverse primer concentrations. Run reactions in duplicate using a three-step protocol (see Table 1). Use the instrument software to plot Ct against log target concentration and determine the slope and R^2 values. Confirm that the slope lies between −3.2 and −3.5 and that R^2 value is greater than 0.98 (33, 34).
14. For the two-step cycling protocol, which is compatible with the Taqman detection system, perform the denaturation step for 15 s at 95°C and annealing/extension stage for 1 min at 60°C using an appropriate annealing temperature for the primer set used. See Table 1 for three-step cycling programme suitable for Molecular Beacon and SYBR green chemistries.
15. To determine assay sensitivity and reproducibility, prepare a tenfold serial dilution of cDNA from a sample which represents

the calibrator sample. Reactions should use optimal primer pair and probe concentrations to derive PCR efficiencies from at least three standard curves. The PCR efficiencies of amplification for each target is defined according to the relationship, $E = 10^{[-1/\text{slope}]}$, where E = efficiency and range from 90 to 110%. The R^2 value of the standard curve should exceed 0.9998 (33, 34).

16. Do not assume that Ct values *within* or *between* each treatment will have a Gaussian distribution with equal variance. These assumptions will not be valid in small sample sizes, so a paired parametric test cannot be performed. Use a Jarque–Bera or Shapiro–Wilk W test in Excel to check for non-normality.
17. Alternatively, use the relative expression software tool (REST) to calculate Ratio values (35). REST enables pairwise comparisons of up to 16 values in a given control or sample group, using the fixed reallocation randomisation test to obtain the significance level. It has the advantage of considering several interactions between gene, treatment, or sample number, making no distributional assumptions about the data organisation and is therefore highly suitable for gene comparisons between several targets and multiple housekeeping genes. Additionally, the purpose of the pairwise variation eliminates differences between two genes and therefore removes any non-specific variation in normalised factors. For multiple variables of less than 6 experimental conditions, use REST-MCS. However, for multiple comparisons greater than 6, log-transform each ratio value in Excel to compress the variance prior to relevant parametric tests. For example, in Figs. 3 and 4, a two-way ANOVA coupled to the post hoc *t*-test, with Bonferroni correction factor, allows us to further establish differences among multiple experimental groups, with no limit on test conditions or data points following log transformation. However, to date, there are no software tools which consider the effect of more than two factors in multiple treatment groups greater than 6 in a statistically relevant manner. Thus, the normalisation strategies and problems used to analyse quantitative PCR data are a subject of much debate (36–38).

Acknowledgements

Dr Chowdhury would like to thank Drs. Kerry Elliot, Lindsay Ramage, and Ying Zhou for their excellent support at the Queens Medical Research Institute, Edinburgh University. The protocols were developed with funding by the Wellcome Trust (project grant: 073972).

References

1. Grodzinsky, A.J., Levenston, M.E., Jin, M., and Frank, E.H. (2000) Cartilage tissue remodeling in response to mechanical forces. *Annu. Rev. Biomed. Eng.* **2,** 691–713.
2. Millward-Sadler, S.J. and Salter, D.M. (2004) Integrin-dependent signal cascades in chondrocyte mechanotransduction. *Ann. Biomed. Eng.* **32,** 435–446.
3. Griffin, T.M. and Guilak, F. (2005) The role of mechanical loading in the onset and progression of osteoarthritis. *Exerc. Sport Sci. Rev.* **33,** 195–200.
4. Sah, R., Kim, Y.J., Doong, J.Y.H., Grodzinsky, A.J., Plaas, A.H.K., and Sandy, J.D. (1989) Biosynthetic response of cartilage explants to dynamic compression. *J. Orthop. Res.* **7,** 619–636.
5. Bachrach, N.M., Valhmu, W.B., Stazzone, E., Ratcliffe, A., Lai, W.M., and Mow, V.C. (1995) Changes in proteoglycan synthesis of chondrocytes in articular cartilage are associated with the time-dependent changes in their mechanical environment. *J. Biomech.* **28,** 1561–1569.
6. Valhmu, W.B., Stazzone, E.J., Bachrach, N.M., Saed-Nejad, F., Fischer, S.G., Mow, V.C., and Ratcliffe, A. (1998) Load-controlled compression of articular cartilage induces a transient stimulation of aggrecan gene expression. *Arch. Biochem. Biophys.* **353,** 29–36.
7. Ragan, P.M., Badger, A.M., Cook, M., Chin, V.I., Gowen, M., Grodzinsky, A.J., and Lark, M.W. (1999) Down-regulation of chondrocyte aggrecan and type-II collagen gene expression correlates with increases in static compression magnitude and duration. *J. Orthop. Res.* **17,** 836–842.
8. Blain, E.J., Mason, D.J., and Duance, V.C. (2001) The effect of cyclical compressive loading on gene expression in articular cartilage. *Biorheology* **40,** 111–117.
9. Lee, J.H., Fitzgerald, J.B., Dimicco, M.A., and Grodzinsky, A.J. (2005) Mechanical injury of cartilage explants causes specific time-dependent changes in chondrocyte gene expression. *Arthritis Rheum.* **52,** 2386–2395.
10. De Croos, J.N., Dhaliwal, S.S., Grynpas, M.D., Pilliar, R.M., and Kandel, R.A. (2006) Cyclic compressive mechanical stimulation induces sequential catabolic and anabolic gene changes in chondrocytes resulting in increased extracellular matrix accumulation. *Matrix Biol.* **25,** 323–331.
11. Madhavan, S., Anghelina, M., Rath-Deschner B., Wypasek, E., John, A., Deschner, J., Piesco, N., and Agarwal, S. (2006) Biomechanical signals exert sustained attenuation of proinflammatory gene induction in articular chondrocytes. *Osteoarthr. cartil.* **14,** 1023–1032.
12. Mio, K., Saito, S., Tomatsu, T., and Toyama, Y. (2005) Intermittent compressive strain may reduce aggrecanase expression in cartilage: a study of chondrocytes in agarose gel. *Clin. Orthop. Relat. Res.* **433,** 225–232.
13. Chowdhury, T.T., Bader, D.L., and Lee, D.A. (2001) Dynamic compression inhibits the synthesis of nitric oxide and PGE_2 by IL-1β stimulated chondrocytes cultured in agarose constructs. *Biochem. Biophys. Res. Commun.* **285,** 1168–1174.
14. Chowdhury, T.T., Bader, D.L., and Lee, D.A. (2003) Dynamic compression counteracts IL-1β induced release of nitric oxide and PGE_2 by superficial zone chondrocytes cultured in agarose constructs. *Osteoarthr. Cartil.* **11,** 688–696.
15. Murata, M., Bonassar, L.J., Wright, M., Mankin, H.J., and Towle, C.A. (2003) A role for the interleukin-1 receptor in the pathway linking static mechanical compression to decreased proteoglycan synthesis in surface articular cartilage. *Arch. Biochem. Biophys.* **413,** 229–235.
16. Fanning, P.J., Emkey, G., Smith, R.J., Grodzinsky, A.J., Szasz, N., and Trippel, S.B. (2003) Mechanical regulation of MAPK signalling in articular cartilage. *J. Biol. Chem.* **278,** 50940–50948.
17. Fitzgerald, J.B., Jin, M., Dean, D., Wood, D.J., Zheng, M.H., and Grodzinsky, A.J. (2004) Mechanical compression of cartilage explants induces multiple time-dependent gene expression patterns and involves intracellular calcium and cyclic AMP. *J. Biol. Chem.* **279,** 19502–19511.
18. Hung, C.T., Henshaw, D.R., Wang, C.C., Mauck, R.L., Raia, F., Palmer, G., Chao, P.H., Mow, V.C., Ratcliffe, A., and Valhmu, W.B. (2000) Mitogen-activated protein kinase signaling in bovine articular chondrocytes in response to fluid flow does not require calcium mobilization. *J. Biomech.* **33,** 73–80.
19. Li, K.W., Wang, A.S., and Sah, R.L. (2003) Microenvironment regulation of extracellular signal-regulated kinase activity in chondrocytes: effects of culture configuration, interleukin-1, and compressive stress. *Arthritis Rheum.* **48,** 689–699.
20. Zhou, Y., Millward-Sadler, S.J., Lin, H., Robinson, H., Goldring, M., Salter, D.M., and Nuki, G. (2007) Evidence for JNK-dependent up-regulation of proteoglycan synthesis and for activation of JNK1 following cyclical mechanical stimulation in a human

chondrocyte culture model. *Osteoarthr. Cartil.* **15,** 884–893.

21. Agarwal, S., Deschner, J., Long, P., Verma, A., Hofman, C., Evans, C.H., and Piesco, N. (2004) Role of NF-kappaB transcription factors in anti-inflammatory and pro-inflammatory actions of mechanical signals. *Arthritis Rheum.* **50,** 3541–3548.
22. Guilak, F., Fermor, B., Keefe, F.J., Kraus, V.B., Olson, S.A., Pisetsky, D.S., Setton, L.A., and Weinberg, J.B. (2004) The role of biomechanics and inflammation in cartilage repair and injury. *Clin. Orthop. Relat. Res.* **423,** 17–26.
23. Knight, M.M., Ghori, S.A., Lee, D.A., and Bader, D.L. (1998) Measurement of the deformation of isolated chondrocytes in agarose subjected to cyclic compression. *J. Med. Eng. Phys.* **20,** 684–688.
24. Lee, D.A., Knight, M.M., Bolton, J.F., Idowu, B.D., Kayser, M.V., and Bader, D.L. (2000) Chondrocyte deformation within compressed agarose constructs at the cellular and sub-cellular levels. *J. Biomech.* **33,** 81–95.
25. Lee, D.A., Noguchi, T., Frean, S.P., Lees, P., and Bader, D.L. (2000) The influence of mechanical loading on isolated chondrocytes seeded in agarose constructs. *Biorheology* **37,** 149–161.
26. Lee, D.A. and Bader, D.L. (1997) Compressive strains at physiological frequencies influence the metabolism of chondrocytes seeded in agarose. *J. Orthop. Res.* **15,** 181–188.
27. Lee, D.A. and Knight, M.M. (2004) Mechanical loading of chondrocytes embedded in 3D constructs: in vitro methods for assessment of morphological and metabolic response to compressive strain. *Methods Mol. Med.* **100,** 307–324.
28. Chowdhury, T.T., Arghandawi, S., Brand, J., Akanji, O.O., Salter, D.M., Bader, D.L., and Lee, D.A. (2008) Dynamic compression counteracts IL-1β induced iNOS and COX-2 expression in chondrocyte/agarose constructs. *Arthritis Res. Ther.* **10,** R35.
29. Akanji, O.O., Sakthithasan, P., Salter, D.M., and Chowdhury, T.T. (2009) Dynamic compression alters NFκB activation and Iκ-α expression in IL-1β stimulated chondrocyte/agarose constructs. *Inflamm. Res.* **59(1),** 41–52.
30. Raveenthiran, S.P. and Chowdhury, T.T. (2009) Dynamic compression inhibits fibronectin fragment induced iNOS and COX-2 expression in chondrocyte/agarose constructs. *Biomech. Model. Mechanobiol.* **8(4),** 273–283.
31. Schroeder, A., Mueller, O., Stocker, S., Salowsky, R., Leiber, M., Gassmann, M., Lightfoot, S., Menzel, W., Granzow, M., and Ragg, T. (2006) The RIN: an RNA integrity number for assigning integrity values to RNA measurements. *BMC Mol. Biol.* **7,** 3.
32. Auger, H., Lyianarachchi, S., Newsom, D., Klisovic, M.I., Marcucci, G., and Kornacker, K. (2003) Chipping away at the chip bias: RNA degradation in microarray analysis. *Nat. Genet.* **35,** 262–293.
33. Bustin, S.A. and Nolan, T. (2004) Pitfalls of quantitative real-time reverse-transcription polymerase chain reaction. *J. Biomol. Tech.* **15,** 155–166.
34. Nolan, T., Hands, R.E., and Bustin, S.A. (2006) Quantification of mRNA using real-time RT-PCR. *Nat. Protoc.* **1,** 1559–1582.
35. Pfaffl, M.W., Horgan, G.W., and Dempfle, L. (2002) Relative expression software tool (REST) for group wise comparison and statistical analysis of relative expression results in real time PCR. *Nucleic Acids Res.* **30(9),** e36.
36. Yuan, J.S., Wang, D., and Stewart, C.N. (2007) Statistical methods for efficiency adjusted real-time PCR quantification. *Biotechnol. J.* **3,** 112–123.
37. Karlen, Y., McNair, A., Perseguers, S., Mazza, C., and Mermod, N. (2007) Statistical significance of quantitative PCR. *BMC Bioinformatics* **8,** 131.
38. Fundel, K., Haag, J., Gebhard, P.M., Zimmer, R., and Aigner, T. (2008) Normalization strategies for mRNA expression data in cartilage research. *Osteoarthr. Cartil.* **16,** 947–955.

Chapter 7

Genetic Modification of Chondrocytes Using Viral Vectors

Teresa Coughlan, Aileen Crawford, Paul Hatton, and Michael Barker

Abstract

The use of isolated cells to construct engineered tissues provides the opportunity to genetically modify those cells prior to the formation of tissue. This should make it possible to create transgenic human model tissues that can be used to determine gene function as well as to identify or validate potential therapeutic targets. As proof of principle, we have used RNA interference to selectively suppress the expression of aggrecanase genes in human chondrocytes, in an attempt to determine which of these key enzymes have roles in arthritic cartilage destruction. This combination of gene targeting and tissue engineering we are using should be equally applicable to the identification of gene function in other biological systems.

Key words: RNAi, Aggrecanase, ADAMTS, Cartilage, Arthritis, Transgenic, Tissue engineering, Retrovirus

1. Introduction

There are two major structural components in articular cartilage: type II collagen and aggrecan. The former is a fibrillar protein that gives cartilage its tensile strength, while the compressive resistance of the tissue is provided by the proteoglycan, aggrecan. Both of these molecules are cleaved in arthritic disease – leading to cartilage loss, by the members of the matrix metalloproteinase (MMP) family of enzymes, though aggrecan is also cleaved at a second, distinct site by another family of enzymes. These are known as "aggrecanases," and the genes encoding them were recently identified (1).

Aggrecanases belong to the ADAMTS (a-disintegrin and metalloproteinase with thrombospondin type I motifs) family of proteinases, and although most work has focused on ADAMTS-4 and -5, ADAMTS-1, -8, -9, and -15 also have aggrecanase activity

John W. Haycock (ed.), *3D Cell Culture: Methods and Protocols*, Methods in Molecular Biology, vol. 695,
DOI 10.1007/978-1-60761-984-0_7, © Springer Science+Business Media, LLC 2011

in vitro (2–6). In transgenic mice, it has recently been demonstrated that ADAMTS-5 rather than ADAMTS-4 or -1 is the enzyme responsible for aggrecan breakdown in models of arthritis (7–9), whereas in other species, ADAMTS-4 or -9 may be more important (10–12). In human cartilage explants, a recent report implicated both ADAMTS-4 and -5 in tissue breakdown (13). Here, we demonstrate knockdown of ADAMTS-5; by the use of RNA interference (RNAi) and by using these cells to engineer de novo transgenic human cartilage, we aim to show definitively which of them are responsible for aggrecan loss in arthritic cartilage.

RNAi technology has been developing for the past 10 years and is now a widely utilised tool for studying gene function (14). Initial and current methods have used short RNA fragments, or "short interfering RNA" (siRNA) introduced directly into the cells, though it is now also possible to generate transgenic viruses that can infect the cells, whereupon they integrate and generate multiple copies of the "short hairpin RNA" (shRNA) within the cell. Unlike the siRNA methods, transduction of shRNA virus can generate long-term knockdown of expression, has the ability to produce stable "knockdown" cell-lines, can be used to infect primary cells which are otherwise difficult to transfect, and can be used in vivo.

We have used retrovirus expressing targeted shRNAs to knockdown aggrecanase gene expression, though the methods described herein can also be adapted to the production of transgenic tissue overexpressing a gene of interest, as we have previously shown (15).

2. Materials

2.1. Retroviral Constructs

1. RNAi-Ready pSIREN-RetroQ vector (BD Biosciences, Oxford, UK) was used to generate retrovirus. Control shRNAs and full instructions are supplied with the plasmid.

2.2. Molecular Biological Reagents

1. Oligonucleotides, real-time PCR probes, and 2× QPCR buffer (Eurogentec, Southampton, UK).
2. Enzymes, BSA, random hexamers, and dNTPs (Promega, Southampton, UK).
3. TRI-reagent, solvents, and plasmid preparation kits (Sigma).
4. Maxiprep kits (Qiagen, Crawley, UK).

2.3. Plastic Ware

1. Pipette tips (CLP, Northampton, UK) and Eppendorf tubes (Fisher Scientific, Loughborough, UK) for molecular biological procedures must be RNAse/DNAse free, and for bacterial/mammalian cell work must be sterile.

2. Plastic tissue culture vessels (Nunc, Loughborough, UK and Iwaki, SLS, Nottingham, UK).
3. 386-Well plates (Greiner, Stonehouse, UK) and lids (Biorad, Hemel Hempstead, UK).

2.4. Bacterial Culture

1. α-Select competent cells (Bioline, London, UK): any of bronze (10^7 cfu/μg DNA), silver (10^8 cfu/μg DNA), or gold (10^9 cfu/μg DNA) efficiency can be used. Store at −80°C and never use past the use-by date or if accidentally thawed.
2. Luria Broth (LB): 1% tryptone, 5% yeast extract, 1% NaCl. Autoclave.
3. LB-agar: LB plus 1.5% agar added prior to autoclaving. Once cooled to about 50°C, carbenicillin can be added.
4. Carbenicillin: make up a 100 mg/mL solution in 70% ethanol, and use at 1/1,000× final concentration.
5. Agar plates: pour a layer of LB-agar into as many 10-cm plastic culture dishes as required and leave with the lids off to set and dry. These can be stored and used for up to a month at 4°C.

2.5. Mammalian Cell Culture

1. C28/I2 chondrocytes were kindly provided by Dr. Mary Goldring (New York, USA).
2. GP2-293 cells can be obtained with the pSIREN vector.
3. Complete medium: Dulbecco's Modified Eagle Medium (DMEM) containing Glutamax (Invitrogen, Paisley, UK), 10% heat-inactivated foetal calf serum (FCS), and penicillin (100 U/mL)/streptomycin (0.1 mg/mL).
4. Packaging plasmid (pVSV-G) (Clontech-Takara Bio Europe, France).
5. Collagen I, polybrene (hexadimethrine bromide), and puromycin (Sigma-Aldrich, Poole, UK).
6. Cytokines (Peprotech, London, UK).

3. Methods

3.1. Selecting RNAi Target Sequences

1. The sequence of the mRNA transcript of the gene of interest can be obtained from the "Entrez" database (http://www.ncbi.nlm.nih.gov/sites/entrez?db=nuccore&itool=toolbar). Identify "AA" dimers within the sequence and locate the 19 nucleotides downstream of them, avoiding the regions within the 5′- and 3′-untranslated regions or within 75 bases of the start codon. These 19 nucleotide sequences are potential candidate target sites for shRNA. Reject sequences with repeated

nucleotide runs of three or more and ensure that the GC content of selected sequences is between 30 and 70%, ideally 50%. Use the "Blast" comparison tool (http://blast.ncbi.nlm.nih.gov/Blast.cgi) to check each candidate against the genome of the species of interest, to ensure no exact match is found with any other mRNA transcript. Ideally, 4–5 candidate sequences per gene should be pursued, preferably distributed throughout the length of the transcript.

2. For each target sequence, two complementary oligonucleotides are designed and synthesised, which will be annealed together and ligated into an expression vector to form a template for transcription of the shRNA. In the case of cloning into the pSIREN vector, the oligonucleotide sequences include *Bam*HI (5′-GGATCC-3′) and *Eco*RI (3′-CTTAAG-5′) restriction enzyme recognition sites, enabling directional cloning into the vector. Other inclusions are a purine residue downstream of the *Bam*HI site to enable RNA polymerase binding, a 7–9 nucleotide hairpin loop sequence such as 5′-TTCAAGAGA-3′ and a poly-T tract for the termination of transcription. A *Mlu*I restriction site (5′-ACGCGT-3′) can be included downstream of the terminator sequence for confirmation of the presence of the insert in ligated plasmid clones. Oligonucleotides typically consist of 5 bases of 5′ restriction site, the purine residue, 19 bases of sense target sequence, 9 bases of hairpin loop, 19 bases of antisense target sequence, 6 bases of terminator, and 6 bases of diagnostic restriction site. Table 1 shows the two complementary sequences used to generate each of four shRNAs for the ADAMTS-5 gene plus control shRNA.

3.2. Cloning shRNA Oligonucleotides into Vectors

1. Each of the two oligonucleotides is redissolved in molecular biology grade water, to a concentration of 100 μM. The two complementary strands are mixed in a 1:1 ratio, giving a 50 μM concentration of double-stranded oligonucleotide. Using a PCR machine, heat the mixture to 95°C for 30 s followed by 72°C for 2 min, 37°C for 2 min, and 25°C for 2 min. The double-stranded oligonucleotide can be used immediately in a ligation reaction or stored at −20°C.

2. Dilute each pair of annealed oligonucleotides to 0.5 μM, and set up ligation reactions as follows, using the components that accompany the pSIREN vector:

   2 μL linearised pSIREN vector (25 ng/mL)
   1 μL annealed oligonucleotide (0.5 μM)
   1.5 μL 10× ligase buffer
   0.5 μL BSA (10 mg/mL)
   9.5 μL H_2O
   0.5 μL T4 ligase (200 U)

Table 1
shRNA oligonucleotide sequences

Contfor	5′-GATCCGTGCGTTGCTAGTACCAACTTCAAGAGATTTTTTACGCGTG
Contrev	5′-AATTCACGCGTAAAAAATCTCTTGAAGTTGGTACTAGCAACGCACG
1for	5′-GATCCGGCACGCGCGCTACACCCTATTCAAGAGATAGGGTGTAGCG CGCGTGCCTTTTTTACGCGTG
1rev	5′-AATTCACGCGTAAAAAAGGCACGCGCGCTACACCCTATCTCTTGAATA GGGTGTAGCGCGCGTGCCG
3for	5′-GATCCGGGACAAGAGCCTGGAAGTGTTCAAGAGACACTTCCAGGCTC TTGTCCCTTTTTTACGCGTG
3rev	5′-AATTCACGCGTAAAAAAGGGACAAGAGCCTGGAAGTGTCTCTTGAACAC TTCCAGGCTCTTGTCCCG
11for	5′-GATCCGGAAGCTGCCTGCGGTGGAATTCAAGAGATTCCACCGCAGGCAG CTTCCTTTTTTACGCGTG
11rev	5′-AATTCACGCGTAAAAAAGGAAGCTGCCTGCGGTGGAATCTCTTGAATT CCACCGCAGGCAGCTTCCG
19for	5′-GATCCGCTATAGCGGTTGGAGCCACTTCAAGAGAGTGGCTCCAACCG CTATAGCTTTTTTACGCGTG
19rev	5′-AATTCACGCGTAAAAAAGCTATAGCGGTTGGAGCCACTCTCTTGAAGT GGCTCCAACCGCTATAGCG

The two complementary DNA strands designed to form each ADAMTS-5 shRNA template are shown, along with the sequences of control (non-targeted) shRNA template. The sequences include *Bam*HI and *Eco*RI restriction enzyme sites to enable cloning into the pSIREN retroviral vector, as well as a *Mlu*I site to confirm the presence of insert in ligated constructs

Set up additional reactions using the control-annealed oligonucleotide as well as a ligation negative control substituting water for oligonucleotide. Flick tubes to mix and incubate at room temperature for 3 h then store at −20°C until use.

3. Competent cells should be stored at −80°C and never allowed to thaw until use. Thaw sufficient tubes of α-select competent cells (Bioline) were used for each ligated construct (N.B. competent cells are fragile therefore always keep tubes on ice and never vortex). If desired, a tube with the addition of a control plasmid such as PUC19 can also be included as a positive control for transformation as well as a tube lacking DNA to control for the effectiveness of the antibiotic plates. Add 2 μL of the relevant ligation mixture to each tube of cells or 10 ng of PUC19, using the pipette tip to gently mix. Incubate the tubes on ice for 30 min. Heat tubes either in a water bath or heat block to 42°C for precisely 45 s and immediately return tubes to ice for 2 min. Using aseptic technique, add

250 μL LB and spread 50 or 150 μL onto LB-agar plates containing carbenicillin (see Note 1). Allow plates to dry and incubate upside down overnight. Check plates for isolated bacterial colonies, indicating successful transformation and ligation. There should be no bacterial growth on the plate of cells that lacked DNA, and generally be less or no colonies on the no insert control ligation plate. This, however, is not always the case; therefore, it is worth in any case of screening colonies on the positive ligation plates for the presence of shRNA insert.

3.3. Identification of Correct Plasmid Clones and Isolation of Plasmids

1. Each bacterial colony represents the growth of a single clone of cells containing one type of plasmid harbouring the ampicillin (carbenicillin) resistance gene. Several clones of each potential shRNA construct should be assessed for the presence of ligated shRNA. Select, for example, three colonies for screening from each type of shRNA ligation/transformation. Using a sterile pipette tip, carefully collect cells from an individual isolated colony and drop the tip into 5 mL LB-carbenicillin in a sterile universal tube. Incubate the tubes at 37°C with shaking overnight (see Note 2).
2. According to the Sigma protocol using their "Genelute" miniprep columns and reagents, centrifuge a total of 3 mL of the bacterial cultures sequentially in 1.5-mL Eppendorfs to pellet the cells (pSIREN plasmids are low copy number which means that much less plasmid is present per amount of bacterial culture than for high copy number plasmids). Follow the rest of the kit protocol as described, with the final elution being with 50 μL water.
3. Quantify the DNA using spectrophotometry (5 μL plasmid preparation diluted in 495 μL water then measure the absorbance in a quartz cuvette at 260 nm. 1 absorbance unit equals 50 μg/mL DNA). Store plasmid preparations at −20°C. Presence of the expected shRNA was verified using the in-house Applied Biosystems 373048-capillary sequencer, priming with the "U6for" primer (sequence: 5′-GGGCAGGAA GAGGGCCTAT-3′).

3.4. Maxiprep of Plasmid DNA for Making Virus

1. All the DNA constructs used are low copy number plasmids; therefore, we used large culture volumes to obtain sufficient DNA for transfection. For each plasmid required, prepare 500 mL Luria Broth (LB) in at least a 2-L volume glass conical flask. Autoclave the flasks and add 500 μL carbenicillin once cooled. Begin with a "starter culture" such as that prepared following transformation of plasmid (see above) and add 1 mL to each flask. Incubate overnight at 37°C with shaking. Centrifuge the culture in large centrifuge bottles at

5,000 × *g* for 15 min and discard the supernatant, recording the wet weight of the pellet. We have found that these cultures generally yield a pellet weight of 3 g, and the Qiagen (Crawley, UK) "Endofree" maxiprep kit suggests using 80 mL volumes of the kit buffers in their protocol for low copy number plasmids and this amount of pellet wet weight. Follow the maxiprep kit protocol except instead of using the cartridges, filter the neutralised lysate through a filter paper (provided with the buffer set) and use two of the columns per prep kit, combining the DNA pellets at the end.

2. Quantify DNA as above, before the isopropanol precipitation step, and ethanol precipitate under sterile conditions to give a final concentration of ≤1 μg/mL (see Note 3). Aliquot the DNA and store at −20°C.

3.5. Transfection to Produce Retrovirus (see Note 4)

1. Virus is produced by transient transfection of particular cell-lines that package recombinant viral RNAs into infectious but replicate incompetent viral particles. They possess in their genome an appropriate combination of some of the required *gag pol* and *env* genes which, combined with the expression of the rest of the required genes plus packaging signals from the transfected DNA, allow the production and secretion of virus from the cell into the medium. For production of retrovirus, GP2-293 cells are used, which possess the *gag* and *pol* genes, while the *env* gene is replaced by the vesicular stomatitis G protein in the pVSV-G plasmid which is co-transfected with the retroviral construct. VSV-G coat protein greatly increases the uptake of retrovirus by chondrocytes (and other cell types) and increases the stability of the viral particles, allowing concentration by centrifugation. It is imperative that only low passage number (<10–20) cells are used to prepare virus in order to gain sufficiently high titre. These cells are prone to detachment of the monolayer from the tissue culture plastic, so are maintained in collagen-coated vessels (see Note 5) and medium changes must be performed carefully (drop-wise). Cells are maintained in complete medium and are split (1 in 3 every 3–4 days) simply by tapping culture flasks to detach cells. Begin with the required number of collagen-coated 10-cm tissue culture dishes, depending on how much of each virus is required (see below). Include an additional dish as positive control for transfection. On the day of transfection, cells should have formed a 50–80% confluent monolayer.

2. Before beginning transfection, remove the medium from all plates and add 9 mL fresh medium (take 10 mL into the pipette to avoid "squirting" the last millilitre onto cells, risking cell detachment). Prepare labelled 1.5-mL Eppendorfs, one for each dish to be transfected. For the positive control,

add 30 μg green fluorescent protein (GFP)-expression plasmid (any expression plasmid which encodes GFP), and for retroviral transfections, to each tube add 15 μg pVSV-G and 15 μg retroviral shRNA construct. Make the volume up to 450 μL in each tube. To each tube then add 50 μL 2.5 M $CaCl_2$ and flick to mix. Prepare labelled 15-mL Falcon tubes, one for each dish of cells. Add 500 μL HEPES-buffered saline (HeBS) to each tube (see Note 6). Using a 1-mL plugged pipette attached to a mechanical pipettor, "bubble" the HeBS continuously with one hand, while adding the DNA-$CaCl_2$ solution drop-wise with the other hand. Flick each tube to mix and leave to stand at room temperature for 20 min. After this time, carefully add the appropriate DNA mix to a labelled dish of cells, drop-wise and evenly dispersing onto the medium, then return the dish to the incubator. After approximately 4 h, remove the transfection medium and replace with 9 mL fresh medium.

3. Virus-containing supernatant should be collected 48 h later and either filtered using a 0.45 μm low protein-binding disc filter, or centrifuged for 5 min at 3,000 × *g* to remove cell debris. Twenty-four hours following transfection, the efficiency of DNA take-up can be assessed by visualising the positive control cells under a fluorescence microscope. Generally, all cells should fluoresce green indicating 100% transfection efficiency.
4. Virus can be concentrated to increase titre by ultra-centrifugation at 50,000 × *g* for 90 min at 4°C. Remove the supernatant and resuspend in culture medium to 0.5–1% of the original volume, by incubating overnight at 4°C. Viral supernatants should be aliquoted and stored at –80°C. Note that excessive freeze–thawing and prolonged exposure to room temperature/handling will reduce viral titre.

3.6. Titration of Virus

1. Viral titre is the number of infectious units of virus present in the supernatant. It is measured in colony forming units (cfu) per millilitre. This means the number of viral particles present, measured as the number of cellular colonies (which result from infection of a single cell by a single antibiotic resistant viral particle), following transduction of virus onto cells and antibiotic selection. Viral supernatants can be titred either by transduction onto the cells of interest (C28/I2 chondrocytes in this case) or onto the standard cell-line for viral titration, NIH-3T3, which is used because of its particular efficiency for taking up virus.

   For each viral supernatant to be titrated, prepare a 15-mL Falcon tube labelled -2, containing 1.98 mL complete medium and four tubes labelled -3, -4, -5 and -6, containing

1.8 mL medium each. To the -2 tube, add 20 μL viral supernatant and invert to mix. This tube now contains a 10^{-2} dilution of virus. Add 200 μL of this suspension to the -3 tube. Continue serially diluting the virus by adding 200 μL of each diluted suspension to each following tube. Label the wells of a six-well plate of 80% confluent cells with *N*, -1, -2, -3, -4, -5, and -6 and remove the medium. Add 1 mL of complete medium to the *N*th well. This well serves as the negative control for the presence of virus. To the other wells, add 1 mL of the appropriate viral dilution. To all wells, add 2 μL of 4 mg/mL polybrene and carefully mix. Incubate the plate overnight as described above.

2. The next day, remove the medium and add 2 mL of fresh medium containing puromycin to each well. The appropriate concentration of puromycin needs to be determined empirically for each cell-line, using a kill-curve (see Note 7), such that the amount used in titrations and further experiments should be one that gives complete death of a dish of non-resistant cells in 48 h. For chondrocytes, this was 800 ng/mL puromycin.
3. Maintain the cultures for about 10 days, changing the medium/puromycin every 3–4 days. After this time, check that all cells are dead in the *N*th well and count the number of cell colonies present in the wells containing the highest viral dilutions. Ideally, there should be at least one colony in the -6 well, which would indicate a 1×10^6 cfu/mL viral titre for the original supernatant.

3.7. Transduction, Stimulation, and Collection of Samples for Expression Analysis

1. The amount of virus needed to achieve knockdown of gene expression can vary and therefore may need to be empirically determined. We found that approximately 70,000 cfu added to a 10-cm dish of cells (approximately six million cells) was sufficient to achieve greater than 90% knockdown of gene expression. The literature generally suggests trying anywhere between 100,000 and 3,000,000 cfu for this number of cells. We have successfully used up to 20,000,000 cfu to transduce this number of cells before encountering toxicity problems.

   As described above, for each gene of interest, there will be 4–5 retroviral gene targeted shRNA constructs plus a control shRNA construct. For analysis of mRNA expression knockdown, a single 10-cm dish of cells per construct should be prepared, plus an extra dish of cells as positive control for antibiotic effectiveness, which will not be transduced. Label the dishes and remove the medium, replacing with an appropriate volume of complete medium. This should allow for the addition of virus, making a total volume of 9 mL. Add 18 μL 4 mg/mL polybrene to each dish followed by the appropriate viral supernatant (or complete medium in the case of the control dish). Tilt dishes to mix and return them to the incubator for 24 h, when the medium was replaced and puromycin added (see above). Once all control cells are dead (typically 48 h) and the transduced cells are confluent, which may take a little longer, the medium can be removed from each dish and 4 mL TRI-reagent (Sigma) added. Process each dish of cells as four RNA samples as described below. Note, if desired, a proportion of cells can be grown on in the medium/puromycin as they are now a stably transduced cell-line. Cells can also be stored, frozen in liquid nitrogen (see Note 8).

2. In addition to assessing the levels of knockdown of selected "mixed populations" of transduced cells, we also cloned lines from single cells of transduced populations by diluting the cells and plating into 96-well plates such that on average one cell per well was expected. Wells were subsequently monitored and only those with the desired single colony of cells were chosen to be grown on. Sometimes it is found to be necessary to work with such clonal lines in order to maximise the degree of knockdown for further functional studies.

3.8. RNA Extraction and Analysis

1. N.B. All plasticware should be RNAse/DNAse free and gloves should be worn to reduce the risk of RNAse contamination. Using the protocol that accompanies TRI-reagent (Sigma), which suggests 1 mL TRI-reagent per sample, treat each 10-cm dish of cells as four samples. Triturate using a pipette tip and transfer to a 1.5-mL Eppendorf tube. This suspension can be stored at −80°C if desired for up to 1 month. Follow the TRI-reagent protocol, adding 1 μL glycogen as carrier at the isopropanol precipitation stage if low amounts of RNA are expected (this allows the visibility of RNA pellet even if there is a little presence). Dissolve the resulting RNA pellet in an appropriate volume of water, for example, 60 μL, and quantify using spectrophotometry (5 μL RNA sample diluted in 495 μL water then measure the absorbance in a quartz cuvette at 260 nm. 1 absorbance unit equals 40 μg/mL RNA). Store RNA samples at −80°C.

2. A DNAse treatment step is necessary in order to remove residual genomic DNA (gDNA) that will inevitably be present in the RNA samples. The presence of contaminating gDNA would confound the quantitation of the nucleic acid, as well as potentially lead to false positives in the real-time PCR due to gDNA annealing of primer/probes. For each sample, set up the following reaction, using the components that accompany the DNAse enzyme:

   10–15 μg RNA
   10 μL 10× Buffer

5 μL DNAse (5 U)
1 μL RNAsin (40 U)
Make up to 100 μL final volume with H_2O and flick to mix.

Incubate the tubes at 37°C for 30 min and add 10 μL stop buffer. Heat the tubes to 65°C for 10 min. Phenol–chloroform extract the RNA, by adding 100 μL water to the reaction mix followed by 200 μL phenol (pH 4.3). Vortex the tubes to form an emulsion, then centrifuge at top speed for 5 min. Carefully remove the upper aqueous layer to a fresh tube containing 200 μL chloroform. As before, vortex to form an emulsion, although this may take longer because chloroform does not form an emulsion so readily as phenol. Centrifuge and remove the upper layer again to a fresh tube and add 20 μL 3 M sodium acetate pH 5.5, as well as glycogen, as above, if desired. Flick tubes to mix and add 500 μL ethanol. After vortexing to mix, centrifuge tubes at 12,000 × *g* for 15 min at 4°C. Wash the pellet with 1 mL 70% ethanol as above and remove all excess. Redissolve the resulting pellet in a volume of water, aiming for a solution of ≤1 μg/μL RNA, and quantify using spectrophotometry as above. It is important that all nucleic acid solutions are used at not greater than approximately 1 μg/μL, in order to minimise viscosity and thus avoid inaccuracy in pipetting from the solution.

3. Reverse transcription generates from the RNA, complementary DNA strands, by annealing of DNA primer and use of a viral enzyme to polymerise DNA from an RNA template, hence the term reverse transcription. It is necessary to generate cDNA from the RNA, as the latter cannot be used as template in real-time PCR, the process used for quantification of levels of gene expression. For each sample, make up an annealing mix, including a negative control containing water instead of RNA template, using 0.5-mL thin-walled Eppendorf tubes:

   500 ng to 2 μg RNA
   0.5 μL random hexamers (0.75 μg)
   H_2O (to 14 μL)

   Using a PCR machine, heat the tubes to 70°C for 5 min and cool. Make up a bulk reaction mix as below, using the components that are supplied with the MMLV enzyme. Include sufficient mix for all tubes plus one extra to allow for error.

   5 μL 5× Buffer
   0.7 μL RNAsin (28 U)
   1 μL MMLV reverse transcriptase (200 U)
   5 μL dNTPs (10 mM stock solution containing all four)

Add 11 μL to each annealing mix and incubate the tubes at 37°C for 1 h followed by 70°C for 5 min. Store the resulting cDNA at 4°C until use.

4. We used the standard curve method of real-time PCR to quantify the relative levels of gene expression in our samples. The procedure involves carrying out on samples, one amplification reaction for each transcript of interest, as well as a second reaction for a transcript known to be present in the cells at levels that do not vary under the current experimental conditions. This second reaction controls for sample loading, and levels of this transcript can be used for normalisation of levels of the transcript of interest. So-called "house-keeping" genes, those that are involved in basic cellular processes, are generally chosen as the normalising transcript. We routinely use Cyclophilin-E for this purpose.

   Beginning with a stock of 0.5 μg/μL RNA from a cell sample known to express the gene of interest (we used cytokine-stimulated C28/I2 RNA), create a set of 13 cDNA standards as follows: Initially prepare three serial 1/10 dilutions of the RNA by adding 2 μL of each successive stock to 18 μL of water to give final concentrations of 50, 5, and 0.5 ng/μL. Using these stocks, prepare the standards in 0.5-mL thin-walled tubes as the following amounts: 4, 2, 1, and 0.5 μg (use 8, 4, 2, and 1 μL of the 0.5 μg/μL stock, respectively); 0.25 μg, 100 ng, and 50 ng (use 5, 2, and 1 μL of the 50 ng/μL stock, respectively); 25, 10, and 5 ng (use 5, 2, and 1 μL of the 5 ng/μL stock, respectively); 2.5, 1, and 0.5 ng (use 5, 2, and 1 μL of the 0.5 ng/μL stock, respectively). Include these RNA templates in a reverse transcription reaction as described above.

5. Draw out a "map" of a 96- or 386-well microplate, depending on the model of real-time PCR machine available. Mark on the map the loading positions of cDNAs, including the negative control, the 13 cDNA standards, and triplicates of each sample of interest. If desired, the entire sample set can be loaded in duplicate at opposite ends of the plate; so that both the transcript of interest and the loading control reactions can be run at the same time (it is also acceptable to run these on two separate occasions).

   For the transcript of interest reaction, carefully load 2 μL of each standard and sample into the bottom of the designated wells of the 386-well optical plate, while for the loading control, dilute the cDNAs 1/10 before loading 2 μL in the same way. Make up two separate reaction mixes in bulk for the transcript of interest and the loading control, including the appropriate primer/probe combinations for each (see Note 9). Each mix should be sufficient for the number of

wells required, plus for every six wells, include the mix for an extra well. The reaction mix is as follows:
5 μL 2× QPCR buffer (Eurogentec)
0.2 μL 10 μM forward primer (200 nM final)
0.2 μL reverse primer (200 nM final)
0.05 μL 20 μM FAM/TAMRA-labelled probe (100 nM final)
Make up to 10 μL with H_2O, allowing for the volume of the cDNAs.

Add 8 μL of the appropriate mix to each well, being very careful not to create air bubbles, though if necessary these can usually be removed with the pipette tip. Attach the sticky heat-sealable lid to the plate by rubbing over it with a plastic card, and insert into the real-time PCR machine (we use ABI model 7900). Using the computer software, label the wells on the "map" as required and run a standard reaction, which takes approximately 2 h.

6. After the machine has analysed the data, it will output "Ct" values for each sample, which means the amplification cycle at which the threshold level of fluorescence was reached. These values can be used to determine the relative starting amount of each transcript in samples, by plotting a straight line graph of the standards with Ct value against the amount of RNA. Although the actual numbers are obviously arbitrary, the relative amounts of transcript of interest or loading control can then be interpolated using the equation that defines each curve. To represent data, choose a sample to act as "calibrator" and set its value at "1," or "100," after which all other samples can be shown relative to it (Fig. 1).

4. Notes

1. Ampicillin is the standard antibiotic for use in selective media; however, we use the carbenicillin derivative, as it is more stable and therefore agar plates and cultures can be stored and remain usable for longer (up to 1 month).
2. If desired, once a plasmid with the correct insert has been identified, the transformed bacterial culture can be stored as a glycerol stock. Add 700 μL culture to 300 μL 50% autoclaved sterile glycerol in a 1.5-mL Eppendorf tube, mix, and store at −80°C. Plasmid can be extracted from this stock simply by stabbing the frozen culture with a sterile tip and dropping it into 5 mL LB-carbenicillin, before incubating overnight with shaking and extracting plasmid as described.

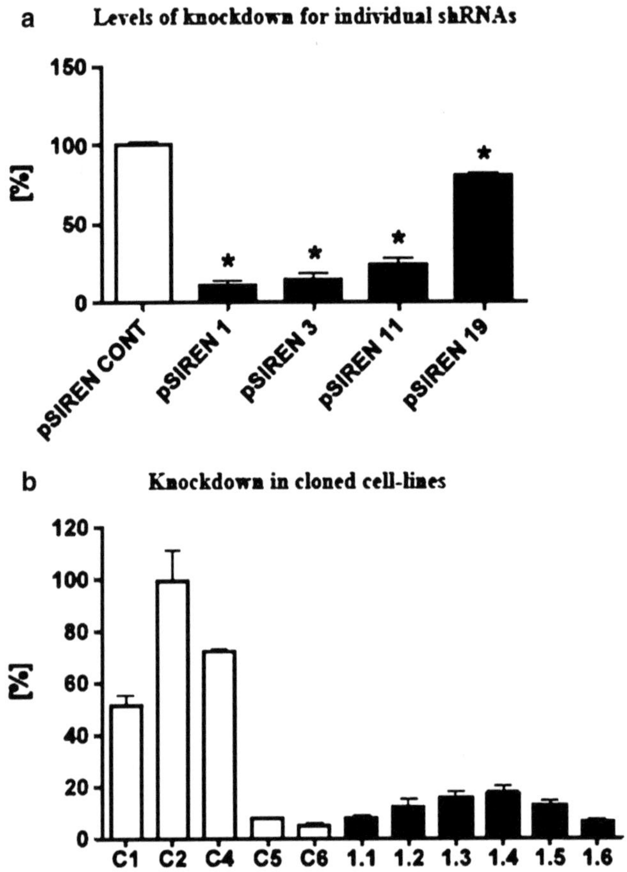

Fig. 1. Knockdown of ADAMTS-5 expression using shRNA-retrovirus. (**a**) C28/I2 chondrocytes were infected with pSIREN retrovirus expressing either of shRNA 1, 3, 11, 19 or control shRNA and selected using 800 ng/mL puromycin. Real-time PCR data for ADAMTS-5 transcript levels are shown as mean ± SEM and relative to control mean = 100%. Data were analysed using Student's *t*-test; $*p<0.05$ vs. control. The degree of knockdown for each shRNA was 1:89%, 3:86%, 11:76%, and 19:19%. (**b**) By dilution, six individual cells each from the pSIRENCONT (C1-6) and pSIREN1 (1.1–1.6) populations were isolated and grown up into clonal populations (population C3 did not survive). Data shown represents real-time PCR analysis of ADAMTS-5 expression in these cell-lines, relative to C2 mean = 100%. Levels were normalised to those of the cyclophilin-E transcript. Clones C2 and 1.6 exhibit maximal difference in expression levels for ADAMTS-5. Primers for ADAMTS-5 detection were "ADAMTS-5FOR" (5′-ggctcacgaaatcggacatt) and "ADAMTS-5REV" (5′-ggaaccaaaggtctcttcacaga) and probe was 5′-tggcctctcccatgacgattccaa, while for cyclophilin-E detection, primers were "Cyclo-EFOR" (5′-ggagggagagcccattgc) and "Cyclo-EREV" (5′-ggcttgttcccaatcttgatg) and probe was 5′-cccgctcaaatcctcagg tgtacatgg.

3. Once the maxiprep DNA has been washed in 70% ethanol, it will be sterile. Roll the ethanol around the insides of the tube before centrifuging. From this point, the tube should only be opened in a laminar flow hood under sterile conditions. Note that it is important to aliquot DNA for transfection as repeated freeze–thawing will cause degradation. DNA for transfection must be of the highest quality.

4. Virus work requires Class 2 containment and appropriate regulatory approval. All plasticware, gloves, etc. that have come into contact with virus should be disposed of with bleaching or autoclaving. All manipulations involving mammalian cells need to be carried out with sterility in a laminar flow hood, and incubations are in humidified 37°C incubators with an atmosphere of 5% CO_2. Confluent cell monolayers are split twice a week 1 in 5 (C28/I2) or 1 in 3 (GP2-293) by either tapping the cell monolayer off the plastic and diluting (GP2-293) or removal of medium, washing with phosphate-buffered saline (PBS), and incubating with trypsin–EDTA solution (0.05% trypsin:0.02% EDTA) for approximately 5 min at 37°C. Once cells have detached (tap flasks if necessary), neutralise the trypsin by adding an equal volume of medium, centrifuge at 500 × *g* for 5 min, and resuspend the resulting cell pellet in an appropriate volume of medium for replating.
5. Dilute a 1 mg/mL solution of collagen I tenfold with sterile water or PBS (use a commercial solution specifically intended for coating plastic). Coat 10-cm dishes with 5 mL of this solution and incubate either in a 37°C incubator for several hours or overnight at 4°C. Remove the excess and rinse with sterile PBS.
6. HeBS is made by adding 16.4 g NaCl, 11.9 g HEPES, and 0.21 g Na_2HPO_4 to 800 mL distilled H_2O, pH 7.05 using 5 N NaOH and make up to 1 L with distilled water. Filter sterilise using a 0.45-μm disc filter and store in 50 mL aliquots at −20°C. Transfection efficiency will vary with each new batch of HeBS, so it is important to verify the function of new batches.
7. Using a stock solution of puromycin at 10 mg/mL in methanol, prepare dilutions of 1/10, 1/100, and 1/1,000 by serial dilutions adding 10–90 μL methanol each time. Using a labelled 24-well plate of 80% confluent cells of interest, replace the medium with 1 mL fresh per well. To triplicate wells, add 5 μL 1/1,000× puromycin (50 ng/mL); 1 μL 1/100× (100 ng/mL); 2 μL 1/100× (200 ng/mL); 4 μL 1/100× (400 ng/mL); 8 μL 1/100× (800 ng/mL); 2 μL 1/10× (2 μg/mL), and 4 μL 1/10× (4 μg/mL). Incubate cells as normal and check every day for cell-death.
8. For long-term storage of mammalian cells, resuspend a cell pellet (typically, we use the contents of a T25 flask) in 1 mL FCS:10% DMSO. Transfer to a labelled 1.5- to 2-mL screw-top plastic cryo-vial and store no longer than overnight in a polystyrene box at −80°C. Remove to a suitable liquid nitrogen storage container.

9. Ready-made and designed to order real-time PCR primer/probe mixes can be obtained from ABI (Foster City, CA, USA); however, we routinely design our own primer and probe sets, using "Primer-express" software (ABI). We ensure chosen amplification products span the intron–exon boundary of a large intron, in order to prevent false positives (these can occur if primers are able to anneal and generate amplification product from gDNA).

References

1. Arner, E. C. (2002) Aggrecanase-mediated cartilage degradation. *Curr. Opin. Pharmacol.* **2**, 322–329.
2. Rodríguez-Manzaneque, J. C., Westling, J., Thai, S. N., Luque, A., Knauper, V., Murphy, G., Sandy, J. D., and Iruela-Arispe, M. L. (2002) ADAMTS1 cleaves aggrecan at multiple sites and is differentially inhibited by metalloproteinase inhibitors. *Biochem. Biophys. Res. Commun.* **293**, 501–508.
3. Kuno, K., Okada, Y., Kawashima, H., Nakamura, H., Miyasaka, M., Ohno, H., and Matsushima, K. (2000) ADAMTS-1 cleaves a cartilage proteoglycan, aggrecan. *FEBS Lett.* **478**, 241–245.
4. Collins-Racie, L. A., Flannery, C. R., Zeng, W., Corcoran, C., Annis-Freeman, B., Agostino, M. J., Arai, M., DiBlasio-Smith, E., Dorner, A. J., Georgiadis, K. E., Jin, M., Tan, X.-Y., Morris, E. A., and LaVallie, E. R. (2004) ADAMTS-8 exhibits aggrecanase activity and is expressed in human articular cartilage. *Matrix Biol.* **23**, 219–230.
5. Somerville, R. P., Longpre, J. M., Jungers, K. A., Engle, J. M., Ross, M., Evanko, S., Wight, T. N., Leduc, R., and Apte, S. S. (2003) Characterisation of ADAMTS-9 and ADAMTS-20 as a distinct ADAMTS subfamily related to Caenorhabditis elegans GON-1. *J. Biol. Chem.* **278**, 9503–9513.
6. Yamaji, N., Nishimura, K., Abe, K., Ohara, O., Nagase, T., and Nomura, N. (2001) Novel metalloproteinase having aggrecanase activity. Yamanouchi Pharmaceutical Co. Ltd, Japan.
7. Stanton, H., Rogerson, F. M., East, C. J., Golub, S. B., Lawlor, K. E., Meeker, C. T., Little, C. B., Last, K., Farmer, P. J., Campbell, I. K., Fourle, A. M., and Fosang, A. J. (2005) ADAMTS5 is the major aggrecanase in mouse cartilage *in vivo* and *in vitro*. *Nature* **434**, 648–652.
8. Glasson, S. S., Askew, R., Sheppard, B., Carito, B., Blanchet, T., Ma, H.-L., Flannery, C. R., Peluso, D., Kanki, K., Yang, Z., Majumdar, M. K., and Morris, E. A. (2005) Deletion of active ADAMTS5 prevents cartilage degradation in a murine model of osteoarthritis. *Nature* **434**, 644–648.
9. Little, C. B., Mittaz, L., Belluoccio, D., Rogerson, F. M., Campbell, I. K., Meeker, C. T., Bateman, J. F., Pritchard, M. A., and Fosang, A. J. (2005) ADAMTS-1-knockout mice do not exhibit abnormalities in aggrecan turnover *in vitro* or *in vivo*. *Arthritis Rheum.* **52**, 1461–1472.
10. Powell, A. J., Little, C. B., and Hughes, C. E. (2007) Low molecular weight isoforms of the aggrecanases are responsible for the cytokine-induced proteolysis of aggrecan in a porcine chondrocyte culture system. *Arthritis Rheum.* **56**, 3010–3019.
11. Tortorella, M. D., Malfait, A.-M., Deccico, C., and Arner, E. (2001) The role of ADAM-TS4 (aggrecanase-1) and ADAM-TS5 (aggrecanase-2) in a model of cartilage degradation. *Osteoarthr. Cartil.* **9**, 539–552.
12. Demircan, K., Hirohata, S., Nishida, K., Hatipoglu, O. F., Oohashi, T., Yonezawa, T., Apte, S., and Ninomiya, Y. (2005) ADAMTS-9 is synergistically induced by interleukin-1β and tumour-necrosis factor-α in OUMS-27 chondrosarcoma cells and in human chondrocytes. *Arthritis Rheum.* **52**, 1451–1460.
13. Song, R.-H., Tortorella, M. D., Malfait, A.-M., Alston, J. T., Yang, Z., Arner, E. C., and Griggs, D. W. (2007) Aggrecan degradation in human articular cartilage explants is mediated by both ADAMTS-4 and ADAMTS-5. *Arthritis Rheum.* **56**, 575–585.
14. Martin, S. E. and Caplen, N. J. (2007) Applications of RNA interference in mammalian systems. *Annu. Rev. Genomics Hum. Genet.* **8**, 81–108.
15. Kafienah, W., Al-Fayez, F., Hollander, A. P., and Barker, M. D. (2003) Inhibition of cartilage degradation. *Arthritis Rheum.* **48**, 709–718.

Chapter 8

Stem Cell and Neuron Co-cultures for the Study of Nerve Regeneration

Paul J. Kingham, Cristina Mantovani, and Giorgio Terenghi

Abstract

Many experimental in vivo studies have indicated that Schwann cells are key facilitators of peripheral nerve regeneration but their clinical therapeutic potential may be limited. Recent advances suggest that stem cell therapy could one day be used to treat nerve traumas. We have shown how adult stem cells can be differentiated into a Schwann cell phenotype, characterised by expression of glial cell proteins and promotion of neurite outgrowth. The development of new cell culture models which mimic the in vivo regeneration environment will help us to better understand the functional benefits of these cells. Here, we describe a stepwise approach towards this, moving from traditional two-dimensional non-contact co-cultures to new three-dimensional models utilising fibrin matrices.

Key words: Adult stem cell, Dorsal root ganglia, Fibrin, Glia, Peripheral nerve, Schwann cell, Regeneration

1. Introduction

The hallmark of the peripheral nervous system is the ability of the nerve to regenerate following injury, a process which involves a series of complex cell–cell and cell–extracellular matrix interactions (1). Animal models of nerve trauma have shown how nerve axotomy results in macrophage recruitment, degradation of axons and their myelin components, and subsequent proliferation of the resident Schwann cells, which align within the original basal lamina to form the bands of Büngner. Activated Schwann cells produce extracellular matrix molecules such as laminins and secrete a range of nerve growth factors and chemokines which act to direct axons towards the distal stump (2–4). Once axonal contact is re-established, Schwann cells return to their differentiated phenotype to ensheath and myelinate the regenerating axons (5).

John W. Haycock (ed.), *3D Cell Culture: Methods and Protocols*, Methods in Molecular Biology, vol. 695,
DOI 10.1007/978-1-60761-984-0_8, © Springer Science+Business Media, LLC 2011

Despite this knowledge, many of the underlying cell–cell signalling interactions remain to be elucidated, and much work is required to improve the clinical outcome for patients suffering from nerve trauma. In vitro models are required which mimic the three-dimensional environment through which the nerve regenerates. Furthermore, this approach can be used with the aim of developing a tissue-engineered construct for the injured peripheral nervous system. Despite the obvious advantages of using Schwann cells in any such construct, their use is limited by the need for a patient nerve biopsy with associated donor site morbidity, and difficulty in culturing the required quantities of cells quickly and efficiently (6). Recent advances in stem cell biology suggest that it might be possible to generate Schwann cells using defined differentiation protocols, which could ultimately be used to treat peripheral nerve injuries.

2. Materials

2.1. Stem Cell Harvest and Culture

1. Stem cell growth medium: 45 mL of MEM-α, 5 mL of foetal bovine serum (FBS), 0.5 mL of penicillin–streptomycin solution, all from Invitrogen, Paisley, UK.
2. Hanks' Balanced Saline Solution (HBSS; Invitrogen).
3. Trypsin (0.25%) – EDTA solution from Invitrogen.
4. Collagenase type I (Invitrogen): prepared fresh on day of stem cell isolation by dissolving 30 mg of collagenase type I in 15 mL of HBSS and filter-sterilised.
5. 70-μm Cell strainer (BD Biosciences, Oxford, UK).
6. Mesenchymal stem cell osteogenesis and adipogenesis kits (Millipore, Watford, UK).

2.2. Differentiation of Stem Cells to a Schwann Cell Phenotype

1. β-Mercaptoethanol (Sigma, Poole, UK).
2. All *trans*-retinoic acid (Sigma, Poole, UK): stock solution of 35 μg/mL diluted in dimethyl sulphoxide (DMSO).
3. Forskolin (Sigma, Poole, UK): stock solution of 10 mM dissolved in DMSO.
4. Basic fibroblast growth factor (bFGF; Peprotech EC Ltd, London, UK): stock solution of 100 μg/mL dissolved in 5 mM Tris–HCl buffer pH 7.6.
5. Platelet-derived growth factor (PDGF-AA; Peprotech EC Ltd): stock solution of 100 μg/mL dissolved in 10 mM acetic acid.
6. Glial growth factor-2 (GGF-2): stock solution of 1.18 mg/mL (Acorda Therapeutics, NJ, USA) (see Note 1).

7. Stem cell differentiation medium: 50 mL of stem cell growth medium, 70 μL of forskolin, 5 μL of bFGF, 2.5 μL of PDGF-AA, and 5 μL/10 μL of GGF-2 (for bone marrow or adipose stem cells, respectively) (see Note 2).

2.3. Immunocytochemistry for Schwann Cell Phenotypic Markers

1. Nunc slide flasks from Thermo Fisher Scientific, Loughborough, UK.
2. Phosphate-buffered saline (PBS): a 10× stock is made with 1.5 M NaCl, 20 mM KH_2PO_4, and 80 mM Na_2HPO_4 with adjustment to pH 7.4 with HCl if necessary. A working solution is made by dilution of one part with nine parts distilled water (see Note 3).
3. Paraformaldehyde (Sigma): a 4% (w/v) paraformaldehyde solution is made by dissolving in 1× PBS and heating to a maximum of 60°C with addition of a few drops of NaOH to clear the solution. Aliquots are frozen at –20°C (see Note 4).
4. Antibody diluent: 0.03% (v/v) Triton X-100, 0.1% (w/v) bovine serum albumin (BSA), and 0.1% (w/v) sodium azide (all dissolved in 1× PBS). This can be stored at 4°C.
5. Primary antibodies: rabbit anti-S100 (1:500 dilution; Dako, Ely, UK), mouse anti-GFAP (1:200; Millipore), and rabbit anti-p75NTR (1:500; Promega, Southampton, UK).
6. Fluorescently coupled secondary antibodies: goat anti-rabbit FITC (1:100; Vector Labs, Peterborough, UK) and goat anti-mouse CY3 (1:200; GE Healthcare, Bucks, UK).
7. Vectashield solution with DAPI (4′,6-diamidino-2-phenylindole) (Vector Labs).

2.4. Adult Schwann Cell Harvest and Culture

1. Poly-D-lysine (PDL; Sigma): 1 mg/mL dissolved in tissue culture grade water. Flasks are coated at a concentration of 100 μg/mL for a minimum of 15 min prior to one wash with 5 mL of tissue culture grade water and air drying (see Note 5).
2. Schwann cell growth medium: 45 mL of DMEM (low glucose), 5 mL of FBS, 0.5 mL of penicillin/streptomycin solution, 50 μL of forskolin, and 2.5 μL of GGF-2.
3. Collagenase type IV (2.5% w/v): 100-μL stocks dissolved in DMEM and frozen.
4. Dispase (Invitrogen): 1 mg/mL solution in DMEM prepared fresh on day of enzyme digestion.
5. Mouse anti-rat Thy 1.1 and rabbit complement (Serotec, Kidlington, UK).

2.5. DRG Neuron and Stem/Schwann Cell Co-culture

1. For effective DRG attachment and neurite outgrowth, 25-mm glass coverslips or tissue culture plasticware is coated with laminin (2 μg/cm^2, Sigma) for 2 h at 37°C (see Note 6).

2. Collagenase type IV (1.25% w/v): 200-μL stocks dissolved in F12 medium and frozen.
3. Trypsin (bovine pancreatic; Worthington Biochemicals): 2.5% (w/v) stocks in F12 medium, filtered-sterilised, and stored frozen as 200-μL aliquots.
4. Bottenstein and Sato (BS) medium for plating of DRG: 20 mL of F12 medium, 100 μg/mL BSA, 100 μg/mL transferrin, 100 μM putrescine, 30 nM sodium selenite, 20 nM progesterone, 10 μM cytosine arabinoside, and 10 nM insulin. Each component is made as 100× stock and frozen.
5. 2.5S beta-NGF (Sigma): 5 μg/mL stock dissolved in 1 mg/mL fatty acid-free BSA solution. Add 8 mL of F12 medium to give a final volume of 10 mL solution of NGF, which can be frozen in 200-μL aliquots.
6. For non-contact cell co-cultures, 6-well tissue culture inserts (1 μm pore size, polyethylene terephthalate membrane) from BD Biosciences.
7. 6-Well and 24-well tissue culture plates from BD Biosciences.
8. Primary antibody for staining neurite outgrowth: mouse anti-β III tubulin (1:1,000 dilution, Sigma).

2.6. Fibrin Matrices

1. Tissucol Kit (Baxter Healthcare) containing individual fibrinogen and thrombin components.
2. Fibrinogen component: 90 mg of fibrinogen, 10 IU of factor XIII, 5.5 mg of fibronectin, 80 μg of human plasminogen, 3.4 UPE of bovine aprotinin are diluted in 1 mL of solution A: 3,000 UIK/mL aprotinin, 25 mM sodium citrate, 29 mM NaCl, 333 mM glycine, and 15 mg/mL human serum albumin. Aliquot and store frozen at −40°C until required.
3. Human thrombin: 500 IU is diluted in 1 mL of buffer B: 40 mM $CaCl_2$, 149 mM sodium chloride, 40 mM glycine, and 50 mg/mL human serum albumin. Aliquot and store frozen at −40°C until required.

3. Methods

Adult stem cells are a multi-potent population of cells, which are now believed to have the capacity to differentiate into cells of mesodermal, endodermal, and ectodermal origins, irrespective of their tissue of origin (7). Methods are described to isolate adult stem cells from bone marrow and adipose tissue based on their relative adherence to tissue culture plastic, followed by their directed differentiation towards a Schwann cell phenotype (Fig. 1).

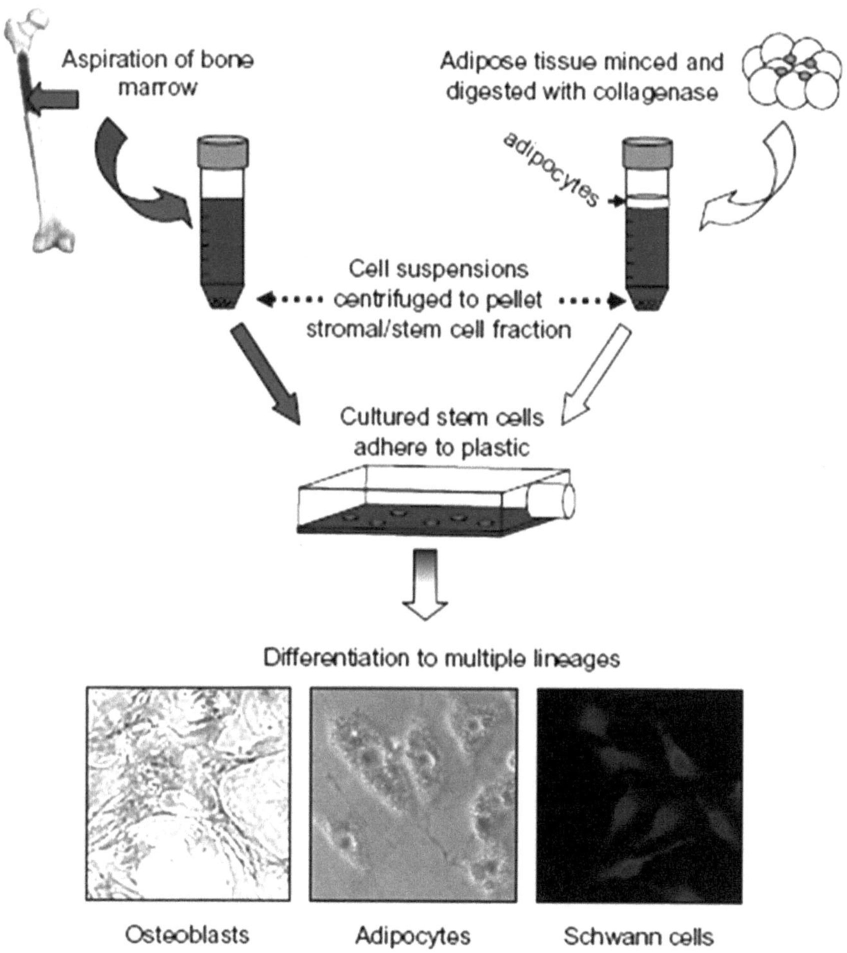

Fig. 1. Scheme for isolating and culturing adult stem cells isolated from bone marrow and adipose tissues. Cells can differentiate along multiple lineages, e.g. osteoblasts (alizarin red stained), adipocytes (Oil Red O stained), and Schwann cells (anti-GFAP antibody stained).

A protocol is also provided for isolation of adult Schwann cells with which to make comparisons of the functional phenotype of these differentiated adult stem cells. The use of fresh nerve tissue generates low yields of Schwann cells, so an in vitro pre-degeneration method is used to mimic Wallerian degeneration and encourage fibroblast outgrowth, thereby generating highly pure Schwann cell cultures. Dissociated sensory neurons are among the most widely used neuronal cell type for primary culture and have the advantage that their isolation from the dorsal root ganglia (DRG) requires axotomy, so they share many of the features of regenerating neurons in vivo. A stepwise approach to explore the interactions between DRG neurons and the differentiated stem/Schwann cells is provided, moving from traditional two-dimensional non-contact co-cultures (Fig. 2) to the development of new three-dimensional models utilising fibrin matrices

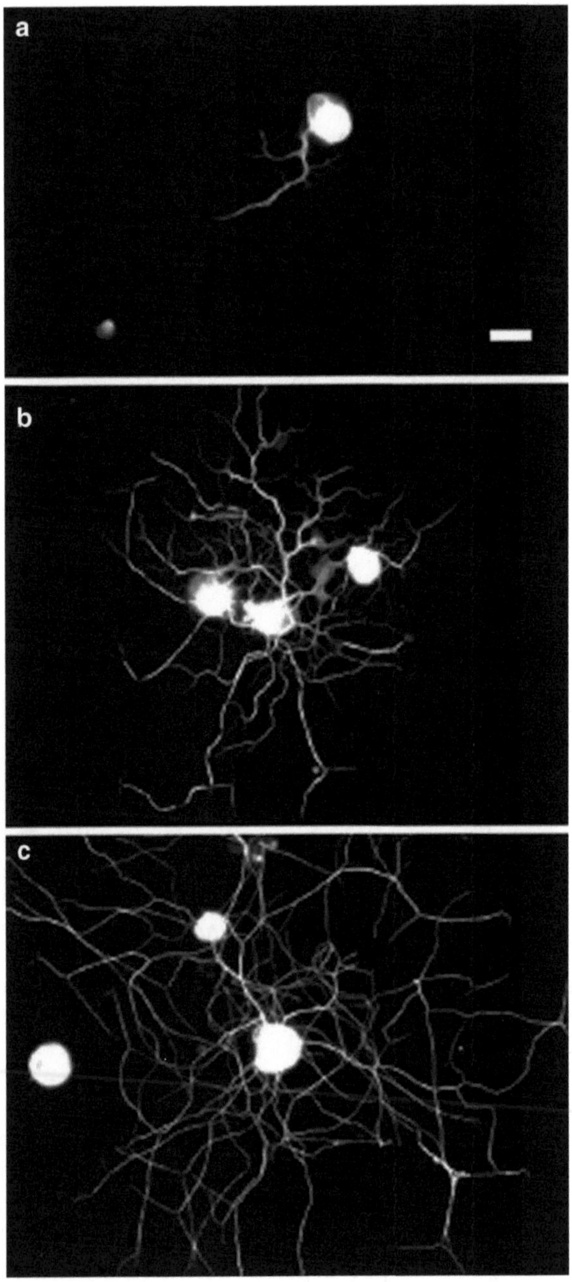

Fig. 2. Schwann-cell-differentiated stem cells promote neurite outgrowth of DRG neurons via released factors. After 48 h plating, DRG neurons alone (**a**), treated with 10 ng/mL NGF (**b**) or co-cultured with differentiated stem cells seeded in tissue culture inserts (**c**) were stained with anti-β III tubulin antibody. Scale bar = 40 μm.

(Figs. 3 and 4). Fibrin is chosen since one of the first reactions of the injured peripheral nervous system involves the formation of a fibrin clot (8). Moreover, fibrin glue is a clinically accepted material which can be used for peripheral nerve repair (9). In the

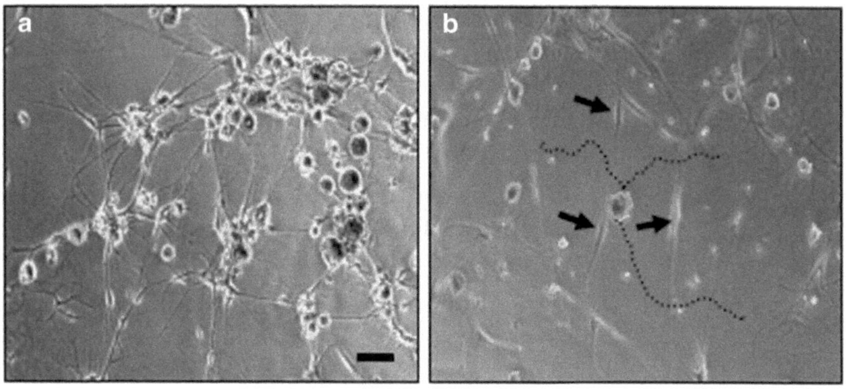

Fig. 3. DRG neurons adhere to and extend neurites on fibrin matrices (**a**) and can interact with differentiated stem cells (*arrows* in **b**). For clarity, some neurites are highlighted with a dashed line. Scale bar = 40 μm.

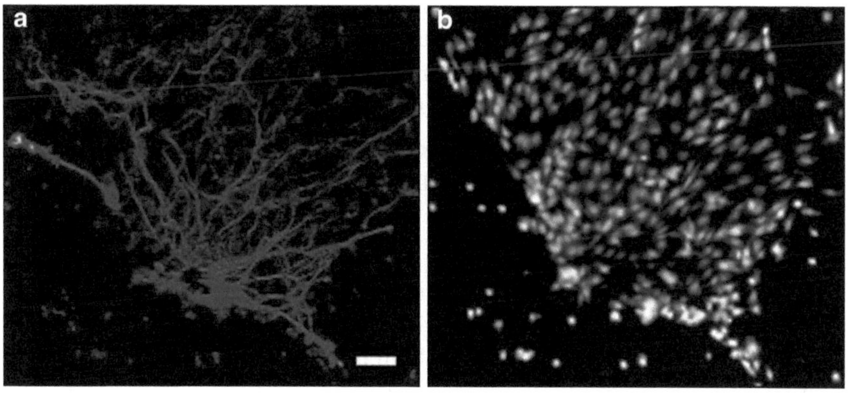

Fig. 4. Co-cultures of DRG neurons and differentiated stem cells within fibrin matrix. (**a**) DRG neurons stained with anti-β III tubulin antibody and (**b**) the corresponding region stained with anti-S100 antibody indicates proliferating differentiated stem cells. Scale bar = 40 μm.

future, it is envisaged that these model systems will be used to identify the signalling mechanisms mediating nerve regeneration and lead to the discovery of new therapeutic options for the treatment of peripheral nerve injuries.

3.1. Stem Cell Harvest and Culture

1. Remove adult rat long bones and dissect out visceral fat encasing the stomach and intestines. Store the fat in HBSS on ice and begin processing the bone by removing proximal and distal ends to reveal the marrow cavity.
2. Inject 10 mL of stem cell growth medium through each marrow cavity using a 21-gauge syringe needle and collect the cell suspension in a centrifuge tube.
3. Triturate the resulting cell suspension using a fresh needle and filter through a 70-μm filter. Centrifuge for 5 min at 600 × *g*.

4. Aspirate the supernatant and resuspend the cell pellet in stem cell growth medium. For each rat, plate 10 mL of suspension in a 75-cm^2 tissue culture flask and incubate in 5% CO_2 at 37°C and wash the flasks daily with HBSS to eliminate non-adherent haematopoietic cells.
5. Begin processing of the fat tissue by removing from HBSS and chop to a fine consistency using a sterile razor blade (see Note 7).
6. For each rat, transfer tissue to 15 mL of freshly made collagenase type I solution and place in a water bath at 37°C for 1–2 h per rat.
7. Neutralise the enzymes by the addition of an equal volume of stem cell growth medium and centrifuge the solution at 800 × *g* for 5 min – at this stage, an upper layer of floating adipose cells should be observed, and a pellet of cells constituting the stromal fraction and containing the stem cells will have formed. Carefully aspirate the upper layer and then the remaining medium to leave the pellet.
8. Resuspend the pellet(s) in 10 mL of stem cell growth medium and pass the solution through a 70-μm filter to remove any large pieces of undissociated tissue. Transfer solution to 15-mL tubes.
9. Centrifuge solutions at 800 × *g* for 5 min and resuspend each pellet in 10 mL of stem cell growth medium and transfer to 75-cm^2 flasks. Maintain cultures at sub-confluent levels in a 37°C incubator with 5% CO_2 and passage with trypsin/EDTA when required.
10. The presence of stem cells in cultures of both bone marrow and adipose tissue can be confirmed by using commercially available kits to differentiate the osteoblast and adipocyte lineages (Fig. 1).

3.2. Differentiation of Stem Cells to a Schwann Cell Phenotype

1. Remove growth medium from sub-confluent cultures at passage 2 and replace with medium supplemented with 0.8 μL of β-mercaptoethanol per 10 mL for 24 h (see Note 8).
2. Wash cells with HBSS and replace with fresh stem cell growth medium containing all-*trans*-retinoic acid (1:1,000 dilution of stock).
3. Three days later, wash the cells with HBSS and replace with stem cell differentiation medium (10 mL). Incubate the cells for 2 weeks under these conditions with addition of fresh differentiation medium approximately every 72 h.

3.3. Immunocytochemistry for Schwann Cell Phenotypic Markers

1. Trypsinise cells from cell culture flasks and re-plate on slide flasks at a density of 10,000 cells. For comparison of staining, control cultures of adult Schwann cells (see Subheading 3.4) can be stained in parallel.

2. After 24–48 h, aspirate medium from cell cultures and then rinse twice in PBS.
3. Fix cells in 4% (w/v) paraformaldehyde for 20 min.
4. Wash samples 3 × 5 min in PBS.
5. Permeabilise using 0.2% (v/v) Triton X-100 dissolved in PBS for 20 min (see Note 9).
6. Wash samples 2 × 5 min in PBS.
7. Block with the suitable serum directed against the species in which the secondary antibody is raised, e.g. 5% (v/v) normal goat serum for 1 h at room temperature.
8. Incubate samples with primary antibodies at 4°C overnight (see Subheading 2.3) in a humidified chamber.
9. Wash samples 3 × 10 min in PBS.
10. Incubate samples with secondary antibodies (see Subheading 2.3) for 1 h at room temperature, with protection from the light to prevent fading of fluorescence.
11. Wash samples 3 × 10 min in PBS.
12. Mount samples with an anti-fading Vectashield solution with DAPI.
13. The slides are viewed through a fluorescence microscope with suitable filters for detection of FITC and CY3 conjugated antibodies.

3.4. Adult Schwann Cell Harvest and Culture

1. Aseptically remove sciatic nerves from adult rats and place in ice-cold DMEM.
2. Using a dissecting microscope remove the epineurium and cut each nerve into 1-mm pieces and place in a 35-mm^2 dish containing 1.5 mL of Schwann cell growth medium.
3. Incubate the nerves for 2 weeks with fresh medium changes every 3–4 days, taking care not to aspirate nerve pieces.
4. After 2 weeks, transfer the pieces of nerve to new 35-mm^2 dishes and add 900 μL of Schwann cell growth medium plus 100 μL of collagenase type 4 and 1 mL of dispase and incubate for 24 h.
5. The next day, triturate the nerves using a 1-mL pipette tip and filter through a 70-μm cell strainer and centrifuge at 600 × *g* for 5 min.
6. Resuspend the pellet in 5 mL of Schwann cell growth medium and seed in one 25-cm^2 PDL-coated flask per rat nerve. Incubate the cells at 37°C with 5% CO_2 until they reach confluence. Upon confluence, the contaminating fibroblasts can be removed by immunodepletion (see Note 10).
7. Trypsinise cultures, centrifuge at 600 × *g* for 5 min, and resuspend the cell pellet in 500 μL of mouse anti-rat Thy 1.1 antibody

at 1:500 dilution in DMEM. Incubate in a water bath at 37°C for 10 min.

8. Add 250 μL of freshly prepared rabbit complement to the cells for 30 min at 37°C and gently agitate the suspension by hand every 10 min to mix.
9. Add 10 mL of Schwann cell growth medium to the mixture and centrifuge at 600 × *g* for 5 min. Resuspend the pellet in Schwann cell growth medium and seed the cells on a new PDL-coated flask and place in incubator.

3.5. DRG Neuron and Stem/Schwann Cell Co-culture

1. Using adult rats excise the spinal column and remove dorsal part to expose spinal cord.
2. Divide the spinal cord in half in the sagittal plane and remove cord tissue to expose the dorsal root ganglia and roots within the vertebral canals.
3. Gently pull individual roots with DRG attached from the vertebral canal and collect the DRG into a 35-mm^2 dish containing F12 medium. Repeat in a separate dish for each animal.
4. Using a dissecting microscope remove nerve roots, taking care to leave DRG intact.
5. Transfer DRG to a small Petri dish containing 1.8 mL of F12 medium.
6. Add 200 μL of collagenase type 4 and incubate for 1 h at 37°C.
7. Remove medium gently and repeat step 6 for a further 45 min.
8. Remove medium and gently wash DRG in fresh F12 medium.
9. Add 1.8 mL of F12 and 200 μL of trypsin and incubate at 37°C for 30 min.
10. Remove trypsin and add 1 mL of F12 medium containing 500 μL of FBS to arrest further digestion.
11. Remove this and wash the DRG gently three times in F12 medium to remove all traces of serum.
12. Add 2 mL of fresh F12 medium to each dish of DRG and transfer to 15-mL centrifuge tubes.
13. Gently mechanically dissociate DRG with a glass pipette. Leave to settle for a couple of minutes, then aspirate the supernatant and place this in a second 15-mL falcon tube.
14. Add a further 2 mL of F12 medium to the remaining cell bolus and repeat the gentle mechanical dissociation as above. Repeat this cycle a total of four times to produce 8 mL of cell suspension (see Note 11).

15. Filter the cell suspension into a fresh 15-mL falcon tube through a 70-μm filter. Combine the tubes to generate a concentrated cell suspension.
16. Centrifuge at 300 × *g* for 5 min.
17. Pipette a 15% (w/v) BSA solution in F12 medium slowly into a 15-mL tube held tilted at an angle to create a trail of viscous BSA down one side of the tube.
18. Remove all but the last 500 μL of the supernatant from the centrifuged cells and resuspend the cell bolus in the remaining 500 μL. Slowly pipette this cell suspension into the pre-prepared tube down the BSA track.
19. Centrifuge at 500 × *g* for 10 min. This creates two layers of liquid with the DRG neurons at the base of the tube and unwanted debris material suspended in a second layer above. Gently aspirate all supernatant and unwanted material to leave only the neuron cell pellet.
20. Resuspend neurons in modified Bottenstein and Sato medium and pipette 400 μL of solution onto laminin-coated coverslips in a 6-well plate. One rat generates enough cells for approximately 6–9 coverslips. Allow cells to adhere for 2 h and then carefully add 2 mL of BS medium to the wells.
21. Transfer tissue culture inserts seeded with stem or Schwann (50,000–1,00,000) cells, 24 h previously, to the 6-well plate and begin co-culture. DRG can be cultured in the absence or presence of NGF. In each well of the plate, add 200 μL of the stock NGF to give a final concentration of 10 ng/mL.

3.6. Fibrin Matrices

1. Dilute the thrombin stock solution 100-fold in solution B and add 200 μL to each well in a 24-well plate.
2. Dilute the fibrinogen component tenfold with solution A and add 200 μL to each well. Allow to clot. Alternatively proceed to step 4 (see Note 12).
3. Combine 5,000-10,000 stem/Schwann cells with DRG in just enough BS medium to cover the surface of the fibrin (approximately 100 μL) and allow to attach for 2 h, followed by addition of a further 500 μL of BS medium (with or without NGF).
4. Dilute the fibrinogen component tenfold with solution A and resuspend DRG/stem/Schwann cell mixture in 200 μL and carefully add to each well. Allow to clot and then add 500 μL of BS medium.
5. Cells on coverslips and on/within fibrin matrices can be immunostained using the protocol described previously (see Subheading 3.3) with rabbit anti-S100 antibody to label differentiated stem/Schwann cells and mouse anti-β III tubulin

for neurite outgrowth of DRG. Image analysis can be performed using software such as Image-Pro Plus (MediaCybernetics, MD, USA) to measure total number of neurites, and branching and length of the longest neurite (10).

4. Notes

1. As an alternative to GGF-2, commercially available recombinant NRG1-β1 (R & D Systems, Abingdon, UK) can be used to differentiate stem cells to a Schwann cell phenotype (11, 12).
2. Prepare only the quantity of differentiation medium as is required at the time of the experiment to minimise loss of growth factor activity. Treatment of the tubes, in which the stock growth factors are stored, with SigmaCote® (Sigma) is also helpful.
3. Unless otherwise stated, solutions are prepared with water which has a resistance of 18.2 MΩcm and total organic content of less than five parts per billion.
4. Paraformaldehyde powder is highly toxic; prepare the solution in a ventilated fume hood and do not heat above 60°C to prevent its molecular breakdown. Once thawed, do not refreeze an aliquot.
5. Tissue culture plastic treated with PDL should be used within 1 week; otherwise, there is noticeable reduction in cell attachment.
6. Coating the glass coverslip first with poly-lysine or poly-ornithine and then laminin helps to increase the concentration of laminin applied. Always thaw laminin on ice to prevent gelling at room temperature.
7. A similar protocol for isolation of rat adipose-derived stem cells from inguinal fat pads has been reported (12).
8. The induction of Schwann cell phenotype is most efficient when the cells do not have to be trypsinised and replated before the second addition of glial growth factors (day 8). To minimise this possibility (especially with respect to adipose stem cells which proliferate at a fast rate), differentiation should be initiated with a maximum of 10,000 cells/mL.
9. Permeabilisation is not required for cell surface markers such as p75NTR, and antigen binding is maximised by using 0.2% EDTA/PBS to detach cells for re-plating on slide flasks.
10. Do not try to immunodeplete Schwann cell cultures prior to reaching confluence as this significantly reduces the viable

yield of cells. Alternative approaches to reduce fibroblast contamination include flushing flasks with a jet stream of ice-cold PBS to detach Schwann cells followed by re-plating (13).

11. The key to good cell yields with minimal contamination is the trituration cycles. Excessive trituration results in low cell yields with excessive cell debris, and under-trituration leaves chunks of tissue intact.
12. If plating multiple wells, use a fresh pipette tip each time to prevent uneven clotting.

Acknowledgements

We would like to thank Acorda Therapeutics for their continuing supply of GGF-2 used to maintain Schwann cell cultures and differentiate stem cells.

References

1. Chen, Z. L., Yu, W. M., and Strickland, S. (2007) Peripheral regeneration. *Annu. Rev. Neurosci.* **30**, 209–233.
2. Ide, C. (1996) Peripheral nerve regeneration. *Neurosci. Res.* **25**, 101–121.
3. Hall, S. (2005) The response to injury in the peripheral nervous system. *J. Bone Joint Surg. Br.* **87**, 1309–1319.
4. Jessen, K. R., and Mirsky, R. (2008) Negative regulation of myelination: relevance for development, injury, and demyelinating disease. *Glia* **56**, 1552–1565.
5. Svaren, J., and Meijer, D. (2008) The molecular machinery of myelin gene transcription in Schwann cells. *Glia* **56**, 1541–1551.
6. Vroemen, M., and Weidner, N. (2003) Purification of Schwann cells by selection of p75 low affinity nerve growth factor receptor expressing cells from adult peripheral nerve. *J. Neurosci. Methods* **124**, 135–143.
7. Phinney, D. G., and Prockop, D. J. (2007) Concise review: mesenchymal stem/multipotent stromal cells: the state of transdifferentiation and modes of tissue repair – current views. *Stem Cells* **25**, 2896–2902.
8. Akassoglou, K., Akpinar, P., Murray, S., and Strickland, S. (2003) Fibrin is a regulator of Schwann cell migration after sciatic nerve injury in mice. *Neurosci. Lett.* **338**, 185–188.
9. Jubran, M., and Widenfalk, J. (2003) Repair of peripheral nerve transections with fibrin sealant containing neurotrophic factors. *Exp. Neurol.* **181**, 204–212.
10. Kingham, P. J., Kalbermatten, D. F., Mahay, D., Armstrong, S. J., Wiberg, M., and Terenghi, G. (2007) Adipose-derived stem cells differentiate into a Schwann cell phenotype and promote neurite outgrowth in vitro. *Exp. Neurol.* **207**, 267–274.
11. Dezawa, M., Takahashi, I., Esaki, M., Takano, M., and Sawada, H. (2001) Sciatic nerve regeneration in rats induced by transplantation of *in vitro* differentiated bone-marrow stromal cells. *Eur. J. Neurosci.* **14**, 1771–1776.
12. Xu, Y., Liu, L., Li, Y., Zhou, C., Xiong, F., Liu, Z., Gu, R., Hou, X., and Zhang, C. (2008) Myelin-forming ability of Schwann cell-like cells induced from rat adipose-derived stem cells *in vitro*. *Brain Res.* **1239**, 49–55.
13. Haastert, K., Mauritz, C., Chaturvedi, S., and Grothe, C. (2007) Human and rat adult Schwann cell cultures: fast and efficient enrichment and highly effective non-viral transfection protocol. *Nat. Protoc.* **2**, 99–104.

Chapter 9

Production of Tissue-Engineered Skin and Oral Mucosa for Clinical and Experimental Use

Sheila MacNeil, Joanna Shepherd, and Louise Smith

Abstract

Since the early 1990s, our understanding of how epithelial and stromal cells interact in 3D tissue-engineered constructs has led to tissue-engineered skin and oral mucosa models, which are beginning to deliver benefit in the clinic (usually in small-scale reconstructive surgery procedures) but have a great deal to offer for in vitro investigations. These 3D tissue-engineered models can be used for a wide variety of purposes such as dermato- and mucotoxicity, wound healing, examination of pigmentation and melanoma biology, and in particular, a recent development from this laboratory, as a model of bacterially infected skin. Models can also be used to investigate specific skin disease processes. In this chapter, we describe the basic methodology for producing 3D tissue-engineered skin and oral mucosa based on de-epidermised acellular human dermis, and we give examples of how these models can be used for a variety of applications.

Key words: Tissue-engineered skin, Tissue-engineered oral mucosa, Pigmentation, Melanoma invasion, Infected skin, Wound healing

1. Introduction

Cultured autologous skin cells have been used to benefit patients with severe burns since the early 1980s, so the production of cultured cells and tissue-engineered skin is probably the most mature of all of the tissue-engineered tissues (recently reviewed in (1, 2)). In this chapter, we focus on producing epidermal dermal tissue-engineered skin and oral mucosa based on de-epidermised acellular human dermis. These tissue-engineered constructs can be used in the clinic, and here, issues of sterilisation and the use of accredited clean rooms must be tackled. However, the majority of the chapter looks at the production of these physiologically relevant tissue-engineered skin and oral mucosa models for in vitro research.

John W. Haycock (ed.), *3D Cell Culture: Methods and Protocols*, Methods in Molecular Biology, vol. 695,
DOI 10.1007/978-1-60761-984-0_9, © Springer Science+Business Media, LLC 2011

There are an increasing number of areas where these 3D physiologically relevant tissue-engineered models come into their own. The examples described in this chapter are the use of tissue-engineered skin to study wound healing, skin graft contracture, bacterial infection of skin, studies of normal skin cell pigmentation and of the pathology of how transformed melanocytes – melanoma cells – interact with skin. For oral mucosa, we describe its production and how this has been used clinically and indicate in brief how it can be used for a wide range of experimental studies.

A major concern when grafting back split-thickness skin (STS) or oral mucosa is the failure to achieve rapid neovascularisation. From our experience and that of others in the area, it is necessary to make sure that the tissue-engineered grafts are not too thick and are grafted onto well-vascularised wound beds. Under these circumstances, neovascularisation will occur within about 5–7 days, and the graft will survive. Grafting tissue-engineered skin onto a poorly vascularised wound bed has resulted in delayed angiogenesis and loss of grafts in our experience (3). Tissue-engineered skin or oral mucosa which becomes vascularised will generally survive long term (see ref. 4).

There are many applications for tissue-engineered skin and oral mucosa. This is a rapidly growing area. There is a need to reduce or replace animal experimentation in the development of products that are used on human skin and hair, but equally many of the more interesting things that happen in skin happen through a dialogue between different skin cells and are rarely seen when individual cells are studied in monolayers. Table 1 lists some of the in vivo and in vitro applications of tissue-engineered skin. With respect to clinical applications, the clinical applications for epidermal–dermal reconstructed skin and oral mucosa are still relatively in their infancy (compared to the expansion and clinical use of cultured keratinocytes or fibroblasts on their own). As it takes a certain amount of time to produce these tissue-engineered constructs, they are entering the clinic for small-scale reconstructive surgery applications. Thus, although theoretically tissue-engineered skin could be used to replace full thickness skin loss in burn patients, in practice, the time taken to produce it (at least 6 weeks in our laboratories) takes it outside of the acute window (within 3 weeks) in which one would like to graft patients with extensive skin loss.

However, there are many reconstructive surgery applications where tissue-engineered skin and oral mucosa will be useful as listed in Table 1. With respect to in vitro applications, the table lists the ones that we are aware of. There will be many others since this area of research is rapidly expanding.

Thus, we have a basic recipe for the production of 3D tissue-engineered skin and oral mucosa which can be flexibly adapted for clinical use and for a wide range of investigations of skin and mucosa biology.

Table 1
Applications of tissue-engineered skin and oral mucosa

Clinical applications
Tissue-engineered skin can be used in reconstructive surgery to replace
Areas of skin contraction
Pigmented lesions
Scar tissue
Tissue loss due to burn injuries
Tissue loss due to skin cancer
Tissue-engineered oral mucosa can be used in reconstructive surgery to replace
Loss of oral tissue due to trauma or cancer
Scarred tissue of the urethra
In vitro uses for tissue-engineered skin
As models of dermatotoxicity and skin irritancy
To study the effects of agents on the skin barrier
To study penetration of agents into skin
As wound healing models
As a model of skin bacterial infection
As models to study angiogenesis
As models to study skin contraction
As models of skin pigmentation
As models to investigate diseases processes
Melanoma
Vitiligo
Psoriasis
Blistering diseases
In vitro uses for tissue-engineered oral mucosa
As models of mucotoxicity
To study penetration of agents into oral mucosa
As models to study oral cancer
As a model of the host–*Candida* relationship
As a model of the host–bacteria relationship in periodontal disease
As a model of radiation/chemotherapy induced mucositis
As a model to study lichen planus

A general issue in obtaining tissue for culture of cells, whether for clinical use or experimental use, is acquiring written informed consent from the patient. In the case of experimental use, this may be relatively simple, confirming that tissue removed at, say for example, elective surgery such as breast reduction or abdominoplasties, can be used for research purposes. The extent to which the research purposes will need to be defined and the nature of the paperwork to be completed will vary from institute to institute and with time, but the essence remains the same – that the patient is fully aware that you are taking tissue that belongs to them and that they have consented freely that you may do this. Implicit in this is the understanding that you do not use it for other purposes and that it is acknowledged as a gift. The ethical consent procedure may exclude use of donated tissue for commercial purposes for example, but this may vary from place to place.

In culturing cells for clinical use, there will be a more extensive review process in which the protocol will have been reviewed by an ethical committee so that, for example, patients will donate a biopsy of skin to be used in their treatment or less commonly in the treatment of other patients. Again, the essence of informed consent is that the patient has a full understanding of the intentions of the researcher and that tissue is freely given for these purposes. Signed documentation is kept within the patient's notes.

We begin by looking at the basic culture of cutaneous and oral keratinocytes and fibroblasts, then the preparation of the acellular de-epidermised dermis (DED), including sterilisation of this where it is required for clinical use, prior to describing the production of tissue-engineered skin and oral mucosa. We then describe ways in which the basic model can be modified to investigate wound healing, pigmentation, melanoma invasion, and wound infection.

2. Materials

2.1. Medium for the Production of Tissue-Engineered Skin

1. Fibroblast culture medium: DMEM high glucose (4,500 mg/L glucose), 10% v/v foetal calf serum (FCS), 2 mM l-glutamine, 0.625 μg/mL amphotericin B, 100 IU/mL penicillin, and 100 μg/mL streptomycin.
2. 3T3 Fibroblast culture medium: DMEM high glucose (4,500 mg/L glucose), 10% v/v newborn calf serum, 2 mM l-glutamine, 0.625 μg/mL amphotericin B, 100 IU/mL penicillin, and 100 μg/mL streptomycin.
3. Green's medium: DMEM high glucose (4,500 mg/L glucose) and Ham's F12 medium in a 3:1 ratio, 10% v/v FCS (UK), 10 ng/mL human recombinant epidermal growth factor, 0.4 μg/mL hydrocortisone, 10^{-10} M cholera toxin, 18 mM adenine, 5 mg/mL insulin, 5 μg/mL apo-transferrin, 20 μM 3,3,5-tri-idothyronine, 2 mM glutamine, 0.625 μg/mL amphotericin B, 100 IU/mL penicillin, and 1,000 μg/mL streptomycin.
4. To make all culture media, the ingredients are mixed at room temperature in a class II laminar flow hood to make a total volume of 500 mL. Medium is stored at <4°C for a maximum of 6 weeks prior to use. Warm an aliquot of medium to 37°C before use.

2.2. Miscellaneous Solutions

1. Cryopreservation medium: 1 mL dimethyl sulphoxide (DMSO) is added to 9 mL of FCS to produce a 10% solution of DMSO in FCS. This solution is made up fresh each time it is needed.
2. "Difco-Trypsin": 0.5 g of Difco Trypsin powder, 0.5 g of D-glucose, and 0.5 mL of phenol red to 500 mL of PBS. Adjust to pH 7.4 using 2 M NaOH and filter-sterilise. Store in 10-mL aliquots at –20°C until needed.
3. Collagenase A: 0.05% (w/v) Solution of collagenase A powder in serum-free fibroblast culture medium. Filter-sterilise this solution and store in 10-mL aliquots at –20°C until needed.
4. Skin collection solution: 500 mL of sterile phosphate-buffered saline, 0.625 μg/mL amphotericin B, 100 IU/mL penicillin, and 1,000 μg/mL streptomycin.

3. Methods

3.1. Sourcing of Skin and Oral Mucosa Samples

1. Our biopsies come from two sources: skin removed at breast reductions and abdominoplasties where there is no shortage of tissue, or smaller pieces of skin often obtained from the trimmed edges of skin grafts (3) in the treatment of major burn patients or from STS biopsies (usually of around 2 cm^2) taken under local anaesthetic for the expansion of keratinocytes for the treatment of patients with non-healing ulcers (5).
2. The smaller the biopsy, the longer it will take to get a large expansion of keratinocytes, but the processing of the biopsies remain the same.
3. In making tissue-engineered skin for clinical use, we usually allow 6 weeks between obtaining the initial biopsy and having the material ready to graft (3).
4. When producing tissue-engineered oral mucosa for grafting, where the cells grow much more rapidly, the grafts can be produced within 3 weeks (see refs. 4, 6).

3.2. Processing of Skin or Oral Mucosa for Cell Isolation

1. When the skin biopsy is transported to the lab, it travels in skin collection solution in a sealed container, enters the laboratory, and is stored in a dedicated refrigerator until use.
2. Ideally, cells are cultured from skin and oral mucosa biopsies within 48 h. However, viable cells can be isolated from skin biopsies up to 4 days after excision.

We next describe the methods for isolation of skin cells. In all cases, the fastest and most reliable expansion of epithelial cells occurs when these are cultured on a growth-arrested layer of 3T3 murine fibroblasts; hence, culture of 3T3s is also described.

3.3. Keratinocyte and Fibroblast Isolation

3.3.1. Keratinocyte Isolation

1. Using sterile forceps and a sterile #22 scalpel cut STS into thin pieces (approximately 0.5 cm × 1 cm) (see Note 1).
2. Place the cut skin in 10 mL of "Difco-Trypsin" and incubate overnight at 4°C (typically 12–18 h).
3. Using sterile forceps check whether the epidermis is easily coming away from the dermis. If not, place the skin back in the fridge. Keep checking until the epidermis comes away easily.
4. Stop enzymatic activity by adding 5 mL of FCS.
5. Place the skin strips (epidermis uppermost) into a Petri dish containing Green's medium.
6. Using sterile forceps gently separate the epidermis from the dermis.
7. Try to place the epidermis so that the two newly exposed surfaces are uppermost.
8. Using a scalpel gently scrape the two newly exposed surfaces (i.e. the top of the dermis and the bottom of the epidermis. This is where the more proliferative keratinocytes are. Melanocytes are also found here).
9. Place the resulting cell suspension into a 25-mL centrifuge tube containing 5 mL of Green's medium.
10. Centrifuge the cells at 200 × *g* for 5 min, remove the supernatant, break the pellet, and then re-suspend the cells in Green's medium. At this point, try to remove any large pieces of epidermis that have been carried over.
11. Perform a cell count using Trypan blue to highlight non-viable cells.
12. Seed cells at a density of ≈4×10^6 in 10–13 mL of Green's medium in T75 flasks (seeded approximately 1 h earlier with 1×10^6 iT3T fibroblast cells). For production of i3T3 cells, see Subheading 3.4 below.
13. Culture cells at 37°C, 5% CO_2 in a humidified atmosphere.
14. Change the medium after 24 h and then every 2–3 days. The cells will generally reach 70–80% confluency in 5–7 days.

3.3.2. Keratinocyte Sub-culture

1. Remove the medium and gently wash the cells with PBS.
2. To remove the i3T3 cells, incubate with 5 mL of 0.02% w/v EDTA at 37°C.
3. Examine by phase contrast microscopy every 5 min and gently tap to encourage fibroblast detachment, ensuring that the keratinocytes are still attached.
4. Remove the i3T3 containing EDTA solution and rinse again with PBS.
5. Add 2 mL of Trypsin/EDTA to the flask and incubate at 37°C.

6. Encourage keratinocyte detachment by gentle tapping and confirm by phase microscopy after 5 min. If the cells have not detached after 5 min, place the flask back into the incubator and check again after another 5 min. Continue this until the cells have detached.
7. Add the cell suspension to 10 mL of Green's medium to neutralise the trypsin and centrifuge at 200 × *g* for 5 min.
8. Remove the supernatant, break the pellet (gentle tapping works well), and re-suspend the cells in a known volume of Green's medium and count the cells prior to use, again using Trypan blue.
9. If sub-culturing for expansion, add ~1×10^6 cells to a T75 flask seeded approximately 1 h earlier with 1×10^6 i3T3 cells.
10. Use primary keratinocytes between passages 1 and 3.

3.3.3. Fibroblast Isolation

1. Take the dermis left over from the keratinocyte isolation step and using sterile forceps and a sterile #22 scalpel mince the dermis (into pieces approximately 1 mm× 1 mm).
2. Place this mince into a Petri dish containing collagenase A solution.
3. Incubate this solution at 37°C overnight.
4. Add 10 mL of fibroblast culture medium to the Petri dish to stop the action of the collagenase.
5. Place the resulting cell suspension, minus any lumps of undigested dermis, into a universal.
6. Spin down the cell solution at 400 × *g* for 10 min.
7. Remove the supernatant (be careful not to lose the pellet – the supernatant is extremely viscous). Break the pellet and re-suspend the cells in fibroblast culture medium. Ideally, add a mycoplasma removal agent.
8. Seed the cells into a T25 flask containing fibroblast culture medium and incubate at 37°C, 5% CO_2 in a humidified atmosphere.
9. Change the medium after 24 h, then every 3–4 days until the cells are 80% confluent (see Note 2).

3.3.4. Fibroblast Sub-culture

1. Remove the medium and gently wash the cells with PBS.
2. Add 2 mL of Trypsin/EDTA to the flask and incubate at 37°C.
3. Encourage cell detachment by gentle tapping and confirm by phase microscopy after 5 min. If the cells have not detached after 5 min, place the flask back in the incubator and check again after another 2–3 min. Continue this until the cells have detached.

4. Add the cell suspension to 10 mL of fibroblast culture medium to neutralise the trypsin and centrifuge at $200 \times g$ for 5 min.
5. Remove the supernatant, break the pellet, and re-suspend the cells in a known volume of fibroblast culture medium and count the cells prior to use.
6. If sub-culturing for expansion, seed $\sim 1 \times 10^5$ cells into a T75 flask.
7. Use primary dermal fibroblasts between passages 3 and 9.

3.4. Production of Irradiated 3T3 Fibroblasts for Keratinocyte Culture

1. Irradiated 3T3 (i3T3) murine fibroblasts are used as a feeder layer during keratinocyte culture. A known number of proliferative 3T3s are stored at passage 13 in cryovials containing 1 mL of cryopreservation medium in liquid nitrogen (-196°C).
2. For production of i3T3, passage 13 3T3s are thawed and expanded using standard fibroblast sub-culture protocol (see Subheading 3.3.4).
3. Once sufficient cell numbers have been achieved (this is usually achieved at passage 17), sub-culture cells again.
4. Expose cells (known concentration suspended in 3T3 culture medium in 25-mL Universal containers) to a cobalt-60 source. They are then exposed to γ-irradiation and receive a total radiation dose of 60 Grays.
5. Count and freeze down at 4×10^6 cells/mL of cryopreservation medium.
6. Place the cryovials in a Nalgene Freezing Container and place in a -80°C freezer overnight.
7. Remove the cryovials, now containing frozen cells, and transfer to a Dewar bucket containing liquid nitrogen (−196°C) for long-term storage.

3.5. Culture of Oral Keratinocytes and Fibroblasts

Culture of oral keratinocytes and fibroblasts is identical to that of culture of cutaneous keratinocytes and fibroblasts with respect to processing and choice of medium (see ref. 6). The only essential differences are that oral keratinocytes and fibroblasts proliferate more rapidly and in broad terms expand approximately twice as fast as cells derived from skin.

3.6. Preparation of DED

Here, the key issue is whether the tissue-engineered skin or oral mucosa has been prepared for clinical use or for experimental use. For clinical use, our policy is to take strenuous steps to reduce the risk of infection for patients receiving tissue-engineered skin (see ref. 7–9) or oral mucosa. This was done by sourcing skin initially from registered tissue banks (where donors are screened for viral and bacterial contamination). In addition to working with skin from accredited tissue banks, skin is then terminally sterilised prior to further processing and for clinical use was also produced under clean room conditions according to protocols approved by the Human Tissue Authority for the UK.

For non-clinical use, DED from elective plastic surgery operations could be used, and there is no absolute necessity to sterilise this dermis for experimental use. Omitting sterilisation results in a better basement membrane and hence good keratinocyte organisation, but there is obviously an increased risk of losing experiments to infection.

The desirable characteristics of the de-epidermised acellular dermis are that it should be flexible, have a normal dermal morphology without damage, retain good mechanical strength, be completely de-cellularised, and have at least a partial retention of basement membrane proteins as these play a major role in attachment of keratinocytes and subsequent organisation of keratinocytes (10).

Our protocol for sterilisation of DED for clinical use is described in (8) and is as follows.

3.6.1. Glycerol- and Ethylene-Oxide-Sterilised DED

1. Store the STS from theatre in skin collection solution at 4°C for up to 14 days.
2. Immerse the skin in a sterile mixture of 50% glycerol and 50% PBS (v/v) for 4 h.
3. Then, immerse in 85% glycerol and 15% PBS mixture for 18 h.
4. Finally, place in 100% glycerol for 26 h.
5. Prior to ethylene oxide sterilisation, remove the skin from 100% glycerol and allow to drip dry.
6. Remove any excess glycerol by gently dabbing the skin with absorbent paper towel.
7. Place the skin into autoclave bags, label, and seal with autoclave tape.
8. Ethylene-oxide-treated skin is then stored in sealed autoclave bags at room temperature until needed.

3.6.2. "Fresh" DED

1. Store the STS from theatre in skin collection solution at 4°C for up to 14 days.
2. Follow the steps for de-epidermisation.

3.6.3. De-epidermisation of STS

1. Using sterile forceps, place the sterilised STS into 100-mL sterile plastic containers containing sterile PBS and incubate at 37°C for at least 2 days.
2. Remove the sterile skin (or "fresh skin") from the PBS solution and place into 100-mL sterile plastic containers containing sterile 1 M NaCl solution.

3. Incubate overnight at 37°C (typically 14–18 h) or until there is visible separation of the dermis from the epidermis (see Note 3).
4. Now, wash the DED twice with PBS and incubate in Green's medium or fibroblast culture medium at 37°C, 5% CO_2 in a humidified atmosphere for a minimum of 48 h. This has a threefold purpose (1) ensuring that any NaCl solution remaining in the DED is washed out, (2) allowing the DED to become saturated with culture medium, and (3) providing a rough sterility check for the DED (in the presence of infection, the medium becomes more acidic, and therefore, the phenol red pH indicator in the medium changes colour from crimson red to yellow. The medium also becomes cloudy).

3.7. Assembly of Tissue-Engineered Skin and Oral Mucosa

Figure 1 shows in diagrammatic form how the tissue-engineered skin is assembled (see Note 4). We have found over the years that we can get good results by introducing epithelial cells and stromal cells together to the papillary surface. This compares well to introducing stromal cells first, either to the reticular surface or to the papillary surface, followed by epithelial cells. In brief, the epithelial cells appear to tell the stromal cells where to go through the basement membrane and into the dermis.

We find that fibroblasts on their own rarely enter into the dermis, particularly if they are placed above an intact basement membrane. Only when keratinocytes are present will the fibro-

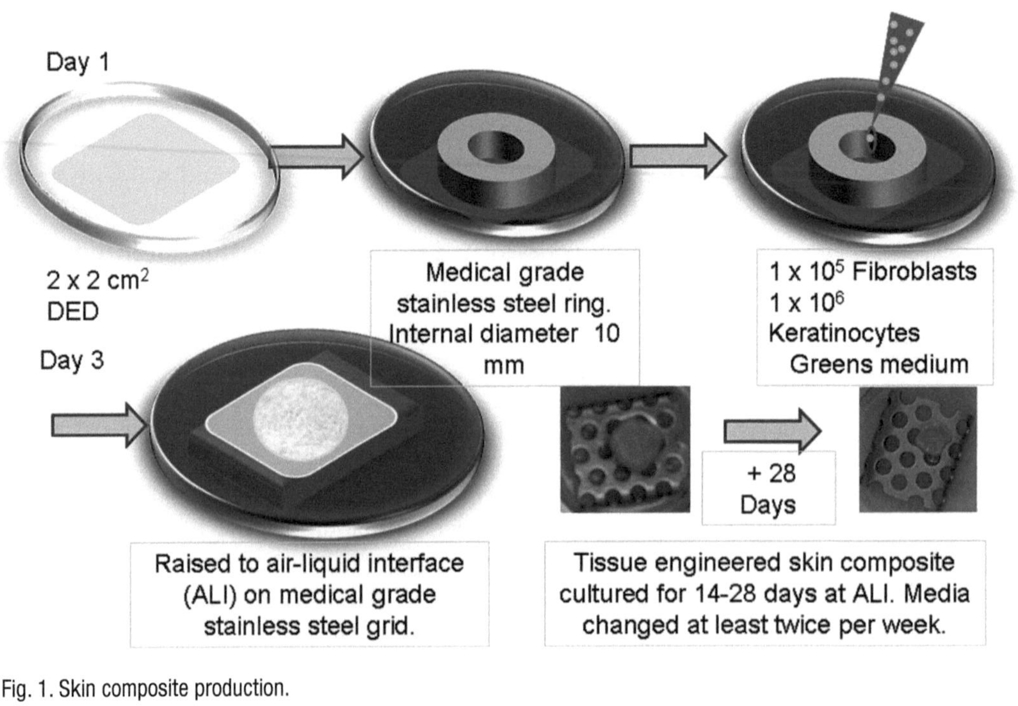

Fig. 1. Skin composite production.

blasts enter the dermis through the basement membrane. This conclusion comes from the work of Ghosh et al. (7) and has been confirmed in a number of unpublished studies in our laboratory since, such that routinely we would now co-culture epithelial cells and fibroblasts and place them together on the upper papillary surface, usually for 1 or 2 days submerged, prior to raising the DED with cells on it to an air–liquid interface. The air–liquid interface has been confirmed, in many studies, to provide a strong stimulus for keratinocyte differentiation.

3.8. Production of Tissue Engineered Skin

The protocol we use was established within our laboratory by Ghosh (7). It was subsequently modified and has been used extensively in our laboratory (7, 8, 10–20).

1. Cut the DED with a sterile scalpel into squares of approximately 2×2 cm and place into the wells of a six-well plate with the papillary dermis facing upwards (see Note 5).
2. Place a chamfered metal (medical-grade stainless steel) ring (internal diameter 1 cm) into the centre of each piece of DED and gently press down with sterile forceps to ensure a watertight seal.
3. Flood the surrounding dermis (outside the ring) with 10% Green's medium, and then inside each ring to check for leakage.
4. Once satisfied that a seal has been obtained, remove the medium from the centre of each ring.
5. Then, seed the DED inside each ring with 1×10^5 fibroblasts and 1×10^6 freshly isolated (P0) skin keratinocytes or 3×10^5 P1-3 skin keratinocytes (or oral keratinocytes for the oral mucosa model), each in 250 μL of 10% Green's medium.
6. After 24 h at 37°C, replace the seeding medium (which should have changed to a more yellowish colour) inside the rings with fresh Green's medium.
7. After a further 24 h, remove the medium from inside the rings with a pipette. The surface of the constructs within the rings should have a slightly yellowish appearance.
8. Then, remove the steel ring using sterile forceps and raise the skin constructs onto stainless steel grids in new tissue culture wells.
9. Add fresh 10% Green's medium to the level of the base of the skin constructs so that it laps the edges of, but does not cover, the DED, forming an air–liquid interface.
10. Replace the medium every 2–3 days.
11. Use skin or oral mucosa constructs for experimentation after >10 days at the air–liquid interface (see Fig. 1).

3.9. Maturation of Tissue Engineered Skin and Oral Mucosa

As these tissue-engineered models develop in the laboratory, assurance that all is well can be obtained by seeing a yellowish tinge on them as they grow. The question of when are they ready to be used will depend on their application.

Figures 2 and 3 show the development of tissue-engineered skin and oral mucosa, respectively, over time. From this, it can be seen that tissue-engineered skin (Fig. 2) achieves a normal-looking gross morphology at around 10–14 days. This can be cultured for up to 28 days as in Harrison et al. 2007 (15, 21). (This is the longest period we have cultured this model at present). With buccal mucosa (see Fig. 3) by 5 days, there is a multilayered epithelium well attached to the underlying dermis, and this continues to mature throughout 14 days of investigation.

Confirmation of the maturation of the epidermis can be achieved by measuring electrical resistance across it (22) or by looking at the expression of specific cytokeratins (21). The most convincing evidence, however, is that these tissue-engineered

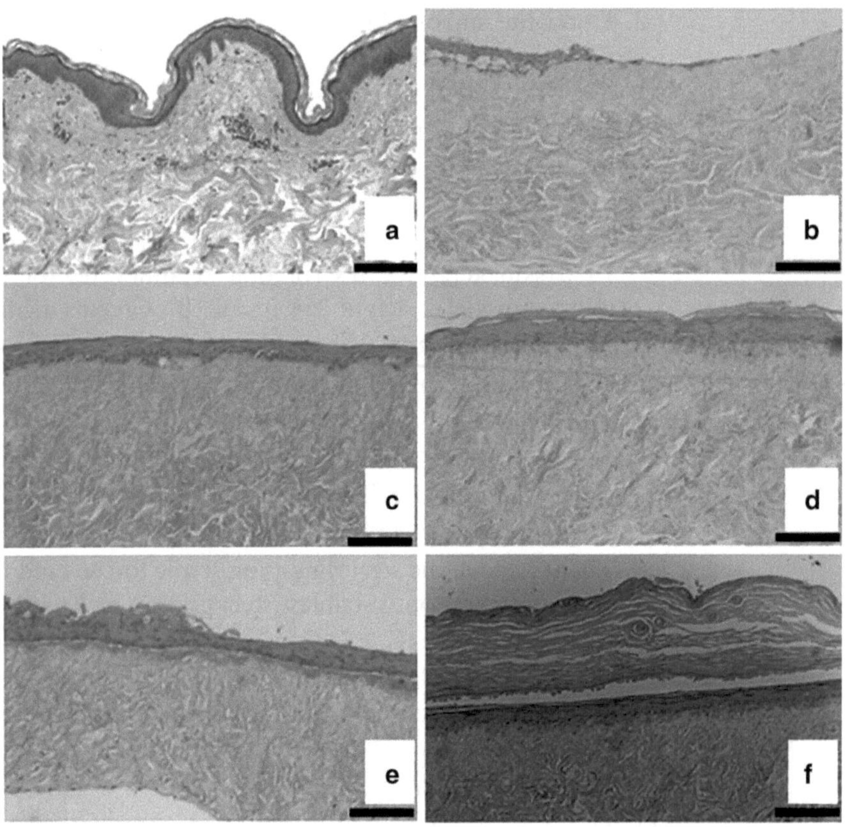

Fig. 2. H&E histology of (**a**) human skin, (**b–f**) tissue-engineered skin raised to an air–liquid interface and sacrificed at (**b**) Day 1, (**c**) Day 7, (**d**) Day 14, (**e**) Day 21, and (**f**) Day 28. Scale bar = 250 mm.

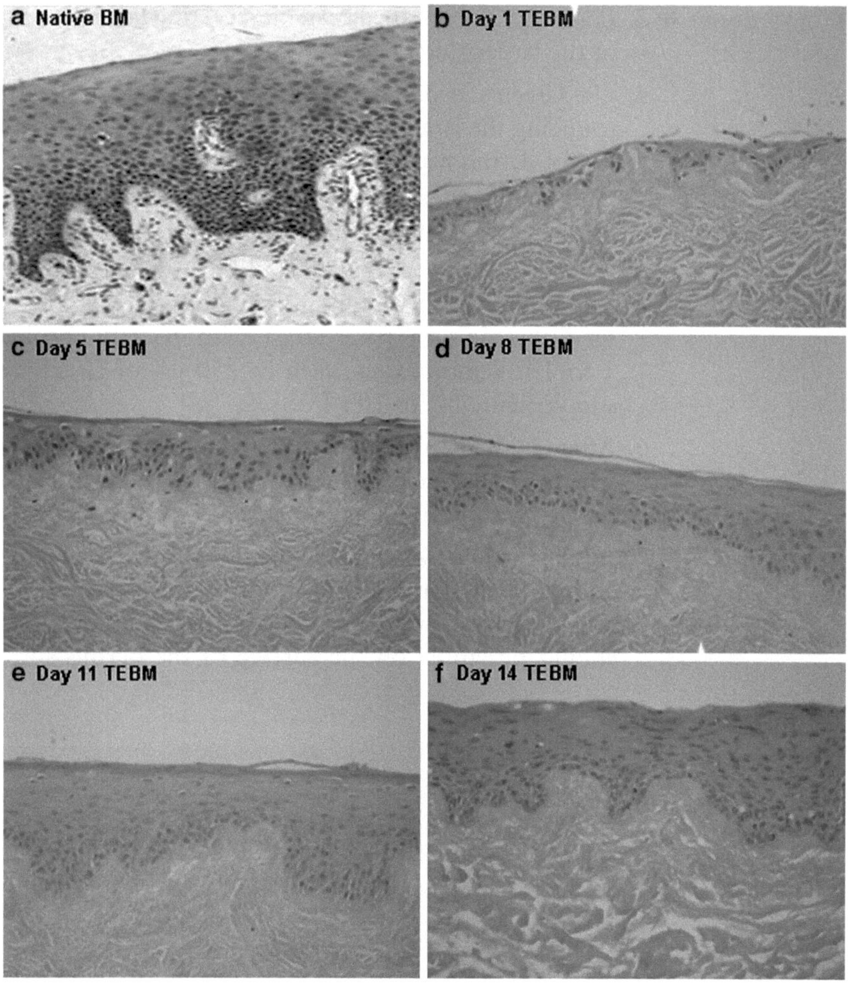

Fig. 3. Tissue-engineered oral mucosa at different time points post ALI, H&E stained. (**a**) Native oral (buccal) mucosa. (**b**) 1 day post ALI, (**c**) 5 days post ALI, (**d**) 8 days post ALI, (**e**) 11 days post ALI, and (**f**) 14 days post ALI.

constructs can be grafted back clinically and survive long term as we have reported in (3) and (4).

3.10. Wound Healing Models

Tissue-engineered skin can be used to produce a wound in which to assess the effect of soluble or insoluble materials on wound healing.

3.11. Method for Preparing a Circular Epidermis Skin Construct for Wound Healing Assays

1. Place DED in tissue culture wells with the papillary surface being uppermost.
2. Place sterile 27-mm diameter steel rings on each piece of dermis and obtain a seal by applying gentle pressure.

3. Position a second, smaller sterile steel ring (1 cm^2) in the middle of the larger ring (Fig. 4a).
4. Use Green's medium (10% FCS, PS/F) to flood the dermis surrounding the larger ring and the dermis between the two rings. To the dermis within the centre of the smaller ring, add 0.5 mL of dispase (2 mg/mL) to disrupt the basement membrane and incubate for 4 h at 37°C. Then, wash the central area three times in Green's medium and flood with fresh Green's medium.
5. Then, remove the medium from between the two rings and seed the ring of bare DED with 6×10^5 fibroblasts and 1.8×10^6 keratinocytes, each in 250 μL of Green's medium supplemented with 10% FCS.
6. After 24-h incubation at 37°C, remove seeding medium, which should have turned to a more yellowish colour, and replace with fresh Green's medium.
7. After a further 24 h, remove the outer ring using sterile forceps and raise the skin constructs onto stainless steel grids in new tissue culture wells. A yellowish "doughnut"-like shape should be visible on the surface of the dermis surrounding

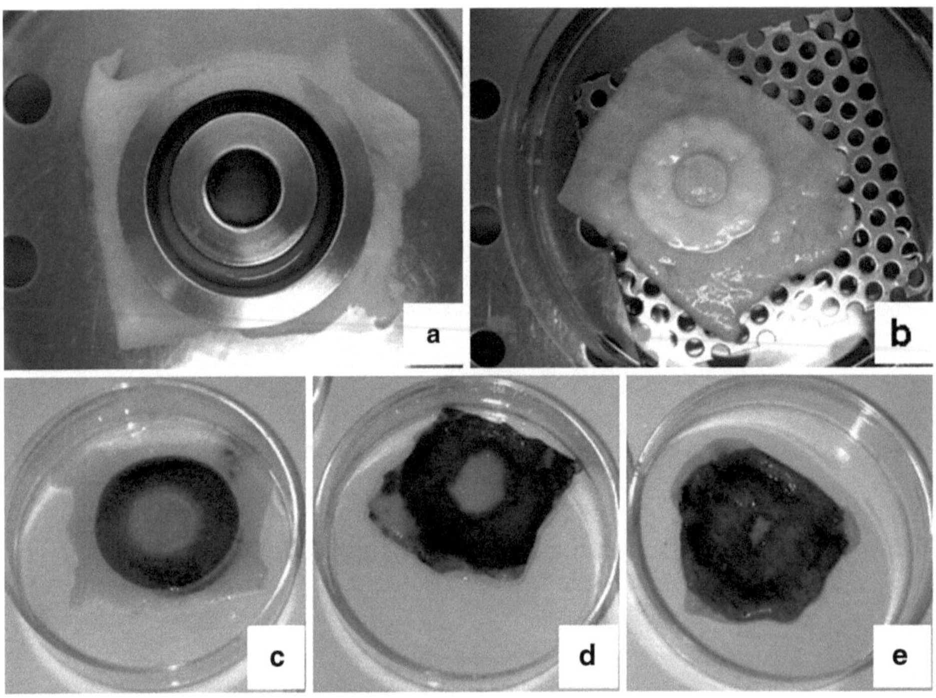

Fig. 4. Production of wound healing model. (**a**) A large piece of DED is cut (approximately 3 × 3 cm). A large (internal diameter = 2 cm) stainless steel ring is placed onto the DED and into this ring is placed a smaller (internal diameter = 1 cm) ring. The cells are seeded into the gap between the two rings to form a donut shape as shown in (**b**) when the construct is raised to an air–liquid interface. Migration of keratinocytes into the centre of the construct was shown by staining the construct with MTT at days (**c**) 1, (**d**) 10, and (**e**) 20.

the inner ring still in position; this is the new epidermis which will have a ring of bare dermis in the centre (Fig. 4b).

8. Add fresh 10% Green's medium to the level of the base of the skin constructs so that it laps the edges of, but does not cover, the DED, forming an air–liquid interface.
9. Change the medium every 2–3 days.
10. After 14 days culture at air–liquid interface, remove the smaller inner ring using sterile forceps.
11. Use skin constructs for experimentation after removal of the smaller inner ring.
12. Stain composites at the required time points with Alamar Blue, diluted 1:10 in PBS, to monitor the rate of keratinocyte migration.
13. Examples of migration at day 0 (day of removal of the central ring), day 7, and day 13 at ALI can be seen in Fig. 4c–e.

3.12. Contraction Models

Pathological skin contraction occurs following burn injuries such as superficial scalds where as many as one-third of the patients may go on to suffer disfiguring skin contraction. It also occurs quite commonly following extensive skin grafting. While there is a good clinical understanding of which patients, which sites of the body, and what circumstances give increased clinical risk of skin contraction, the mechanism of how fibroblasts and keratinocytes can both contribute to skin contraction is not fully understood as discussed in detail in (23).

The 3D tissue-engineered skin model provides an excellent model in which to study aspects of skin contraction, as the keratinocytes and fibroblasts can be added separately or together and the grafts can be studied under a variety of conditions. Figure 5 illustrates the type of results that can be obtained using the tissue-engineered model to study skin contraction, and readers are referred to previous studies on skin contraction from this laboratory using this model (10, 11, 15, 20, 23).

The methodology for producing tissue-engineered skin for contraction studies is as follows.

3.13. Skin Composite for Contraction Studies

1. Produce the skin composite as above, but before raising to the ALI, excise the seeded area by cutting around the seeded area with a sterile #22 scalpel. Use the marks left by the seeding ring as a guide.
2. Photograph the composites against a scale bar. It is best to photograph the composite inside the culture hood but without the lid on the plate. This maintains the sterility of the plate but minimises distortion of the image.
3. Import the still photograph into image analysis software such as Image J (National Institutes of Health, Bethesda, MD).

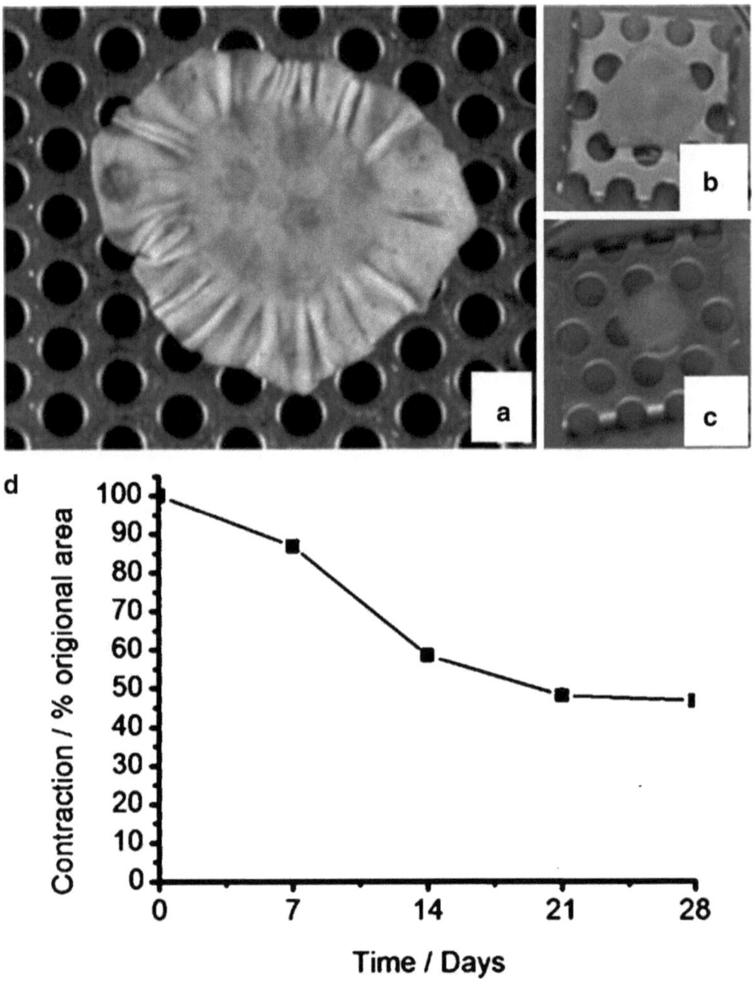

Fig. 5. Contraction of composites. (**a**) Shows the composite after being cultured at an air–liquid interface for 28 days. The *darker area* in the centre of the composite is where the cells are. The bare dermis has had to pleat to accommodate the contraction of the composite over this time period. If the seeded area is excised ((**b**) excised composite at day 1 and (**c**) excised composite at day 28), the contraction is more noticeable, and it is easier to measure the change in area over time using image analysis software such as Image J giving data as shown in (**d**).

4. Use the scale bar to calibrate the image.
5. Draw freehand around the edges of the composite and record the area of the composite.
6. Photograph the composite at least every 7 days.

3.14. Bacterial Infected Skin Wound Model

Bacterial infection of skin is one of the most common contributors to failure of skin wounds to heal. However, despite the vast number of chronic infected skin wounds, it is difficult to undertake investigational, mechanistic, or even therapeutic studies in patients, and most animal models do not duplicate the conditions that prevail in a chronic infected skin wound in man.

We have recently developed a tissue-engineered skin model (24) which, following wounding, can be infected with bacteria, and this is illustrated in Fig. 6. This model is now providing a valuable 3D model for investigating approaches to reduce bacterial infection of skin.

3.14.1. Skin Constructs for Studying Bacterial Infection

The intact epidermis of the skin presents an incredibly efficient barrier to microbiological infection. For this reason, skin constructs must be either damaged, as occurs in vivo, or provide a "built-in" route

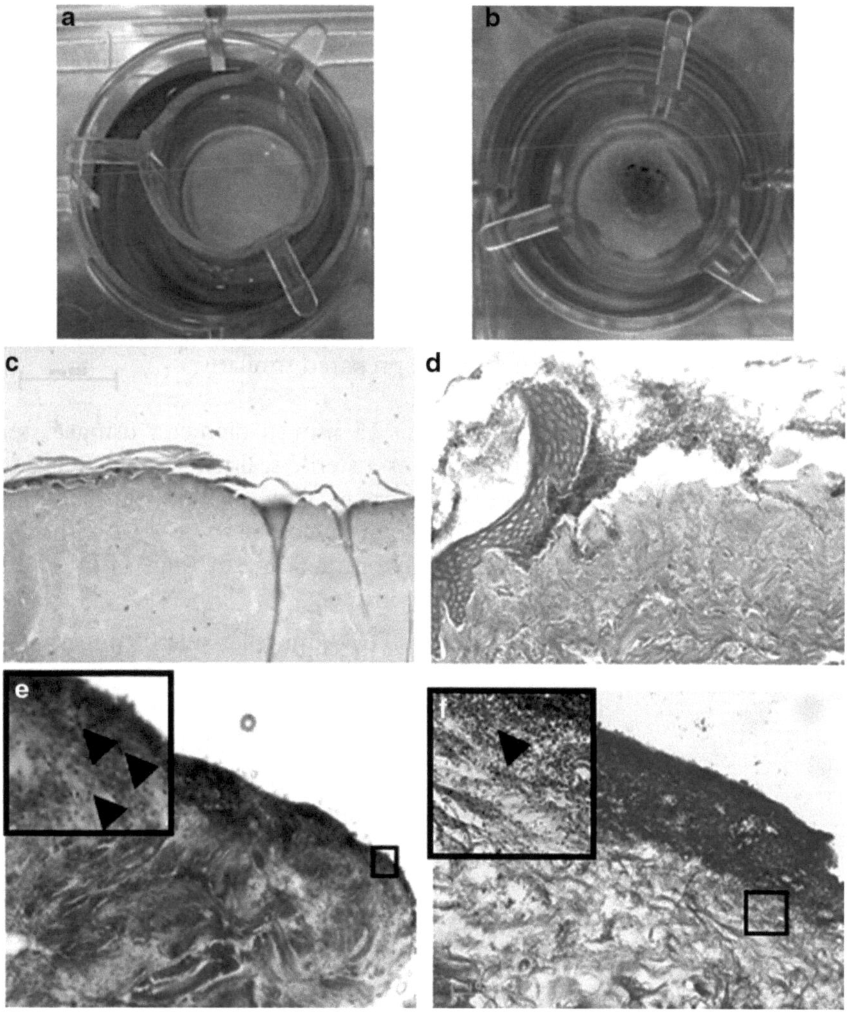

Fig. 6. Bacterial infection of composites. (**a**) Tissue-engineered skin cultured in cell culture inserts. (**b**) Appearance of skin immediately after burning. (**c**) H&E staining of uninfected, burnt skin showing both normal epithelium and damaged area with exposed dermis. (**d**) H&E stain of *P. aeruginosa* infected skin showing bacteria invading beneath epidermis. (**e**, **f**) Burnt skin infected with *S. aureus* and *P. aeruginosa*, respectively for 72 h (Gram stain). Reprinted with permission from Mary-Ann Liebert Publications.

for bacteria to invade into the dermis. There are two methods currently in use to provide a mode of entry; the first is to use skin constructs lacking epidermal cover over a defined area of dermis, and the second is to cause a burn injury to the epidermis.

For the first type, constructs such as those used on the wound-healing assays (see Fig. 4, Subheading 3.9), which are made using the double ring method, provide a circular area of unprotected dermis surrounded by intact epidermis. Bacteria can be introduced directly onto this dermal area. There is one difference in the protocol for wound-healing assays; when constructing the composites with the double rings, it is not necessary to incubate the dermis in the centre with dispase. For bacterial infection assays, the centre ring is flooded with fresh Green's medium at the same time as the DED surrounding the outer ring. In this case, the area between the two rings is seeded with keratinocytes and fibroblasts at this time point. Otherwise, the protocol is as for the wound-healing assays. Composites are used >14 days at ALI, and the skin is infected directly via the central area of dermis (see bacterial infection, below).

For the second type, smaller constructs are prepared as below within tissue culture inserts. The smaller size of these constructs makes maximum use of DED available and can be used for any purpose, not only as an infection model. The engineered oral mucosa model can be prepared similarly.

3.14.2. Method for Preparation of the Skin or Oral Mucosa Model Within Tissue Culture Inserts

1. Cut rings of DED 15 mm in diameter using a sterile cork borer if possible, or a sterile scalpel. The DED is cut slightly larger than the inserts to allow for skin contraction on burning; if the skin or oral mucosa is to be used without burning or other causes of contraction, cut rings of DED 12 mm in diameter to fit the inserts.
2. Place the circles of DED, reticular surface uppermost, within 12-mm tissue culture inserts with 4 μM pores in the base (Greiner). The pores allow the medium to bathe the dermis from below. Ensure that the DED is pushed into the bottom of the inserts by pressing gently with sterile forceps.
3. Suspend the inserts from the edges of 12-well plates into the wells.
4. Add 10% Green's medium to the bottom of the wells, surrounding the suspended inserts, so that it laps the under surface of the DED.
5. Seed the DED with 1×10^5 fibroblasts and 5×10^5 skin keratinocytes (or oral keratinocytes for the oral mucosa model), each in 250 μL of 10% Green's medium.
6. After 24 h at 37°C, replace the seeding medium (which should have changed to a more yellowish colour) inside the inserts and the bathing medium in the wells with fresh Green's medium.

7. After a further 24 h, remove the medium from inside the inserts with a pipette. This allows the skin constructs to be at the air–liquid interface. The surface of the constructs within the inserts should have a slightly yellowish appearance. Replace the Green's medium in the wells with fresh medium.
8. Replace Green's medium every 24 h. Constructs within tissue culture inserts in 12-well plates have access to a smaller volume of Green's medium than constructs grown in steel rings on grids, so the medium needs to be replaced more frequently.
9. Use skin or oral mucosa constructs for experimentation after >10 days at the air–liquid interface (see Fig. 6a for prepared composite within an insert).
10. After 14 days at ALI, the skin constructs within the inserts are burnt in the centre by applying a heated metal rod, 4 mm in diameter, to the surface of the skin for 6 s (Fig. 6b) (see Notes 5 and 6).

3.14.3. Bacterial Infection of Skin Constructs

1. Culture bacterial species of interest in appropriate broth at 37°C for 24 h from stock plates prior to use. Centrifuge broths, then wash the resulting pellet in PBS and re-suspend bacteria in PBS to a concentration of 1×10^{10} cfu/mL. Bacteria can be stored at this concentration in PBS at 4°C for several days.
2. Wash skin constructs (by removing and replacing the Green's medium at least every 24 h) in antibiotic-free Green's medium for 72 h prior to infection. To infect the larger "double ring" constructs, pipette bacteria at the desired concentration (e.g., 1×10^7 bacteria) in 20 μL of appropriate broth per composite directly onto the exposed dermis in the centre of the skin constructs.
3. For constructs in inserts, burn as above immediately prior to infection to provide the bacteria a mode of entry into the skin. Pipette bacteria of interest at the test concentration (e.g., 1×10^7 bacteria) in 100 μL of appropriate broth per construct into the inserts, covering the epidermal surface.
4. Incubate infected skin constructs and non-infected controls in antibiotic-free Green's medium at 37°C/5% CO_2 for the desired amount of time, then sacrifice for analysis.

3.14.4. Analysis of Infected Skin Constructs

1. For histological analysis, bisect skin constructs with a sterile scalpel. Fix the tissue in 10% formalin for >24 h, then process and embed in paraffin using standard techniques. Alternatively, for frozen sections, after bisecting the skin, snap-freeze in liquid nitrogen and store samples at −80°C prior to sectioning on a cryostat.
2. Gram staining: To visualise bacteria in paraffin sections of skin, first dewax the sections through xylene and alcohol to

water as standard (as in for H&E staining). Then, with the slides still wet, perform a Gram stain on a rack over a sink.

3. A standard Gram stain is: 1 min crystal violet, wash in running water, fix in Gram's iodine (1 min), wash in running water, destain in iodine–acetone, 30 s, wash in running water, carbol fuschin (1 min).
4. At each step, pipette enough stain to cover the tissue section. After the final carbol fuschin step, tap the carbol fuschin off the slides, blot sections dry with filter paper, and then mount a coverslip using a non-aqueous mountant. Gram-positive bacteria will appear dark purple and Gram-negative pink (see Fig. 6e–f). Gram-negative bacteria may be more difficult to see since there will be a high level of background staining, also pink, on the tissue sample.
5. In order to quantify the viable bacteria that have invaded into the skin after infection, first weigh the sample. This is important since not all skin constructs will be exactly of the same size or thickness. Then, mince the tissue in a sterile Petri dish using a sterile scalpel into small pieces and homogenise (for example, in a sterile glass homogeniser) in 1 mL of broth. Perform a serial dilution on the resulting homogenates, e.g., $1{:}10^2$, $1{:}10^4$, $1{:}10^6$, etc.
6. Add 10-μL drops of each dilution in triplicate onto agar plates; allow the drops to dry and then incubate the plates at 37°C overnight. Count the number of resulting colonies and deduce the number of viable bacteria in the original sample from the dilution factor. For example, 15 colonies at a $1{:}10^6$ dilution indicates a starting concentration of ~15,000,000 or 1.5×10^7 cfu (colony forming units). It is then possible, using the weights of the tissue samples, to express the numbers of viable bacteria recovered per mg of tissue.

3.15. Skin Pigmentation

Tissue-engineered skin is going to prove very useful in understanding the regulation of skin pigmentation. For example, melanocytes from pale skin donors when introduced to reconstructed skin give rise to barely pigmented skin (25). However, these tissue-engineered models spontaneously pigment if fibroblasts are omitted from the models (26). Exploring this in 2D co-culture models, we demonstrated that while normal fibroblasts will suppress melanocyte pigmentation, stressed fibroblasts (stressed by freezing or gamma radiation) produce soluble factors which induced pigmentation in melanocytes. These results were confirmed for melanocytes derived from skin, hair, and eye, indicating that this may be a generic biology (27).

Figure 7 illustrates that normal human melanocytes when added to these tissue-engineered skin models know where to go, orientating in the basal layer of keratinocytes, and their numbers are tightly

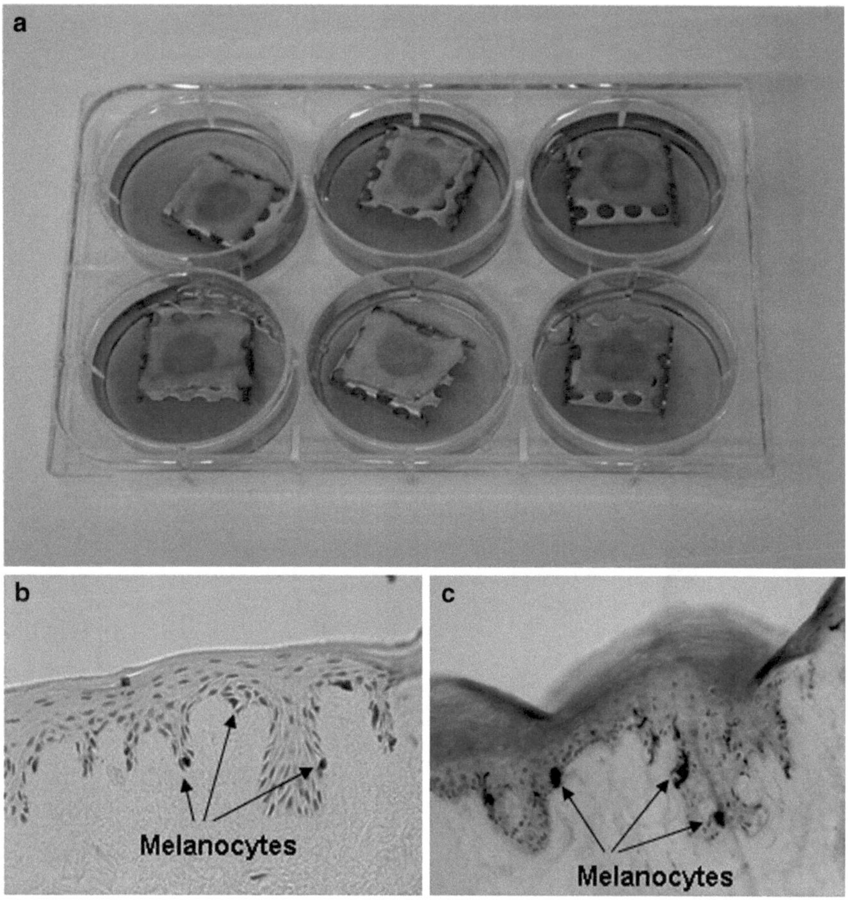

Fig. 7. Tissue-engineered models of pigmentation. (**a**) Shows the appearance of six tissue-engineered skin models containing keratinocytes and melanocytes (but no fibroblasts) pigmenting spontaneously at an air–liquid interface. In the presence of fibroblasts, the same composites had very little pigmentation (details as in Hedley et al. (26)). In (**b**), an H&E stained section shows the location of melanocytes in the basal layer of keratinocytes. Melanocytes are stained with S100. In (**c**), these composites containing melanocytes have been stained for TRP1, one of the key enzymes associated with pigmentation (details as in Eves et al. (25)).

regulated by the keratinocyte population (25, 28). (In contrast, melanoma cells added to the same model do not obey these rules; many cells remain in the upper superficial layers of the epidermis, but some invade through the basement membrane into the dermis).

3.16. Methodology for Producing a Melanocyte Skin Model Is as Follows

1. Produce the skin composite as above (Subheading 3.7), but when seeding with fibroblasts and keratinocytes into the metal ring, add an additional 5×10^5 laboratory expanded melanocytes. The melanocytes will initially be expanded in a melanocyte-specific media, but once added to the composite, all culture then continues in the Green's medium.
2. Culture the composites for at least 10 days at an air–liquid interface.

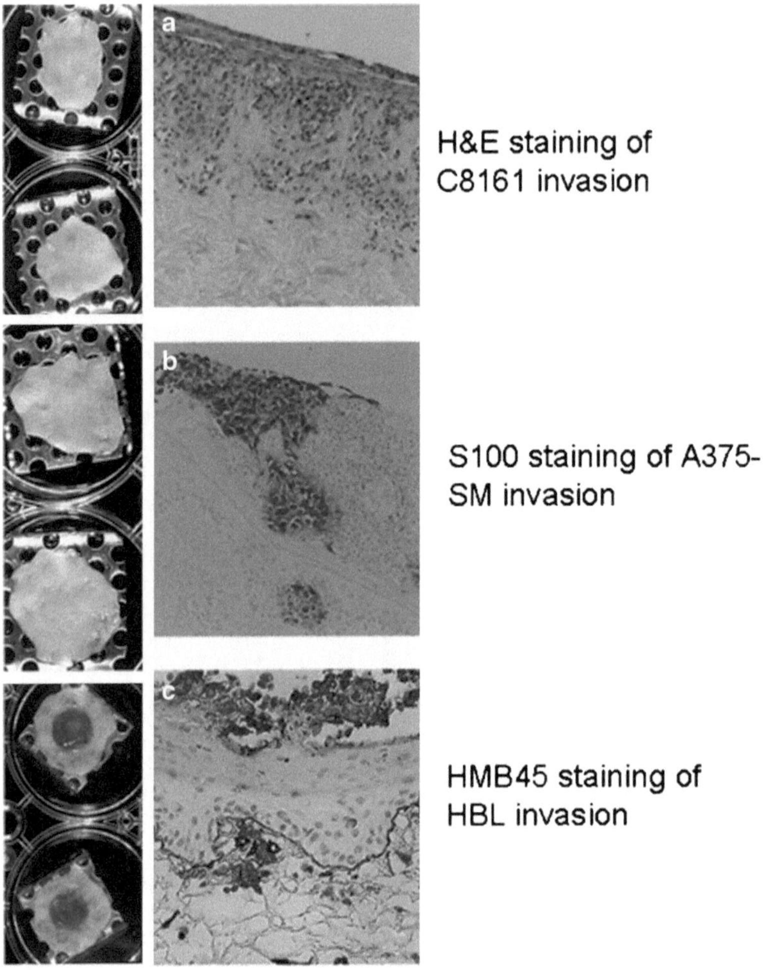

Fig. 8. Melanoma invasion. (**a**) H&E staining of C8161 invasion, (**b**) S100 staining of A375-SM invasion, (**c**) HMB45 staining of HBL invasion. For further details of this model, see Eves et al. (14).

3.17. Skin Composite for Melanoma Invasion Studies

Figure 8 illustrates how the tissue-engineered skin model has been used to study several aspects of melanoma cell invasion. There are many questions to be asked with respect to why some melanomas and not others metastasise so effectively. What is the interaction between the melanoma cell and the extracellular matrix? Do melanoma cells invade via degradative enzyme activity or by migratory activity? What is the evidence that melanoma cells interact with adjacent skin cells and vice versa? For studies on this area from our group, see refs. 13, 14, 18, 29, 30. The methodology for producing a melanoma skin model is as follows:

1. Produce the skin composite as above (Subheading 3.7) and seed fibroblasts (1×10^5) and keratinocytes (1×10^6) into the metal ring with an additional 5×10^4 melanoma cells. Again

these will each initially be expanded in whichever medium is recommended for the particular cell line, but once added to the composite, all culture will be in Green's medium.

2. Culture the composites for at least 10 days at an ALI and for up to 28 days depending on what aspects of melanoma invasion are being studied.

3.18. Oral Mucosa Applications

As mentioned previously, we have made tissue-engineered oral mucosa for clinical use. It has been used to replace scarred tissue in the urethra, and a 3-year follow-up study showed good results for three out of five patients and some recurrent constrictions for two of these five patients. As these patients had previously had a long history of recurrent tissue strictures, this result was not unexpected (4).

For in vitro purposes, we can also use the oral mucosa model to look at the penetration of agents across this barrier. Work ongoing in our laboratory is looking at the penetration of polymersomes being designed to carry drugs and genes into the body through the oral mucosa. Here, the tissue-engineered oral mucosa is an invaluable test bed model in developing the polymersomes.

4. Notes

1. Problems in cell isolation can occur with STS grafts that are too thick (because then it is difficult to separate the epidermis from the dermis) or when the pieces of biopsy are not cut sufficiently small to allow enzyme separation of epidermis and dermis.
2. At this point, the culture contains fibroblasts, keratinocytes, melanocytes, and other skin cells. When the cells have become confluent, subculture the cells into a T75 flask and then when this flask has become confluent, passage cells fairly aggressively (i.e. 1× T75 into 10× T75) to encourage a pure fibroblast population. Only use when you are sure that the culture is not contaminated with other cell types. This is usually around passage 3.
3. The epidermis of the glycerol- and ethylene-oxide-sterilised skin can be difficult to remove. If this proves to be the case, leave in 1 M NaCl longer or gently detach by gentle scraping with a blunt-ended spatula. However, be careful – too much scraping will damage the basement membrane.
4. For any one experiment, use DED from a single patient and wherever possible cut from the same sheet. This is to try to reduce inter-patient variation in skin characteristics and differences in thickness.

5. If unsure which is the papillary and which is the reticular dermis, the papillary dermis tends to look smoother than the reticular dermis which tends to look more ragged and "cotton wool-like."
6. When performing the burn injury, if several samples are to be injured, it is best to hold the heated rod in a thermoprotective glove as it will get very hot.
7. During burn injury, the surface of the skin may stick to the metal rod, so have a pair of sterile forceps to hand to gently tease the skin away from the metal.

Acknowledgements

We gratefully acknowledge the continued support for our research throughout the years of the Burns and Plastic Surgery Consultants of the Northern General Hospital Trust Burns Unit, particularly Mr Eric Freedlander and Mr David Ralston. We also acknowledge the histopathology services provided over the years by Dr Chris Layton – ever helpful.

References

1. MacNeil, S. (2007) Progress and opportunities for tissue-engineered skin. *Nature* **445,** 874–880.
2. MacNeil, S. (2007) Skin tissue engineering, in *Tissue engineering using ceramics and polymers* (Boccaccini, A. R., and Gough, J., Eds.), pp 375–403, Woodhead Publishing Limited, Cambridge.
3. Sahota, P. S., Burn, J. L., Heaton, M., Freedlander, E., Suvarna, S. K., Brown, N. J., and MacNeil, S. (2003) Development of a reconstructed human skin model for angiogenesis. *Wound Repair Regen.* **11,** 275–284.
4. Bhargava, S., Patterson, J. M., Inman, R. D., MacNeil, S., and Chapple, C. R. (2008) Tissue-engineered buccal mucosa urethroplasty-clinical outcomes. *Eur. Urol.* **53,** 1263–1269.
5. Moustafa, M., Simpson, C., Glover, M., Dawson, R. A., Tesfaye, S., Creagh, F. M., Haddow, D., Short, R., Heller, S., and MacNeil, S. (2004) A new autologous keratinocyte dressing treatment for non-healing diabetic neuropathic foot ulcers. *Diabet. Med.* **21,** 786–789.
6. Bhargava, S., Chapple, C. R., Bullock, A. J., Layton, C., and Macneil, S. (2004) Tissue-engineered buccal mucosa for substitution urethroplasty. *BJU Int.* **93,** 807–811.
7. Ghosh, M. M., Boyce, S., Layton, C., Freedlander, E., and MacNeil, S. (1997) A comparison of the methodologies for the preparation of human epidermal-dermal composites. *Ann. Plast. Surg.* **39,** 390–404.
8. Chakrabarty, K. H., Dawson, R. A., Harris, P., Layton, C., Babu, M., Gould, L., Phillips, J., Leigh, I., Green, C., Freedlander, E., and MacNeil, S. (1999) Development of autologous human dermal-epidermal composites based on sterilized human allodermis for clinical use. *Br. J. Dermatol.* **141,** 811–823.
9. Huang, Q., Dawson, R. A., Pegg, D. E., Kearney, J. N., and MacNeil, S. (2004) Use of peracetic acid to sterilize human donor skin for production of acellular dermal matrices for clinical use. *Wound Repair Regen.* **12,** 276–287.
10. Ralston, D. R., Layton, C., Dalley, A. J., Boyce, S., Freedlander, E., and MacNeil, S. (1999) The requirement for basement membrane antigens in the production of human epidermal/dermal composites in vitro. *Br. J. Dermatol.* **140,** 605–615.
11. Chakrabarty, K. H., Heaton, M., Dalley, A. J., Dawson, R. A., Freedlander, E., Khaw, P. T., and MacNeil, S. (2001) Keratinocyte-driven contraction of reconstituted human skin. *Wound Repair Regen.* **9,** 95–106.

12. Dawson, R. A., Goberdhan, N. J., Freedlander, E., and MacNeil, S. (1996) Influence of extracellular matrix proteins on human keratinocyte attachment, proliferation and transfer to a dermal wound model. *Burns* **22,** 93–100.
13. Eves, P., Katerinaki, E., Simpson, C., Layton, C., Dawson, R., Evans, G., and MacNeil, S. (2003) Melanoma invasion in reconstructed human skin is influenced by skin cells – investigation of the role of proteolytic enzymes. *Clin. Exp. Metastasis* **20,** 685–700.
14. Eves, P., Layton, C., Hedley, S., Dawson, R. A., Wagner, M., Morandini, R., Ghanem, G., and MacNeil, S. (2000) Characterization of an in vitro model of human melanoma invasion based on reconstructed human skin. *Br. J. Dermatol.* **142,** 210–222.
15. Harrison, C. A., Gossiel, F., Layton, C. M., Bullock, A. J., Johnson, T., Blumsohn, A., and MacNeil, S. (2006) Use of an *in vitro* model of tissue-engineered skin to investigate the mechanism of skin graft contraction. *Tissue Eng.* **12,** 3119–3133.
16. Harrison, C. A., Heaton, M. J., Layton, C. M., and MacNeil, S. (2006) Use of an in vitro model of tissue-engineered human skin to study keratinocyte attachment and migration in the process of reepithelialization. *Wound Repair Regen.* **14,** 203–209.
17. Hernon, C., Harrison, C. A., Thornton, D. J. A., and MacNeil, S. (2007) Enhancement of keratinocyte performance in production of tissue engineered skin by use of low-calcium medium. *Wound Repair Regen.* **15,** 718–726.
18. MacNeil, S., Eves, P., Richardson, B., Molife, R., Lorigan, P., Wagner, M., Layton, C., Morandini, R., and Ghanem, G. (2000) Oestrogenic steroids and melanoma cell interaction with adjacent skin cells influence invasion of melanoma cells in vitro. *Pigment Cell Res.* **13(8),** 68–72.
19. Ralston, D. R., Layton, C., Dalley, A. J., Boyce, S. G., Freedlander, E., and MacNeil, S. (1997) Keratinocytes contract human dermal extracellular matrix and reduce soluble fibronectin production by fibroblasts in a skin composite model. *Br. J. Plast. Surg.* **50,** 408–415.
20. Thornton, D. J. A., Harrison, C. A., Heaton, M. J., Bullock, A. J., and MacNeil, S. (2008) Inhibition of keratinocyte-driven contraction of tissue-engineered skin in vitro by calcium chelation and early restraint but not submerged culture. *J. Burn Care Res.* **29,** 369–377.
21. Harrison, C. A., Layton, C. M., Hau, Z., Bullock, A. J., Johnson, T. S., and MacNeil, S. (2007) Transglutaminase inhibitors induce hyperproliferation and parakeratosis in tissue-engineered skin. *Br. J. Dermatol.* **156,** 247–257.
22. Bullock, A. J., Barker, A. T., Coulton, L., and Macneil, S. (2007) The effect of induced biphasic pulsed currents on re-epithelialization of a novel wound healing model. *Bioelectromagnetics* **28,** 31–41.
23. Harrison, C. A., and MacNeil, S. (2008) The mechanism of skin graft contraction: an update on current research and potential future therapies. *Burns* **34,** 153–163.
24. Shepherd, J., Douglas, I., Rimmer, S., Swanson, L., and MacNeil, S. (2009) Development of 3-dimensional tissue engineerd models of bacterial infected human skin wounds. *Tissue Eng. Part C Methods* **15(3),** 475–484.
25. Eves, P. C., Bullett, N. A., Haddow, D., Beck, A. J., Layton, C., Way, L., Shard, A. G., Gawkrodger, D. J., and MacNeil, S. (2008) Simplifying the delivery of melanocytes and keratinocytes for the treatment of vitiligo using a chemically defined carrier dressing. *J. Invest. Dermatol.* **128,** 1554–1564.
26. Hedley, S. J., Layton, C., Heaton, M., Chakrabarty, K. H., Dawson, R. A., Gawkrodger, D. J., and MacNeil, S. (2002) Fibroblasts play a regulatory role in the control of pigmentation in reconstructed human skin from skin types I and II. *Pigment Cell Res.* **15,** 49–56.
27. Balafa, C., Smith-Thomas, L., Phillips, J., Moustafa, M., George, E., Blount, M., Nicol, S., Westgate, G., and MacNeil, S. (2005) Dopa oxidase activity in the hair, skin and ocular melanocytes is increased in the presence of stressed fibroblasts. *Exp. Dermatol.* **14,** 363–372.
28. Eves, P. C., Beck, A. J., Shard, A. G., and MacNeil, S. (2005) A chemically defined surface for the co-culture of melanocytes and keratinocytes. *Biomaterials* **26,** 7068–7081.
29. Katerinaki, E., Evans, G. S., Lorigan, P. C., and MacNeil, S. (2003) TNF-alpha increases human melanoma cell invasion and migration in vitro: the role of proteolytic enzymes. *Br. J. Cancer* **89,** 1123–1129.
30. Eves, P., Haycock, J., Layton, C., Wagner, M., Kemp, H., Szabo, M., Morandini, R., Ghanem, G., García-Borrón, J. C., Jiménez-Cervantes, C., and MacNeil, S. (2003) Anti-inflammatory and anti-invasive effects of α-melanocyte-stimulating hormone in human melanoma cells. *Br. J. Cancer* **89,** 2004–2015.

Chapter 10

Three-Dimensional Alignment of Schwann Cells Using Hydrolysable Microfiber Scaffolds: Strategies for Peripheral Nerve Repair

Celia Murray-Dunning, Sally L. McArthur, Tao Sun, Rob McKean, Anthony J. Ryan, and John W. Haycock

Abstract

Injuries to the peripheral nervous system affect 1 in 1,000 individuals each year. The implication of sustaining such an injury is considerable with loss of sensory and/or motor function. The economic implications too are extensive running into millions of pounds (or dollars) annually for provision and support. The natural regrowth of peripheral nerves is possible for small gap injuries (of approximately 1–2 mm). However, patients with larger gap injuries require surgical intervention. The "gold standard" for repairing gap injuries is autografting; however, there are problems associated with this approach, and so, the use of nerve guidance conduits (NGC) is a realistic alternative. We outline in this chapter the development of an NGC that incorporates aligned poly-L-lactide fibres for supporting the growth of organised Schwann cells within a three-dimensional scaffold in vitro. A closed loop bioreactor for growing cells within NGC scaffolds is described together with a method of plasma deposition for modifying the microfibre surface chemistry (which improves the ability of Schwann cells to attach) and confocal microscopy for measuring cell viability and alignment within 3D constructs.

Key words: Schwann cell, Peripheral nerve, Biomaterial scaffold, Nerve guidance conduit, Bioreactor, Poly-L-lactide

1. Introduction

1.1. Peripheral Nerve Injury

Injuries to the peripheral nervous system (PNS) are extremely common, and although axon regeneration is possible, this typically occurs over very short distances of 1–2 mm. However, individuals who receive this type of injury will have continuing discomfort and lifelong disability (1). Where damage to the nerve is too significant for regeneration to take place, surgical intervention will be

John W. Haycock (ed.), *3D Cell Culture: Methods and Protocols*, Methods in Molecular Biology, vol. 695,
DOI 10.1007/978-1-60761-984-0_10, © Springer Science+Business Media, LLC 2011

required, such as end-to-end suturing or autografting. Reconstructive surgery does not result in complete or effective functional recovery, and a major disadvantage of autografting includes a secondary surgical procedure and donor site morbidity. Previous studies have demonstrated that a high number of nerve cells do not survive axatomy, with 35–40% undergoing apoptosis following transection injury (2). This can result in diminished levels of sensitivity despite successful remyelination of established axons by Schwann cells. Strategies to bioengineer peripheral nerves are therefore being developed.

1.2. Nerve Guidance Conduits

By using entubulation devices such as nerve guidance conduits (NGC), axon regrowth from the proximal cone towards the distal stump can be improved. NGCs are designed to provide a favourable microenvironment and a degree of basic physical guidance for improving the rate of reinnervation, the distance of reinnervation, and to a certain extent the orientation of regrowing nerves. Many types of NGC materials have been studied (3), including natural materials such as autologous veins and arteries. However, disadvantages exist with natural materials such as a limited supply and unnecessary surgical procedures. Collagen (4) and chitosan (5) have been used, but concerns of disease transmission continue to exist for these materials (6). Therefore, the use of synthetic NGC materials is increasingly being studied. The advantages of synthetic materials include control over properties such as physical flexibility, porosity, biocompatibility, and biodegradability, while also having control over material purity and quality.

1.3. Schwann Cells and Peripheral Nerve Repair

Schwann cells within the PNS are essential for peripheral nerve repair, and indeed, a lack of these cells can impede regeneration (7). Furthermore, it has previously been shown that incorporation of Schwann cells within an NGC device can improve neuronal cell growth and alignment (8). Following injury, there is a period of rapid proliferation of Schwann cells and the formation of cellular columns known as bands of Büngner (8). These columns act as "guidance pathways" for the regrowth of axons extending from the proximal stump, and it is known that growth cone filaments can extend along regenerative tracts at a rate of up to 1 mm per day (9). However, gap injuries have no spatial information between the proximal and distal axons, and so, we suggest that the provision of a scaffold for Schwann cells in the form of aligned biodegradable fibres will improve subsequent axon alignment. In the present study, we describe the fabrication of aligned microfibre scaffolds made from a hydrolysable polymer, poly-L-lactide. Fibres are surface-coated using plasma polymerisation (10), which improves the ability to support Schwann cell adhesion and growth. Schwann cells can be grown on fibre sheets to rapidly evaluate an ability to support cell growth and alignment.

Additionally, they can be constructed to form experimental conduits for the 3D culture of Schwann cells using a closed loop perfusion bioreactor described herein.

2. Materials

2.1. Poly-L-lactic-acid for Electrospinning Fibres

1. Poly-L-lactic-acid (PLLA) polymer granules (M_n ~99 K) (Fluka, Switzerland) dissolved in dichloromethane (DCM) to 8% (wt/wt).
2. A disposable 22-gauge blunt-end needle, internal diameter of 0.8 mm (I&J Fisnar Inc., Fair Lawn, NJ, USA) and syringe pump (Aladdin 1000).
3. High-voltage power supply (Brandenburg, Alpha series III).
4. An aluminium foil sheet of A5 size (approx. 148 × 210 mm) wrapped around the cylindrical collector.

2.2. Acrylic Acid Plasma Polymer Deposition

1. A glass cylindrical reactor vessel [50 cm long × 8 cm wide (QVF Process Systems Ltd., Stafford, UK)].
2. Power supply for plasma polymerisation = 13.56 MHz from a radio frequency power generator (Coaxial Power Systems Ltd., UK). The power supply should be coupled to the reactor via an impedance matching network (Coaxial Power Systems Ltd., UK).
3. Monomer flow rate should be controlled using a needle valve (BOC Edwards, UK).
4. A pump system is required to generate low pressure within the reactor vessel [Edwards Two Stage 5 Vacuum Pump (BOC Edwards, UK)].

2.3. Maleic Anhydride Plasma Polymer Deposition

1. A stainless steel T-piece reactor (376 × 240 mm).
2. Monomer flow rate should be controlled using a needle valve (BOC Edwards, UK).
3. Power generator for plasma polymerisation = 13.56 MHz from a radio frequency power generator (Coaxial Power Systems Ltd., UK). The power supply should be coupled to the reactor via an impedance matching network (Coaxial Power Systems Ltd., UK).
4. A pulse generator is used to cycle the on and off times of the RF power supply (Thurlby Thandar Instruments, UK).
5. Oscilloscope (Tektronix TDS3014, S.J Electronics, UK) used to minimise reflected power.
6. A pump system is required to generate low pressure within the reactor vessel [Model No. RV8 (BOC Edwards, UK)].

2.4. RN22 Schwann Cell Culture

1. Dulbecco's Modified Eagle's Medium (DMEM): 10% (v/v) foetal calf serum (FCS), 100 units/ml penicillin, 100 μL/mL streptomycin, 2 mM L-glutamine, and 1.25 ml of Fungizone®.
2. RN22 rat Schwann cells (HPACC, Porton Down, Salisbury, UK). Use up to passage 25.
3. 0.25% Trypsin (w/v) and 1 mM ethylenediamine tetraacetic acid (EDTA). Store in 5-mL aliquots at −20°C.
4. RN22 Schwann cells should be passaged using tissue culture polystyrene T75 flasks.

2.5. RN22 Cells for Sheet Culture with PLLA Fibre Scaffolds

1. Medical-grade stainless steel rings with an internal diameter of 12 mm and a depth of 8 mm. (Medical Workshop, Royal Hallamshire Hospital, Sheffield, UK) (see Note 1).
2. Electrospun PLLA fibres on A5 foil sheets cut to 20×20 mm (see Note 2).
3. 6-Well tissue culture plates and phosphate-buffered saline (PBS).

2.6. Closed Looped Bioreactor System Assembly

1. The assembly of a closed system bioreactor should be constructed using a commercially available tissue culture Petri dish, a peristaltic pump (Watson and Marlow, model 205S), a medium bottle (100 ml), silicone tubing, connectors, and three-way valves (Cole-Palmer Instrument Company Ltd., Hanwell, London, UK) (see Fig. 1). The components can all be connected together in series using silicone tubing and connectors for supporting a single conduit within a Petri dish (9 cm diameter).

2.7. Fluorescent Labelling of Schwann Cell Actin Filaments and Nuclei

1. Fixing solution: 3.6% (w/v) paraformaldehyde in PBS.
2. Permeabilising solution: 0.1% (w/v) Triton X-100 in PBS.
3. Actin filament labelling solution: phalloidin-FITC (Invitrogen, UK) (25 μg/mL) in PBS.
4. Nuclei labelling solution: 4′,6-diamidino-2-phenylindole dihydrochloride (DAPI) (300 nM) in PBS.
5. Cells should be imaged using an upright confocal microscope (e.g. LSM 510 Meta, Zeiss, Germany) using water-dipping objective lenses (see Note 3).

2.8. Live/Dead Staining for Cell Viability

1. Live/Dead® BacLight™ kit (Invitrogen, UK) or equivalent can be used to assess cell viability: Syto9 (Component A; 3.34 nM, 300 μL solution in DMSO) (1 μL/mL), propidium iodide (Component B; 20 mM, 300 μL solution in DMSO) (3.5 μL/mL) in 10% DMEM medium.
2. Cells imaged by confocal microscopy (LSM 510 Meta, Zeiss, Germany).

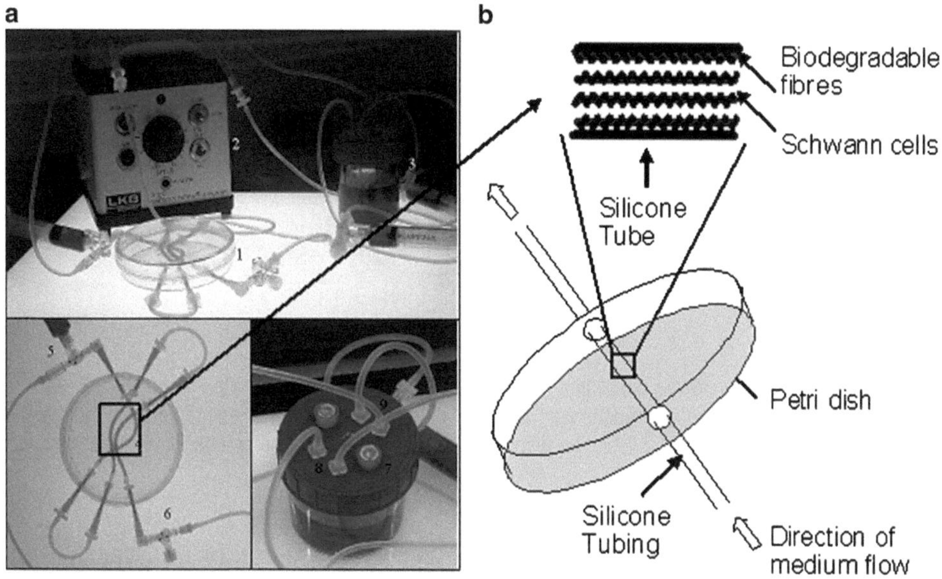

Fig. 1. (**a**) A photograph of the conduit bioreactor. The system consists of (1) a circular polystyrene Petri dish of 9 cm in diameter, (2) a peristaltic pump, and (3) a medium bottle. (4) Inside the Petri dish are three silicone tubes (shown, 40 mm in length and 1.2 mm in diameter connected in series). Contained within each lumen is an aligned microfibre scaffold. (5) Two closable inlets and (6) two closable outlets were present for cell seeding and medium perfusion connected to the cell culture chamber. The medium bottle is used as a reservoir for continuous flow culture. On the lid of the bottle are (7) two vent windows covered with an air-permeable film, (8) two medium inlets, and (9) two medium outlets. (**b**) A schematic diagram of the microfiber alignment within the conduit.

3. Methods

3.1. Fabrication of Biodegradable Aligned Poly-L-lactic-acid Microfibre Scaffolds by Electrospinning

1. Aligned PLLA fibres should be fabricated by electrospinning, using a rotating cylindrical collector.
2. Dissolve PLLA polymer granules in dichloromethane (DCM) under gentle stirring to produce an 8% (wt/wt) viscosity. The solution is then delivered via a disposable 22-gauge blunt-end needle using a syringe pump. The solution must be delivered at a constant flow rate. When a high voltage of 17.5 kV is applied, a fluid jet is ejected towards the earthed rotating collector.
3. The cylindrical collector should be situated at a distance of 20 cm from an ejecting needle (see Note 4).
4. A foil sheet of A5 size (approx. 148×210 mm) wrapped around the cylindrical collector should receive the dried fibres. The rotation speed of the collector will determine whether the fibres are aligned (see Note 5).

3.2. Acrylic Acid Plasma Polymer Deposition

1. PLLA scaffolds can be subjected to a thin acrylic acid plasma polymer surface deposition. Aligned fibre scaffolds should be suspended between two pipette tips and placed upright in a

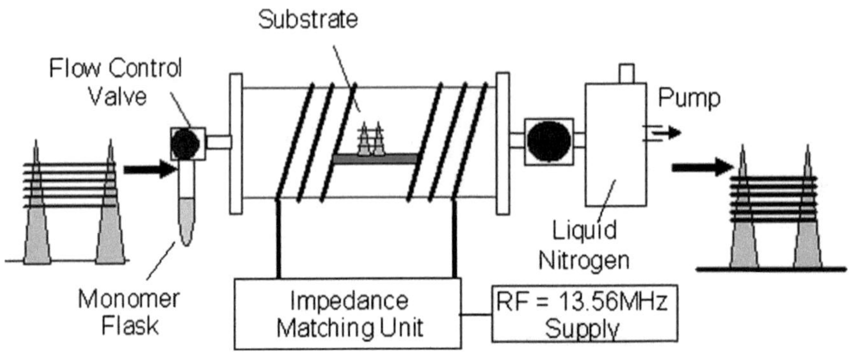

Fig. 2. Schematic diagram of a plasma deposition reactor, showing the connection of the monomer flask, flow control valve, impedance matching unit, liquid nitrogen cold trap, and pump. Aligned polymer microfibers are indicated as the substrate and are shown suspended between two disposable P200 pipette tips. The pipette tips are inserted vertically into a 96-well plate for stability.

96-well plate for stability and inserted into the deposition chamber (see Fig. 2 and Note 6). The vessel should then be pumped down to a base pressure of 1×10^{-3} mbar.

2. Acrylic acid should be used as received from the supplier after several freeze–pump thaw cycles to remove residual gas from the monomer in a vacuum-sealed Young's® monomer flask.
3. Polymerisation takes place in the glass cylindrical reactor vessel (50 cm long × 8 cm wide) with an external helical copper coil electrode.
4. An initial power of 20 W is applied for 5 min before being reduced to 10 W for 15 min. A monomer flow rate of 2.4 cm^3/min is then maintained for a further 5 min after termination of the plasma to quench any remaining radicals.

3.3. Maleic Anhydride Plasma Polymer Deposition

1. PLLA scaffolds can be subjected to a thin maleic anhydride plasma polymer surface deposition. Aligned fibre scaffolds should be suspended between two pipette tips and placed upright in a 96-well plate for stability and inserted into the reaction chamber (see Fig. 2). The vessel should then be pumped down to a base pressure of 1×10^{-3} mbar.
2. Maleic anhydride should be used as received from the supplier, placed into an appropriate flask, and covered in foil.
3. Polymerisation should take place in a stainless steel T-piece reactor vessel (376 × 240 mm) with an internal aluminium disc electrode. The reactor should be heated to 37°C.
4. The monomer flow rate is 2.5 cm^3/min, and the fibres are coated with maleic anhydride for 20 min. The pulse generator should be set to 80 μs of on time and 800 μs off time. The input power is set to 10 W. The total power is then equivalent to 1 W.

5. The TDS3014 oscilloscope is used to adjust the correct plasma wave discharge and minimises reflected power during the polymerisation period.
6. After a 20-min deposition, the monomer flow is maintained for a further 5 min after termination of the plasma to quench any remaining radicals.

3.4. RN22 Schwann Cell Culture

1. RN22 rat Schwann cells should be used up to passage 25 for experimentation and expanded in T75 flasks. Cells should be cultured in Dulbecco's Modified Eagle's Medium (DMEM).
2. Cells should be passaged when approaching 70% confluence using trypsin/EDTA to ensure continuation of the cell line. Cells are recommended to be split at a ratio of 1:10 so that they approach confluence after 72 h in culture.

3.5. RN22 Cells in Sheet Culture with PLLA Micro Fibre Scaffolds

1. Medical-grade stainless steel rings (internal diameter of 12 mm and depth of 8 mm) should be used to weigh down a monolayer of PLLA fibres electrospun onto A5 foil sheets (see Subheading 3.1).
2. For individual scaffold experiments, sheets should be cut to 20×20 mm (see Note 3).
3. Aligned fibre scaffolds should be placed in the bottom of a 6-well tissue culture plate with sterilised stainless steel rings placed on top and fibres washed with 70% ethanol for 5 h to dissolve any residual solvents left from the polymer processing stage and sterilise the scaffolds (see Fig. 3 and Note 7).
4. Ethanol is removed and scaffolds are washed with 1 ml of PBS (3×) and then left in PBS overnight (see Note 8). PBS is removed and 1×10^5 RN22 Schwann cells are seeded onto

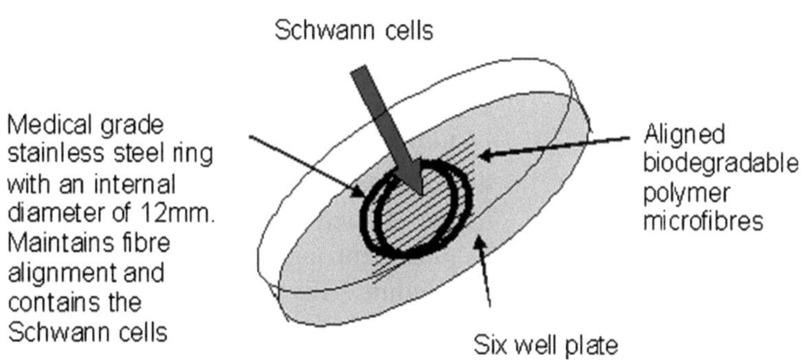

Fig. 3. Schematic diagram illustrating a sheet of aligned biodegradable PLLA microfibers supported by aluminium foil. A medical-grade stainless steel ring with a 12-mm internal diameter is used to keep the fibres aligned and in place throughout the experiment. 1×10^5 Schwann cells are seeded directly into the centre of the ring and onto the fibres for static culture.

scaffolds contained within the steel ring. Cells should be cultured in a humidified incubator at 37°C (95% air/5% CO_2) for 48–96 h according to experimental parameters.

3.6. Closed Loop Bioreactor System Assembly

1. The assembly of the closed system bioreactor is detailed in Subheading 2.6. Components can be connected together in series using silicone tubing. We recommend connecting a single experimental conduit rather than three conduits in series or parallel (although this is possible and is illustrated in Fig. 1).
2. The medium bottle should contain 30 mL of DMEM medium to enable continuous flow.
3. After experimentation, the system can be reused by flushing with 70% ethanol under flow (5 mL/h) for 5 h to sterilise, followed by a PBS wash overnight (5 mL/h).

3.7. Construction of Experimental Nerve Guidance Conduits

1. Aligned PLLA fibres should be produced on an A5 foil sheet.
2. A calculation should be made to attain an approximate 33% bulk volume of fibres per experimental conduit. This should be based on prior knowledge of: (1) the internal diameter of the outer conduit wall, (2) the average microfiber diameter, and (3) the number of fibres. For example, 8,500 fibres with an individual average diameter of 7.5 μm (=44.2 μm^2 cross-sectional area (csa)) and hence a total csa of $3.73 \times 10^5 \mu m^2$ can be introduced into a 1,200-μm diameter conduit (1.13×10^6 μm^2 total csa).
3. The approximate number of required fibres can be counted under low magnification along the foil sheet. To remove the fibres from the foil while keeping them aligned, two small cell culture pipette tips should be used (see Note 6). The pipette tips are placed at either end of the foil sheet and rolled inwards. The fibres attach to these pipette tips and remain aligned. Pulling each pipette upright suspends the aligned fibres between them. Once suspended upright, remove the aluminium foil carefully. Thus, a "tennis-net" structure is formed.
4. A thread of fine cotton can then be looped around one end of the aligned fibres. A needle attached to the thread is then pulled through the 1.2-mm internal diameter outer silicone conduit. This maintains the fibres aligned along the length of the conduit. The conduit can then be cut at either end using a sharp scalpel and connected to the bioreactor system.

3.8. Cell Introduction and 3D Culture in the Closed Bioreactor System

1. The system is sterilised with 70% ethanol for 5 h at a flow rate of 0.8 mL/min and then washed with PBS (0.8 mL/min) overnight. 5×10^5 Schwann cells/mL are introduced into the bioreactor using a 1-mL syringe via the three-way valve,

enabling direct delivery to the scaffold while maintaining sterility (see Fig. 1).

2. The Petri dish supporting the conduit, the medium reservoir, and the peristaltic pump are then placed inside a tissue culture incubator at 37°C in a 95% air/5% CO_2 humidified atmosphere. The pump is switched on briefly (0.8 mL/min) to position the cells within the scaffold and then switched off for 2 h. This is an extremely important step and enables the cells to adhere to the scaffold fibres (see Note 9). The pump can then be switched on with a continuous flow of 0.8 mL/min for experimentation lengths of 24–96 h, depending on experimental parameters.
3. After experimentation, conduits can remain connected to the bioreactor and be processed for confocal microscopy in situ.

3.9. Live/Dead Staining for Cell Viability as Sheet Scaffolds or Within the Bioreactor

1. Remove the culture medium from the samples and wash with PBS for 3 × 15 min.
2. Use a Live/Dead® BacLight™ solution to assess cellular viability. 1 ml is added to the microfiber sheet and incubated for 15 min at 37°C. The solution is then removed and washed with PBS (3×).
3. Cells integrated within scaffold fibres are imaged by confocal microscopy (LSM 510 Meta confocal microscope, Zeiss, Germany). Syto-9 images are taken at $\lambda_{ex} = 480$ nm/$\lambda_{em} = 503$ nm (for live-cell visualisation), and propidium iodide images are taken at $\lambda_{ex} = 490$ nm/$\lambda_{em} = 617$ nm (for dead cell visualisation). Samples should be immersed in PBS and an upright water-dipping lens used for visualisation.
4. When viability staining is performed for conduits, cells can be stained as above in situ, and washed with PBS (3×) using the introduction ports and pump.
5. Conduits are then disassembled from the reactor and carefully prepared for microscopy by removing a longitudinal slice of the silicone wall using a sharp scalpel and visualised by confocal microscopy under PBS as above.

3.10. Immunofluorescence Confocal Microscopy of the Schwann Cell Cytoskeleton

1. For Schwann cells cultured on aligned fibres within 6-well plates – remove culture medium from the wells and wash with PBS (3×).
2. Add fixing solution to the samples for 30 min at room temperature to fix the cells.
3. Remove the fixing solution and wash with PBS (3×).
4. Permeabilise cells by incubation in permeabilising solution for 10 min at room temperature, remove, and wash with PBS (2×) gently.

5. Incubate cells with 10 μg/mL phalloidin-FITC in PBS for 30 min to specifically label actin microfilaments [at room temperature, then remove and wash with PBS (2×)].
6. Incubate cells with 300 nM DAPI (10 μL/10 mL) for 10 min to label nuclei (at room temperature). Remove and wash with PBS (2×), leaving the third PBS wash in the well plate for objective lens immersion.
7. Samples are best viewed immediately – initially by epifluorescence microscopy, to identify approximate cell position and focal plane. Images can then be taken by confocal microscopy. Phalloidin-FITC images are taken at $\lambda_{ex}=495$ nm/$\lambda_{em}=513$ nm (for actin-filament visualisation, green emission) and for DAPI at $\lambda_{ex}=358$ nm/$\lambda_{em}=461$ nm (for nuclei visualisation, blue emission).
8. When cytoskeletal staining is performed using the 3D bioreactor, cells are processed in situ using the same steps as above using the introduction ports and pump.
9. Conduits are then disassembled from the reactor and carefully prepared for microscopy by removing a longitudinal slice of the silicone wall using a sharp scalpel and visualised by confocal microscopy under PBS as above. A typical image of Schwann cells cultured on PLLA alone versus acrylic acid coated PLLA fibres is shown in Fig. 4, where cells have been visualised using a Live/Dead® BacLight™ stain.

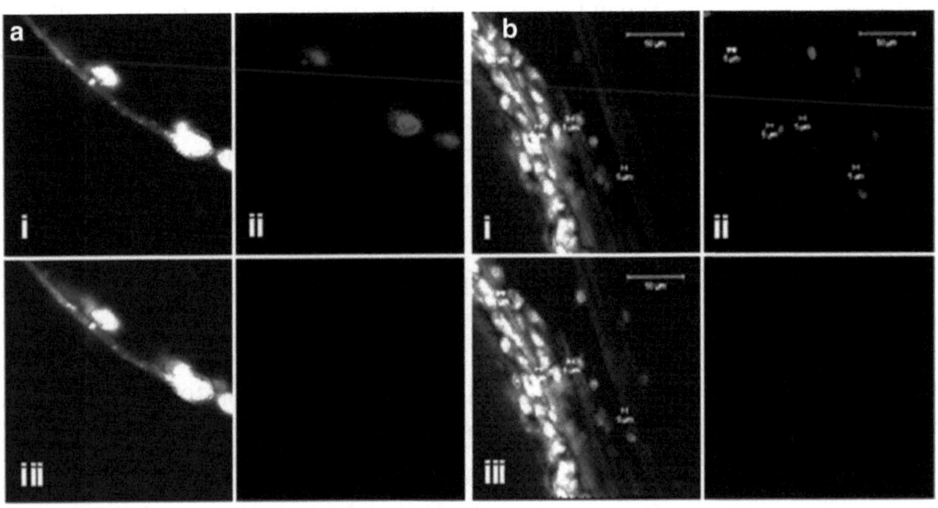

Fig. 4. Live–dead confocal microscopy showing Schwann cells cultured on aligned PLLA fibres for 96 h within the closed loop bioreactor. (**a**) Uncoated PLLA fibres and (**b**) acrylic-acid-coated PLLA fibres. (**a**, **b**): (i) Live-cell staining; (ii) dead-cell staining; (iii) composite images of live and dead cell stains shown in (i) and (ii).

4. Notes

1. The stainless steel rings were manufactured by a hospital medical workshop according to specifications provided by us. An important consideration is the internal diameter – which should permit the insertion of a water-dipping immersion objective lens. The metal rings must be made from medical-grade stainless steel and should be autoclaved before each use.
2. The PLLA fibres should be supported by the aluminium foil throughout the duration of the experiment and not removed from this support.
3. It is very important to obtain fluorescent micrograph images using an upright confocal microscope that employs water-dipping lenses. If a confocal microscope is unavailable, an upright epifluorescence microscope with water-dipping lenses will enable micrographs to be obtained, but will not enable *z*-stacks or three-dimensional imaging.
4. The rotating cylindrical collector was manufactured by a university workshop according to specifications provided by us, where the width = 20 cm and diameter = 8 cm.
5. To achieve aligned fibres, the collector must be rotated at a relatively high speed. We use 600 rpm, which is approximately tenfold faster than that required to collect non-aligned fibres.
6. Transferring the fibres from the foil sheet to the pipette tips will create a "tennis-net" arrangement for plasma polymer deposition. Use some glass paper (or sharp scissors) to roughen the surface of the pipette tips. This makes it easier for the fibres to attach to the tips.
7. The addition of any solutions to the fibres has to be introduced extremely slowly so as to avoid any movement of the aligned fibres.
8. The extended PBS incubation step is extremely important, as without it we find that Schwann cells do not attach to the fibres. It is suspected that residual contaminants from the polymer processing stage are removed at this point.
9. The Schwann cells when in suspension can readily be viewed through the silicone tubing of the bioreactor due to light scattering within the culture medium. This visual method can be used initially upon introduction to ascertain the relative position of the cells – and importantly to identify the point at which they enter the conduit scaffold.

Acknowledgements

We are grateful for financial support from the EPSRC (UK).

References

1. Noble, J., Munro, C.A., Prasad, V.S., and Midha, R. (1998) Analysis of upper and lower extremity peripheral nerve injuries in a population of patients with multiple injuries. *J. Trauma* **45**, 116–122.
2. McKay Hart, A., Brannstrom, T., Wiberg, M., and Terenghi, G. (2002) Primary sensory neurons and satellite cells after peripheral axotomy in the adult rat: timecourse of cell death and elimination. *Exp. Brain Res.* **142**, 308–318.
3. Schmidt, C.E. (2003) Neural tissue engineering: strategies for repair and regeneration. *Annu. Rev. Biomed. Eng.* **5**, 293–347.
4. Yoshii, S. and Oka, M. (2001) Collagen filaments as a scaffold for nerve regeneration. *J. Biomed. Mater. Res.* **56**, 400–405.
5. Suzuki, M., Itoh, S., Yamaguchi, I., Takakuda, K., Kobayashi, H., Shinomiya, K., et al. (2003) Tendon chitosan tubes covalently coupled with synthesized laminin peptides facilitate nerve regeneration *in vivo*. *J. Neurosci.* **72**, 646–659.
6. Mikos, A.G. and Temenoff, J.S. (2000) Formation of highly porous biodegradable scaffolds for tissue engineering. *J. Biotechnol.* **3**, 114–119.
7. Hall, S.M. (1986) The effect of inhibiting Schwann cell mitosis on the re-innervation of acellular autografts in the peripheral nervous system of the mouse. *Neuropathol. Appl. Neurobiol.* **12**, 401–414.
8. Huang, Y.C. and Huang, Y.Y. (2006) Biomaterials and strategies for nerve regeneration. *Artif. Organs* **30**, 514–522.
9. Kingham, P.J. and Terenghi, G. (2006) Bioengineered nerve regeneration and muscle reinnervation. *J. Anat.* **209**, 511–526.
10. Salim, M., Wright, P.C. and McArthur, S.L. (2009) Studies of electroosmotic flow and the effects of protein adsorption in plasma-polymerized microchannel surfaces. *Electrophoresis* **30**, 1877–1887.

Chapter 11

Encapsulation of Human Articular Chondrocytes into 3D Hydrogel: Phenotype and Genotype Characterization

Rui C. Pereira, Chiara Gentili, Ranieri Cancedda, Helena S. Azevedo, and Rui L. Reis

Abstract

This chapter is intended to provide a summary of the current materials used in cell encapsulation technology as well as methods for evaluating the performance of cells encapsulated in a polymeric matrix. In particular, it describes the experimental procedure to prepare a hydrogel matrix based on natural polymers for encapsulating and culturing human articular chondrocytes with the interest in cartilage regeneration. Protocols to evaluate the viability, proliferation, differentiation, and matrix production of embedded cells are also described and include standard protocols such as the MTT and [3H] Thymidine assays, reverse transcription polymerase chain reaction (RT-PCR) technique, histology, and immunohistochemistry analysis. The assessment of cell distribution within the 3D hydrogel construct is also described using APoTome analysis.

Key words: Natural polymers, Hydrogels, Gelation, Chondrocyte encapsulation, Cell culture and survival within 3D matrix, Cell-based technology, Cartilage regeneration

1. Introduction

Cell encapsulation technology holds promise in many areas of medicine and biotechnology. For example, some important applications are treatment of diabetes, production of biologically important chemicals, and controlled release of drugs. The hope that encapsulated cells might be used as a therapeutic strategy is increasingly being realized and is expected to have enormous potential in medicine in the near future. However, to be a viable complement to the current methods of cell transplantation therapy, encapsulation technology has to fulfill the strict requirements applicable to these types of therapeutic strategies, such as

John W. Haycock (ed.), *3D Cell Culture: Methods and Protocols*, Methods in Molecular Biology, vol. 695,
DOI 10.1007/978-1-60761-984-0_11, © Springer Science+Business Media, LLC 2011

performance, biosafety, biocompatibility, stability, availability, purity, characterization, and cost (1). In a typical encapsulation process, cells are suspended in a liquid solution, which is then rapidly solidified or gelled to entrap cells in the matrix. It is widely understood that in order to encapsulate living cells within an artificial matrix, the encapsulating conditions must not damage the living cells and the matrix material must be biocompatible and at the same time provide a natural microenvironment for the cells to sustain their viability, function, and growth or differentiation. Additionally the mechanical strength of the encapsulating matrix is critical, particularly when long-term implantation is envisaged. The mechanical strength is essential for maintaining the matrix integrity and to withstand manipulations associated with in vitro culture, implantation, and in vivo existence. Moreover, it is essential that the matrix should have suitable diffusion properties to ensure sufficient access of nutrients, for encapsulated cells to remain viable and functional, and the removal of secreted metabolic waste products. Thus, there is a need for relatively mild cell encapsulation methods, which offers control over properties of the encapsulating matrix.

An exciting approach for cell delivery is the use of materials that can undergo a gelation process that is cell-compatible and can be injected into the body (i.e., hydrogels). This approach enables the clinician to transplant the cell and polymer combination in a minimally invasive way (2). Hydrogels are appealing for cell delivery because they are highly hydrated three-dimensional networks of polymers that provide a place for cells to adhere, proliferate, and differentiate and they can often be processed under relatively mild conditions. Gelation can occur through a change in temperature (thermoreversible gels), or via chemical crosslinking, ionic crosslinking, or formation of an insoluble complex. Covalent crosslinking is a common method to precisely control the crosslinking density of hydrogels. However, the toxicity of cross-linking molecules must be considered. Many chemical crosslinking agents negatively impact cell viability and cause reactions in vivo, precluding the use of most chemically crosslinked gels for encapsulation. Ionic crosslinking with multivalent counterions is a simple way to form hydrogels. However, those ions could be exchanged with other ionic molecules in aqueous environments, resulting in an uncontrolled deterioration of the original properties of hydrogels.

Multicomponent blends of synthetic, semi-synthetic, and naturally occurring macromolecules have been explored for the encapsulation of a variety of mammalian cell types (3–5). Naturally-derived polymers are abundant, biodegradable, and usually undergo gelation under gentle conditions. In addition, they have shown low toxicity. Natural polyanions include several polysaccharides (e.g., alginate, carboxymethyl cellulose, carrageenan, cellulose sulfate, gellan gum, gum arabic, heparin, hyaluronic acid, xanthan, dextran sulfate) and chitosan, the only naturally occurring polycation.

Although alginate-based materials are still receiving a great deal of attention, a wide variety of polyelectrolyte gels and complexes have now been investigated for cell encapsulation. Alginate, a negatively-charged polysaccharide, can either be used alone, or in conjunction with positively-charged polylysine to form alginate–polylysine polyelectrolyte complexes. Alginate itself may be ionically crosslinked by divalent cations such as calcium and barium. Although the encapsulating method based on ionic crosslinking of alginate (a polyanion) with polylysine or polyomithine (polycation) offers relatively mild encapsulating conditions and quite stable matrices, their mechanical properties and long-term stability are poor. Moreover, these polymers when implanted in vivo, are susceptible to cellular overgrowth, which restricts the permeability of the matrices to nutrients, metabolites, and transport proteins from the surroundings, leading to starvation and death of encapsulated cells.

Other encapsulation matrices have been tested and they were either not sufficiently mechanically stable or suffered from other surface or matrix related deficiencies such as shrinkage in either PBS or in culture media. In order to overcome these limitations, several approaches have been considered and tested, like crosslinking, chemical adjustment of charge density, combination of low and high molecular weight polyelectrolytes, adjustment of osmotic pressure, polymer grafting, polymer blending, among others.

Currently, hydrogels are being used in an attempt to engineer a wide range of tissues, including cartilage, bone, muscle, fat, liver, and neurons. Hydrogels have a similar macromolecular structure to cartilage, which is a highly hydrated tissue composed of chondrocytes embedded in type II collagen and glycosaminoglycans. Thus, cartilage is an obvious tissue to engineer using hydrogel matrices. Numerous hydrogel systems embedded with chondrocytes have been developed and tested both in vitro and in vivo.

Carrageenans are high molecular weight sulfated polygalactans derived from several species of red seaweeds (*Rhodophyceae*). The most common forms of carrageenan are λ, κ and ι. Carrageenan has alternating disaccharide units composed of d-galactose-2-sulfate and d-galactose-2,6-disulfate, being the galactose residues joined by -1,3 and -1,4 linkages. Carrageenans resemble to some extent the naturally occurring glycosaminoglycans owing to their backbone composition of sulfated disaccharides. The different composition and conformation of carrageenans results in a wide range of rheological and functional properties. Carrageenans form thermoreversible gels. The polysaccharide is water soluble when heated and gels upon cooling the solution in presence of electrolytes (Ca^{2+} and K^{+}). κ-carrageenan gels in the presence of K^+ ions to form strong crisp gels, whereas ι-carrageenan gels in the presence of Ca^{2+} ions to form elastic gels (6). Carrageenans have been used extensively in the food industry. In addition to incorporation in foods, carrageenan has been used as an ingredient in pharmaceuticals and personal care products, such as toothpaste and cosmetics (7).

The use of carrageenan has not been fully explored as potential biomaterial and we believe that the structural diversity of carrageenans can provide very interesting rheological and functional properties in cell encapsulation technology. Because of its widespread use in food and pharmaceutical industry, purified carrageenan materials are readily available with reliable, predictable, and chemically defined composition.

The aim of the present chapter is to provide experimental procedures to prepare a hydrogel-based matrix for the encapsulation of chondrocytes, and evaluate its potential as a cell-delivery carrier for cartilage regeneration strategies. We focus on methods for culturing and expanding human chondrocytes obtained from articular cartilage biopsies and evaluating the maintenance of cell viability and function within the 3D hydrogel matrix. Characterization methods to evaluate other properties of the hydrogel system, such as hydrogel strength and stability, gelation kinetics, and hydrogel permeability are also of central importance, but these topics will not be described in this chapter.

2. Materials

2.1. Hydrogel

1. Carrageenan polysaccharides (ι- and κ-carrageenans both from Fluka Biochemika, Denmark).
2. Human fibrinogen and human thrombin (Baxter AG, Vienna, Austria) plus the syringe holder – Duploject. The syringe holder allows (with a single press) for an injectable hydrogel to be formed with an equal distribution of both components.
3. High molecular weight hyaluronic acid from *Bacillus subtilis* (kindly provided by Prof. Hilborn, Uppsala University).
4. Double distilled water, $CaCl_2$ and KCl powders to trigger gel formation.

2.2. Cell Culture: Primary Culture of Human Articular Cartilage Biopsies

1. Coon's F-12 Modified Ham's medium (Biochrom A.G. Berlin, Germany) supplemented with 10% fetal calf serum (FCS) (Invitrogen Life Technologies, Carlsbad, CA).
2. Trypsin solution (0.25%) with ethylenediamine tetraacetic acid (EDTA) (1 mM) (Invitrogen Life Technologies, Carlsbad, CA).

3. Collagenase I, collagenase II (both from Worthing Biochemical, Lakewood, NJ), hyaluronidase (Sigma, St. Louis, MO), and trypsin (Invitrogen Life Technologies, Carlsbad, CA). Store at −20°C.

2.3. Cell Culture: 2D Monolayer Cell Expansion

1. Coon's F-12 Modified Ham's medium supplemented with 10% fetal calf serum.
2. Serum free medium (8) freshly prepared twice a month in accordance with the composition listed in Table 1. Human recombinant growth factors (Austral Biologicals, San Ramon, CA, USA) and insulin (Sigma-Aldrich, Steinheim, Germany).
3. Trypsin solution (0.25%) with ethylenediamine tetraacetic acid (EDTA) (1 mM).

2.4. Cell Culture: 3D Cell Encapsulation and Culture Under Chondrogenic Differentiation

1. Agarose working solution 1% (w/v) in PBS (1×). Transfer this solution to a glass bottle and autoclave to ensure adequate sterilization. Store at room temperature.
2. Plastic ring containers (made, for example, of 2 mL plastic pipettes cut in equal parts, or pre-made plastic rings used in

Table 1
Serum free medium composition (adapted from Malpeli et al. (8))

Component (protein/growth factor)	Final concentration
Apo-transferrin	25 μg/mL
Ascorbic acid	250 μM
Biotin	33 μM
Cholesterol	13 μM
Dexamethasone	10 nM
Epidermal growth factor	5 ng/mL
Fibroblast growth factor 2	5 ng/mL
Holo-transferrin	25 μg/mL
Human serum albumin	1% (w/v)
Insulin	10 μg/mL
Linoleic acid	4.5 μM
N-acetylcysteine	50 μM
Platelet derived growth factor	5 ng/mL
Sodium pantothenate	17 μM
Sodium selenite	30 nM

cell biology laboratories to collect cell clones) to be used as a mold for the injectable hydrogel (to give form and shape). Choose an appropriate size to determine the volume of the hydrogel and sterilize by autoclaving.

3. Chondrogenic medium (9): Coon's F-12 Modified Ham's medium supplemented with 6.25 μg/mL Bovine Insulin, 6.25 μg/mL Human Apotransferrin, 1.25 μg/mL Linoleic Acid, 5.35 μg/mL Bovine Serum Albumin, 1 mM sodium pyruvate, 50 μg/mL Ascorbic acid, 10 ng/mL Transforming Growth Factor β-1 (TGF β-1), and 10^{-7} M dexamethasone.

2.5. Viability and Proliferation of Encapsulated Cells

1. Thiazolyl Blue staining (MTT; Sigma-Aldrich) powder dissolved at concentration of 5 mg/mL in freshly prepared phosphate buffer solution (PBS) (1×). Store aliquots at −20°C.
2. Spectrophotometer (Beckman Du® 640, USA).
3. 1450 Microbeta – Liquid Scintillation & Luminescence Counter (Wallac, Triliux, USA).
4. [Methyl-^{3}H] Thymidine (Amerstam, GE Healthcare, UK) solution. Store at 2°C (see Note 1).
5. EcoLite™ – Scintillation Liquid (JCN, Costa, CA, USA). Store at room temperature and protect from light.
6. DNA buffer solution (Invitrogen, Carlsbad, CA, USA). Store at 4°C.
7. Washing solutions, absolute ethanol, and PBS (1×). Store these solutions at room temperature.

2.6. ApoTome Image Acquisition

1. Axiovert 200M microscope equipped with the ApoTome module and the AxiocamHR camera (Carl Zeiss, Jena, Germany) as well as Axiovision Software (Carl Zeiss, Jena, Germany).
2. Hoechst 33342 at final concentration of 5 μg/mL in PBS (1×). Store at 4°C and protect from light.

2.7. Histology and Immunohistochemistry Analysis

1. Graded ethanol solutions (70, 90, 95, and 100%); xylene and wax.
2. 4% Formaldehyde solution in PBS. Store aliquots at −20°C.
3. Methanol/H_2O_2 solution: methanol/H_2O_2 (49:1, v/v). Store at 4°C.
4. Sodium acetate solution: 50 mM Na Acetate, pH 5.0. Adjust the pH solution with HCl (0.05 M) or with NaOH (0.05 M).
5. Peroxidase substratum solution: 3-amino-9-ethylcarbazole 0.04% in dimethylformamide, 45 mM Na Acetate (pH 5), 0.03% H_2O_2 (100:900:1). Store at room temperature and protect from light.

6. Harris's haematoxylin solution (Bio Optica, Milano, IT).
7. Mounting medium: Gel mount (Biomeda Corp., Foster City, CA, USA).

2.8. mRNA Extraction

1. PCR supplies (microcentrifuge tubes, PCR tubes, PCR plates, pipette tips) and water (diethylpyrocarbonate (DEPC) treated water) free of DNAase/RNAase.
2. Trizol Reagent (Invitrogen, Carlsbad, CA, USA). Store at 4°C.
3. Homogenizer (Ultra Turrax® Model T25, IKA Labortechnik GMbH, Germany).
4. 2-Isopropanol.
5. Solution of chloroform/isoamyl alcohol (49:1) (v/v). Store at room temperature.
6. 70% Ethanol solution. Store at −20°C.

3. Methods

3.1. Hydrogel Preparation

1. Weigh out 0.6 and 0.4 g of ι- and κ-carrageenans, respectively, and place them inside appropriate tubes to be autoclaved (120°C for 20 min) to ensure sterility. Sterilize double distilled water, glassware, and lab tools (e.g., two or three beakers, magnetic stirrers, and forceps) by autoclaving them prior to use.
2. Under sterile conditions (laminar flow cabinet), heat 50 mL of double distilled water (covered with a cap to avoid evaporation) up to 65°C and dissolve the carrageenan powder under constant stirring until forming a homogenous preparation.
3. When the solution is clear, homogenous, and without bubbles, switch off the stirrer and the heater power. Leave it at room temperature. Transfer the polymeric solution to a 50-mL tube. The solution is ready to be used. It can remain at room temperature from days to weeks until further use.

3.2. Human Articular Chondrocytes Primary Culture

1. Clean human articular cartilage biopsy samples from connective and muscular tissues and/or subchondral bone and cut them in small fragments (see Note 2).
2. Wash articular cartilage fragments several times with freshly prepared PBS (1×) and centrifuge in PBS (1×) at 180 ×*g* for 5 min.
3. Discard the supernatant and refill with the enzymatic cocktail solution made of: trypsin 0.25% (v/v), collagenase I (400 U/mL), collagenase II (1,000 U/mL), and hyaluronidase (1 mg/mL). Place in a thermostatic bath at 37°C for 30 min.

4. Collect the supernatant and block the enzymatic activity with Coon's modified ham's medium with 10% fetal calf serum (FCS). Centrifuge for 10 min at 180×*g*.
5. Plate the cell suspension under anchorage-dependent conditions: plastic Petri dishes coated with Coon's modified ham's medium with 10% of FCS.
6. Repeat steps 3–5 until no biopsy material is available. Have in mind that this procedure can take from 2/3 h up to 1 entire day, depending on the amount of biopsy material. Therefore, prepare, at least 1 day before, chemicals and materials (aliquots of the different enzymatic solutions, solutions, several bistouries, and forceps) necessary to perform this procedure.
7. Keep the cells in culture under conditions described in step 5 for 3/4 days. Trypsinize cells and perform an initial count (counting is necessary to report cell duplication number relative to the starting number of primary cells). The next step is to re-plate the cells under serum-free medium conditions.

3.3. 2D Monolayer Cell Expansion

1. Before starting the 2D monolayer expansion of human articular chondrocytes under serum-free medium conditions, it is important to be aware that all growth factors should be previously aliquoted and ready for use. The medium used during all the expansion period of a primary culture should be always from the same batch of aliquots. This is important for reducing variability between independent experiments.
2. Enzymatically detach and count the cells (reported as step 7 in Subheading 3.2). Use Coon's modified Ham's medium with 10% of FCS to block the activity of trypsin.
3. Wash three times with PBS (1×) the Petri dish coated with Coon's modified Ham's medium (plus 10% FCS) to remove any residual traces of serum. Re-plate the cells at a density of $4 \times 10^4/\text{cm}^2$ using serum free medium prepared according to the Table 1. Change the medium three times a week.
4. When cells have a sub-confluent distribution, repeat step 2 using Coon's modified Ham's medium with 1% of FCS instead of 10% FCS. Repeat step 3 until the cell number reaches the number necessary to start the experiment.

3.4. 3D Cell Encapsulation and Culture Under Chondrogenic Differentiation Conditions

1. Prepare aliquots of fibrinogen at a final concentration of 3,000 UIK/mL, as well as thrombin (500 IU/mL). Add hyaluronic acid powder to the latter solution to obtain a concentration of 0.75% w/v, and allow the polymer to dissolve overnight at 4°C under gentle shaking.
2. At the end of in vitro cell expansion, detach enzymatically the cells and count them. Calculate the number of hydrogels to be prepared (with and without cells) based on the final number

of cells available for encapsulation and the cell density within the 3D hydrogel.

3. Prepare a medium that will trigger the gel formation of carrageenan solution. For that, dissolve $CaCl_2$ (0.265 g/L) and KCl (0.4 g/L) in Coon's modified Ham's medium and render this solution sterile by syringe filtration (0.22-μm filter).
4. Add fibrinogen solution 35% (v/v) to the carrageenan polymeric solution. Mix them well with a pipette avoiding the formation air bubbles (solution B). Make the cell suspension in the medium prepared in step 3 and add thrombin solution to a final concentration of 5.9% (v/v) (Solution A) (see Note 3). The cell number depends on the final cell density within the hydrogel (i.e., 2×10^6 cells/mL). See Fig. 1.
5. Load solution A and B in to two separate 1-mL syringes (see Fig. 1). Place them with care in the Duploject double syringe holder and set the connector at the syringe extremity end. Insert the needle and be ready to fill the plastic ring molds for giving shape and size to the injectable hydrogel.
6. Mix well before starting the injection to avoid sinkage of cells to the bottom of the syringe by gravity. Inject the same volume

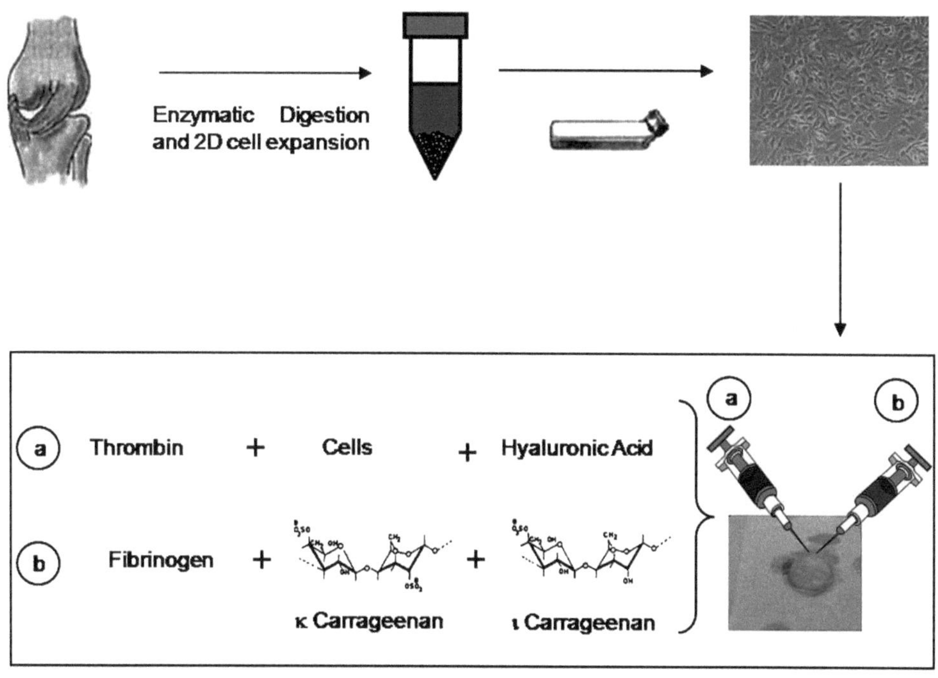

Fig. 1. Schematic representation of the method to form the hydrogel. From the *top left* (following the *arrows*): human articular chondrocytes biopsy and successive digestions; Light microscopy image of primary monolayer culture in serum free medium during the 2D expansion period; biodegradable hydrogel system with the content composition in both syringes (a and b) and the macroscopic morphology of the formed gel. Reproduced with permission from Pereira et al. (14).

of hydrogel into the rings within a 24 multi-well plate (previously coated with agar).

7. Wait 5–10 min and turn the gels upside down to allow a better cell distribution within the hydrogel (to avoid cells sinking to the bottom of the gel by gravity). Wait another 5–10 min. Meanwhile prepare the chondrogenic medium. Calculate the volume necessary to be prepared knowing that the total volume of the hydrogel must be covered with medium.
8. Discard the plastic rings and add medium to each well. Place the 24 multi-well plate in to an incubator (37°C and 5% CO_2). Change the chondrogenic medium three times a week.

3.5. In Vitro Viability and Proliferation of Encapsulated Cells

It is necessary for each time point to evaluate the viability and proliferation rates of the human articular chondrocytes encapsulated within the gel. Repeat the procedure described below: MTT assay corresponding to the viability analysis (10) and [methyl-^{3}H] Thymidine (11) to proliferation.

3.5.1. MTT Assay (Viability)

1. Prepare serum-free medium with MTT at a final concentration of 50 μL/mL. The final volume depends not only on the number of samples to be analyzed (remember to perform the analysis at least in triplicate) but also on the volume occupied by the construct. All individual samples must be completely immersed. The volume used for all the samples must be the same, even if their volume is different.
2. At pre-determined time points, carefully remove the medium from the well in which the 3D construct was kept during the in vitro culture without damaging it.
3. Refill the well with recently prepared medium described in step 1. Incubate (37°C and 5% CO_2) for 2–3 h protected from light.
4. Collect the medium and add 1 mL absolute ethanol. The hydrogel with encapsulated cells will form blue dots (the dots correspond to the cells). Break up the hydrogels by pipetting the construct up and down. The same number of movements should be applied to all the conditions.
5. For each sample, collect 1 mL of liquid into separated tubes and measure the dye absorbance at 570 nm with background subtraction at 670 nm in a spectrophotometer.
6. Repeat this procedure for remaining time points programmed for the experiment and at the end plot the absorbance values versus time.

3.5.2. [Methyl-^{3}H] Thymidine Assay (Proliferation)

1. Prepare serum free medium with [methyl-^{3}H] Thymidine at a final concentration 1 μCi/mL. Before handling radioactive material, ensure you are familiar with good radiation safety

practice (see Note 1). Place all materials in contact with the radioactive material in a closed bag and place in a designated radioactive waste container for disposal.

2. Remove the medium from the sample for analysis. Incubate the 3D constructs (37°C and 5% CO_2) with the medium prepared in step 1 for 16 h ensuring that the hydrogels are completely immersed in the medium.
3. After the incubation period, collect the medium in a closed flask and place it in a container (designated for radioactive solutions). Wash the samples several times with PBS (1×) to remove residual [methyl-^{3}H] Thymidine (i.e., not incorporated into cellular DNA) from the 3D hydrogels. Add these solutions to the flask containing [methyl-^{3}H] Thymidine.
4. Add 500 μL of DNA buffer solution to the 3D hydrogel and break up the hydrogel by pipetting up and down, according to the protocol used for the MTT assay (MTT, step 4). Recover the supernatant and store at −20°C.
5. After collecting the samples for all the experimental time points, thaw the frozen samples and transfer 400 μL of the solubilized-cell solution into 4 mL scintillation vials and count. Specify a volume for each and add 2 mL of scintillation liquid. Transfer the solution to the Beta Counter containers and read. The samples must be performed in triplicate for each condition as a minimum. Plot the values versus time. Cell proliferation is evaluated by [methyl-^{3}H] Thymidine incorporation, which correlates to cell number as a function of culture time.

3.6. APoTome Analysis

This is an epifluorescent illumination technique (12, 13), which allows images to be obtained throughout most of the hydrogel without the need for any manipulation procedures, such as dehydration for histology (which may introduce artifacts). This can be a powerful tool for in vitro cellular characterization (distribution and viability) of hydrogels.

1. Carefully aspirate the medium from the well where the samples to be analyzed are located.
2. Incubate (37°C and 5% CO_2) the samples with a serum-free medium containing Hoechst 33342 at a final concentration of 5 μg/mL for 30 min (and protect from light).
3. Transfer the hydrogel to a small Petri dish (3 mL) and start to acquire images by epifluorescent illumination using an Axiovert 200M microscope equipped with the ApoTome module. Set up the camera and select the correct filter for the required channel. Start the image acquisition stacking optical sections and subsequent 3D reconstruction using the Axiovision Software (see Fig. 2). During this procedure,

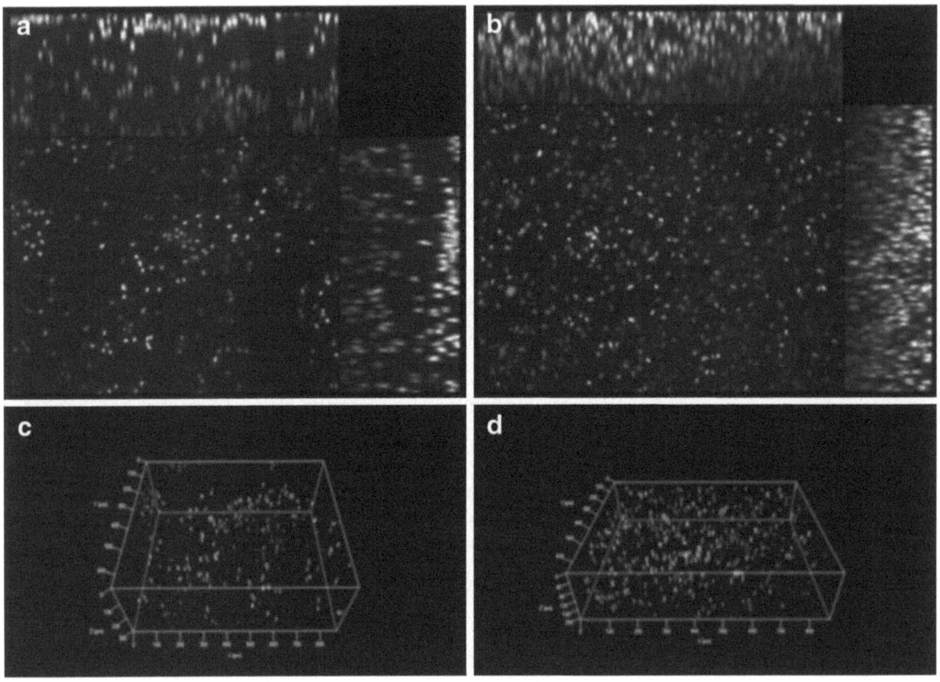

Fig. 2. Representative images of live human articular chondrocytes encapsulated within the hydrogel at 1 (**a**, **c**) and 3 weeks (**b**, **d**) of culture after labeling of the cell nuclei with Hoechst and optical sectioning by structured epifluorescent illumination. An image projection along the *X*, *Y*, and *Z* axes (**a**, **b**), and their respective 3D reconstructed stacks (**c**, **d**). Reproduced with permission from Pereira et al. (14).

try not to overexpose the sample to the fluorescent light (to prevent photo bleaching). Afterwards, samples can be processed for histology.

3.7. Histology and Immunohistochemistry Analysis

1. Discard the medium, rinse the sample three times with PBS (1×) and transfer the sample to a test tube. Add 10 mL of 4% formaldehyde solution and incubate for 6 h under rotation conditions. Make sure that the tube is tightly closed to avoid possible leaks.
2. Dehydrate the samples by immersing them in a graded series of ethanol solutions (70, 90, 95, and 100%) for 30 min each and xylene (10 min). Due to the high water content of the hydrogel, sample shrinkage is normally observed during dehydration (due to removal of the intrinsic water in the sample). Be careful not to lose any sample during the dehydration steps.
3. Embed the hydrogels in paraffin. Perform a series of different melting points on the paraffin specifically at 45, 55, and 60°C (each one for 45 min). Perform the final paraffin inclusion and keep the sample at −20°C.

4. With the help of a microtome, cut sample sections with a thickness of between 4–6 μm. Dewax the sections repeating step 2 in reverse in order to perform any specific staining steps (e.g., Harris's haematoxylin and Eosin, Toluidine Blue) or immunohistochemistry.

3.7.1. Immunohistochemistry

1. After step 4, incubate the cut sections with methanol/H_2O_2 (49/1) solution for 30 min to inhibit endogenous peroxidases.
2. Discard the solution and incubate with hyaluronidase solution for 20–30 min. Wash several times with PBS (be careful to not detach the sample from the glass).
3. To reduce non-specific binding, incubate with goat serum for 1 h. Wash three times with PBS. Incubate with a specific primary antibody for 1 h at room temperature.
4. Wash the sections three times with PBS and incubate with secondary biotinylated anti-mouse IgG and peroxidase-conjugated egg-white avidin. Wash the sections with PBS and add peroxidase substrate solution. Incubate at room temperature (protected from daylight) for a period of 15 min.
5. Counterstain the sample with Harris's haematoxylin. Mount the section with gel mount and observe by optical microscopy.

3.8. mRNA Extraction/Isolation

1. Collect the medium from the wells and wash the 3D hydrogels several times with PBS.
2. Aspirate the hydrogel carefully with the help of a pipette and place in a tube.
3. Add 1 mL of Trizol and place immediately on ice.
4. Homogenize the samples at 15,000 × *g* for 30 s, rinsing thoroughly the probe with DEPC water.
5. Transfer 1 mL of sample to a 1.5-mL tube. Incubate at room temperature for 5 min. Add 200 μL of chloroform/isoamyl alcohol (49:1) and incubate the samples at room temperature for 4 min.
6. Centrifuge at 12,000 × *g* for 10 min at 4°C.
7. Collect the supernatant into a new 1.5-mL microcentrifuge tube. Be very careful not to disturb the lower part of the gradient solution, to minimize possible protein contamination.
8. Add an equal volume of 2-isopropanol per tube (1:1). Hand mix and incubate at room temperature for 10 min.
9. Centrifuge at 12,000 × *g* for 10 min at 4°C. At this stage, a pellet will form in the bottom of the tube. Be aware of possible difficulties for visualizing the pellet due to its small size and irregular shape.

10. Discard the liquid and add 1 mL of 70% ethanol. Centrifuge at 7,500 × *g* for 5 min at room temperature.
11. Carefully remove the ethanol solution and dissolve the RNA pellet in water (50–100 μL RNAase-free water). The RNA of each sample is ready to be quantified and consequently used in a RT-PCR.

4. Notes

1. [Methyl-^{3}H) Thymidine is a radioactive reagent with an approximate 15 day decay period. When working with radioactive chemicals, it is important to follow correct local and national safety guidelines. This will include the use of personal protective equipment (e.g., always work with gloves and use two pairs if possible) and the disposal of radioactive waste (radioactive liquids and other materials in contact with radioactive chemicals), which must be via designated containers for radioactive waste disposal.
2. The importance of protocol design in yielding cells of adequate quantity and quality is paramount for the success of this technology. Therefore, several variables should be considered when performing any type of primary culture (i.e., the total final number of cells enzymatically obtained during the tissue digestion and its biological quality). During the first step, it is very important to clean the biopsy sample from all adherent tissues (muscular, connective, and bone) to avoid possible contamination of cartilage cells. The biological quality of the primary culture will depend on the biological functionality of the tissue (cartilage in this case) and also on the degree of tissue purity.
3. Having a homogeneous distribution of cells in the syringe before the injection of the hydrogel is extremely important. Be aware that by gravity cells in suspension will sink to the bottom of the gel. To avoid that, shake the syringe and try to keep it horizontal before injection. In this way, the cell distribution within the hydrogel construct will be more uniform. Even knowing that this is an easy to handle injectable hydrogel, with in situ gelling properties, try to minimize any spill to avoid wasting cells and polymer solution.

Acknowledgements

The authors would like to thank Recco Orthopedic staff members for helpful discussions, particularly to Dr. Michele Grandizio; patients for biopsy material donation, as well as to Mrs Daniela

Marubbi and Monica Scaranari for technical assistance with histological sample processing. This work was supported by funds from the Italian MUR (FIRB-Tissuenet project), the European Union funded STREP Project HIPPOCRATES (NMP3-CT-2003-505758), and the European NoE EXPERTISSUES (NMP3-CT-2004-500283).

References

1. Orive, G., Hernandez, R.M., Gascon, A.R., Calafiore, R., Chang, T.M.S., de Vos, P., Hortelano, G., Hunkeler, D., Lacik, I., and Pedraz, J.L. (2004) History, challenges and perspectives of cell microencapsulation. *Trends Biotechnol.* **22**, 87–92.
2. Lee, K.Y. and Mooney, D.J. (2001) Hydrogels for tissue engineering. *Chem. Rev.* **101**, 1869–1879.
3. Prokop, A., Hunkeler, D., DiMari, S., Haralson, M.A., and Wang, T.G. (1998) Water soluble polymers for immunoisolation I: Complex coacervation and cytotoxicity. *Microencapsul. Microgels Iniferters* **136**, 1–51.
4. Prokop, A., Hunkeler, D., Powers, A.C., Whitesell, R.R., and Wang, T.G. (1998) Water soluble polymers for immunoisolation II: Evaluation of multicomponent microencapsulation systems. *Microencapsul. Microgels Iniferters* **136**, 53–73.
5. Oliveira, J.T. and Reis, R.L., Hydrogels from polysaccharide-based materials: fundamentals and applications in regenerative medicine, in *Natural-based polymers for biomedical applications*, R.L. Reis, et al., Editors. 2008, Woodhead Publishing Limited: Cambridge. pp. 485–514.
6. Yuguchi, Y., Thuy, T.T.T., Urakawa, H., and Kajiwara, K. (2002) Structural characteristics of carrageenan gels: temperature and concentration dependence. *Food Hydrocoll.* **16**, 515–522.
7. Rinaudo, M. (2008) Main properties and current applications of some polysaccharides as biomaterials. *Polym. Int.* **57**, 397–430.
8. Malpeli, M., Randazzo, N., Cancedda, R., and Dozin, B. (2004) Serum-free growth medium sustains commitment of human articular chondrocyte through maintenance of Sox9 expression. *Tissue Eng.* **10**, 145–155.
9. Johnstone, B., Hering, T.M., Caplan, A.I., Goldberg, V.M., and Yoo, J.U. (1998) In vitro chondrogenesis of bone marrow-derived mesenchymal progenitor cells. *Exp. Cell Res.* **238**, 265–272.
10. Denizot, F. and Lang, R. (1986) Rapid colorimetric assay for cell-growth and survival – modifications to the tetrazolium dye procedure giving improved sensitivity and reliability. *J. Immunol. Methods* **89**, 271–277.
11. Yan, W.Q., Yang, T.S., Hou, L.Z., Suzuki, F., and Kato, Y. (1994) Effect of concanavalin A on morphology and DNA synthesis of resting chondrocyte cultures. *Shi Yan Sheng Wu Xue Bao* **27**, 300–305.
12. Osakada, F., Ikeda, H., Mandai, M., Wataya, T., Watanabe, K., Yoshimura, N., Akaike, A., Sasai, Y., and Takahashi, M. (2008) The generation of rod and cone photoreceptors from mouse, monkey and human embryonic stem cells. *Nat. Biotechnol.* **26**, 352–352.
13. Frey, M.R., Dise, R.S., Edelblum, K.L., and Polk, D.B. (2006) p38 kinase regulates epidermal growth factor receptor downregulation and cellular migration. *EMBO J.* **25**, 5683–5692.
14. Pereira, R.C., Scaranari, M., Castagnola, P., Grandizio, M., Azevedo, H.S., Reis, R.L., Cancedda, R., and Gentili, C. (2009) Novel injectable gel (system) as a vehicle for human articular chondrocytes in cartilage tissue regeneration. *J. Tissue Eng. Regen. Med.* **3**, 97–106.

Chapter 12

Micro-structured Materials and Mechanical Cues in 3D Collagen Gels

James B. Phillips and Robert Brown

Abstract

Collagen gels provide a versatile and widely used substrate for three-dimensional (3D) cell culture. Here we describe how cell-seeded Type-I collagen gels can be adapted to provide powerful 3D models to support a wide range of research applications where cell/substrate alignment, density, stiffness/compliance, and strain are critical factors. In their fully hydrated form, rectangular collagen gels can be tethered such that endogenous forces generated as resident cells attach to and remodel the fibrillar collagen network can align the substrate in a controllable, predictable, and quantifiable manner. By removing water from collagen gels (plastic compression), their density increases towards that of body tissues, facilitating the engineering of a range of biomimetic constructs with controllable mechanical properties. This dense collagen can be used in combination with other components to achieve a range of functional properties from controlled perfusion, or tensile/compressive strength to new micro-structures. Detailed methodology is provided for the assembly of a range of 3D collagen materials including tethered aligned hydrogels and plastic compressed constructs. A range of techniques for analysing cell behaviour within these models, including microscopy and molecular analyses are described. These systems therefore provide a highly controllable mechanical and chemical micro-environment for investigating a wide range of cellular responses.

Key words: Collagen, Three-dimensional cell culture, Hydrogel, Plastic compression, Cell alignment, Neurons, Schwann cells

1. Introduction

This chapter describes the creation of cell culture models in which mechanical cues can be controlled and manipulated. The first part describes systems for *modelling cell function* based on the use of collagen gels and the second part describes how collagen gels can be modified *towards tissue density*. The key cellular parameters that can be controlled using these approaches are the alignment,

John W. Haycock (ed.), *3D Cell Culture: Methods and Protocols*, Methods in Molecular Biology, vol. 695,
DOI 10.1007/978-1-60761-984-0_12, © Springer Science+Business Media, LLC 2011

density, stiffness/compliance, and strain. Whilst a range of three-dimensional (3D) cell culture materials are available, Type-I collagen is the substrate of choice in these systems for a number of reasons. It is the major constituent of many of the tissues which are being modelled and supports the survival of most cell types, making it relevant for many model applications. It can be obtained from animal sources in a reliable and pure form, and unlike other popular 3D gel substrates it is not supplemented with mixtures of other bioactive ECM proteins and growth factors, which can complicate experimental test approaches and thus diminish the usefulness of culture models whose strength lies in their simplicity and focus.

From a mechanical point of view these cell-collagen gel systems normally represent the resultant or sum of two sets of forces, endogenous (cell-origin) and exogenous, including gravity. In practice, many of the substrates (especially unmodified hyperhydrated collagen gels) are so compliant that even very small forces are significant and so only occasional, specially designed, systems can test either component in isolation. Endogenous forces are generated as resident cells attach to and remodel the fibrillar collagen network and this behaviour can be harnessed in collagen gels to allow strain generated by the resident cells to contract and align the substrate in a controllable, predictable, and quantifiable manner (1, 2). Alternatively, exogenous strain can be applied to cells by stretching or compressing the gel (3). Furthermore, collagen gels can be prepared with a wide spectrum of densities (and therefore compliance), ranging from hydrogels with minimal protein content to materials with similar collagen density and mechanical properties to body tissues (4).

Here we describe the production, maintenance, and analysis of versatile rectangular collagen gel models which have been used successfully to investigate the effect of strain generated by and applied to cells, and to create co-culture models in which the alignment of cells was explored (1–3, 5–9, 20). These rectangular gel models are particularly useful since the mechanical microenvironment at distinct regions of the same gel varies in a predictable manner, providing test and control regions (1, 20). This section will use models developed for nervous system research as an exemplar (10–13, 20), but the procedures described can be used to generate models for many other investigations by substituting the nervous system cells with the cells of interest. We will then describe methods by which water can be removed from collagen gels in a controlled manner in order to adjust the density of collagen gels towards that of body tissues, creating a versatile range of biomimetic constructs with controllable mechanical properties (4, 14) for a wide variety of tissue engineering and tissue modelling applications.

The approaches described here facilitate the construction of 3D models in which the micro-structure is controlled such that the mechanical cues perceived by the cells can be predicted and understood. There are no situations in which cells receive no mechanical stimulus (even floating in micro-gravity, attached and motile cells would apply load to each other), and mechanical stimuli have the potential to evoke substantial responses in cells. The value in these approaches is that the mechanical micro-environment of the cells is sufficiently well understood to be taken into account when investigating cellular responses.

2. Materials

2.1. Cell-Seeded Collagen Gels

1. Type 1 collagen solution: typically 2 mg/mL in 0.6% acetic acid.
2. 10× minimum essential medium (MEM).
3. Sodium hydroxide (5 M aqueous solution for initial stages of neutralisation, 1 M for drop-wise, final adjustment).
4. Rectangular culture moulds with aspect ratio 3:1 and appropriate tethering bars (see Fig. 1a, b). Moulds can be constructed from stainless steel, nylon, or other cell-culture-suitable materials. Nylon mesh is particularly useful for fabrication of the tethering bars and can be purchased from suppliers of plastic cross-stitch canvas (routine haberdashery). Such tethering bars should ideally be of overall low density and, in some cases, able to float in medium (1).
5. Culture medium will depend on the cell type and experimental parameters, but it is often useful to supplement medium with 50 μg/mL ascorbic acid, a co-factor in the synthesis of collagen by fibroblasts, which is frequently important for remodelling and may even aid contraction. However, ascorbate oxidises rapidly and gradually becomes ineffective in the medium and so must either be replaced almost daily or be used as the phosphate form which has a far longer effective half-life.
6. Collagenase: (from *Clostridium histolyticum*) 0.125% w/v in DMEM.

2.2. Immunostaining, mRNA Extraction

1. Fixation requires 4% paraformaldehyde, stored and used at 4°C, permeabilisation requires 0.5% Triton X-100. Other immunostaining reagents will depend on the choice of cells and antigens. For neural cultures, antisera against β-III tubulin, S100, and GFAP can be used to detect neurons, Schwann cells, and astrocytes, respectively (10, 20). Normal serum from

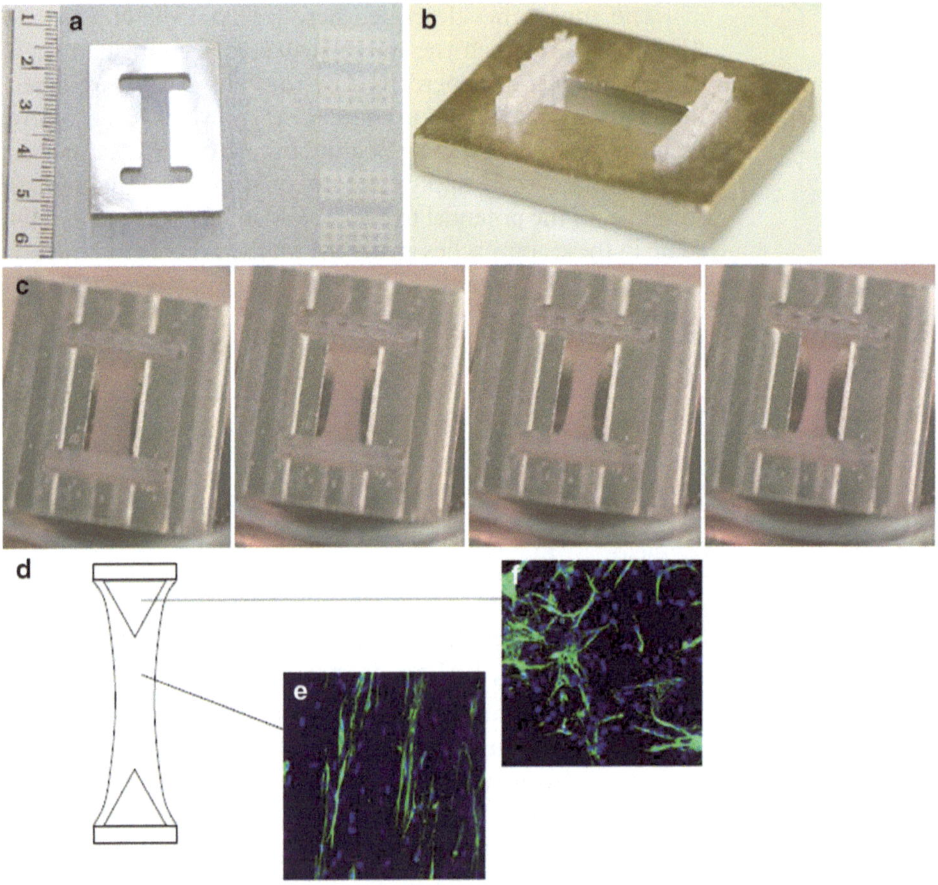

Fig. 1. Typical stainless-steel culture mould for making small tethered aligned cellular gels (**a**). The tethering bars are supported at each end of the rectangular mould (**b**) and the collagen integrates with them before setting. Gels adopt a characteristic shape as cells contract and align (**c**), in this case images were captured at 2, 4, 8, and 12 h after setting. Tethered gels such as this exhibit distinct regions (**d**) containing aligned (**e**) and unaligned cells (**f**).

the species in which the secondary antibody was raised can be used at 5% for blocking after permeabilisation.

2. RNA extraction requires Trizol® (Invitrogen) and liquid nitrogen.

3. Methods

The start point for all of the models described is a standard Type-I collagen hydrogel which can be seeded with cells and transferred to a mould where it sets. Incorporation of stiff porous tethering bars in the mould allows the collagen to integrate, forming a secure anchoring point when required. If required, collagen gels can be connected via the tethering bars to a culture force monitor

(2), or the tethering bars can be anchored to the mould, so that the contraction of the cells causes alignment of the centre portion of the gel (see Fig. 1). Cells in this aligned region can be compared with unaligned counterparts in the stress-shielded delta zones near the tethering bars, or can be subjected to dual cues such as placement of biomaterials providing contact guidance perpendicular to the direction of principal strain (15).

3.1. Modelling Cell Function

3.1.1. Cell-Seeded Collagen Gel

To make a standard (~2 mg/mL) collagen gel of volume 5 mL:

1. Prepare and sterilise moulds and tethering bars (see Fig. 1).
2. Mix 4 mL collagen solution with 0.5 mL 10× MEM and store at 4°C (see Note 1).
3. Prepare the cell suspension in a volume of 0.5 mL culture medium. Cell density can be varied (see Note 2).
4. Neutralise the solution of collagen and MEM by drop-wise addition of NaOH (see Note 3).
5. Immediately add the neutral solution of collagen to the cell suspension and swirl to mix, then transfer to the mould, ensuring solution fully integrates with the tethering bars if present (see Note 4).
6. Place mould assembly into humidified incubator (37°C, 5% CO_2) and allow to set (see Note 5).
7. Gently add culture medium (37°C) on top of the gel; avoid damaging or dislodging it.
8. Maintain in culture, changing medium regularly.
9. Cells will adhere to the collagen matrix and contract the gel. Because the gel is anchored at its ends, the contractile forces will cause the cells to align to the direction of principal strain, parallel to the long axis of the gel. Stress shielding caused by the stiff tethering bars results in a triangular region at each end of the gel in which there is no principal strain and therefore cells are randomly aligned (see Fig. 1). These stress-shielded regions can be useful intrinsic control areas in studies looking at the effects of cell alignment (see Note 6) (10, 20).

3.1.2. Removal of Cells from Collagen Gels

In most cases analysis of cell responses will be conducted within the 3D gel environment, but occasionally it is useful to remove live cells from collagen gels. This can be achieved by digesting the gel using collagenase solution and standard cell culture techniques (see Note 7).

1. Remove medium and wash gels briefly in PBS.
2. Add collagenase (0.125%) and incubate at 37°C for 10–20 min until gel has dissolved and cells are in suspension
3. Collect cells by centrifugation (100 × *g*, 5 min), discard supernatant, and resuspend cells in culture medium.

4. If collected cells are going to be seeded into another collagen gel, repeated washes using large volumes of fresh culture medium are recommended to remove all traces of collagenase.

3.1.3. To Analyse Cell Responses in Collagen Gels

A wide range of microscopic (including time lapse live cell imaging) and molecular analyses can be applied to collagen gel cultures by adapting standard protocols used for monolayer culture and tissue samples. Here the adaptations required for successful immunocytochemistry and extraction of mRNA from cells in gels are described.

3.1.3.1. Immunostaining and Microscopy

Immunofluorescence detection of cell-specific proteins enables identification of cells in mixed populations as well as providing information about changes in protein expression in cultures subjected to experimental treatments.

1. Remove medium, wash briefly with PBS, fix using 4% paraformaldehyde at 4°C for at least 3 h (see Note 8).
2. Immunostaining protocol will depend on the cells and antigens to be detected. In general, the same protocol can be followed that is used for monolayer cultures, but for 3D gels double all incubation times. Wash steps should be thorough, for example 3×5 min washes in 20 mL of PBS.
3. Following immunostaining, gels can be stored in PBS at 4°C.
4. Immunostained cells in collagen gels are easily visualised using standard fluorescence microscopy, but, if possible, a confocal microscope should be used so elaborate 3D structures can be observed and measured and complex cellular interactions can be resolved and quantified (see Fig. 2). Gels can be placed carefully onto a glass slide or coverslip for direct observation on an inverted stage microscope (see Note 9) or bathed in shallow solution for observation using water-immersion objectives with an upright microscope.

3.1.3.2. Extraction of mRNA from Cellular Collagen Gels

Extraction of mRNA from gels facilitates analysis of changes in transcription monitored using RT-PCR (see Fig. 3) or microarray techniques. Techniques were developed for extraction from collagen gels (high protein background) (15) and repeated for dense gels (16).

1. Remove medium and wash briefly in PBS, then carefully transfer gel into a sterile tube containing Trizol® (1 mL Trizol® for 0.5 mL gel).
2. Freeze immediately in liquid nitrogen (samples can be stored at –80°C).
3. After thawing, homogenise gels by trituration using a 19-G hypodermic needle on a syringe until no lumps of collagen gel remain.

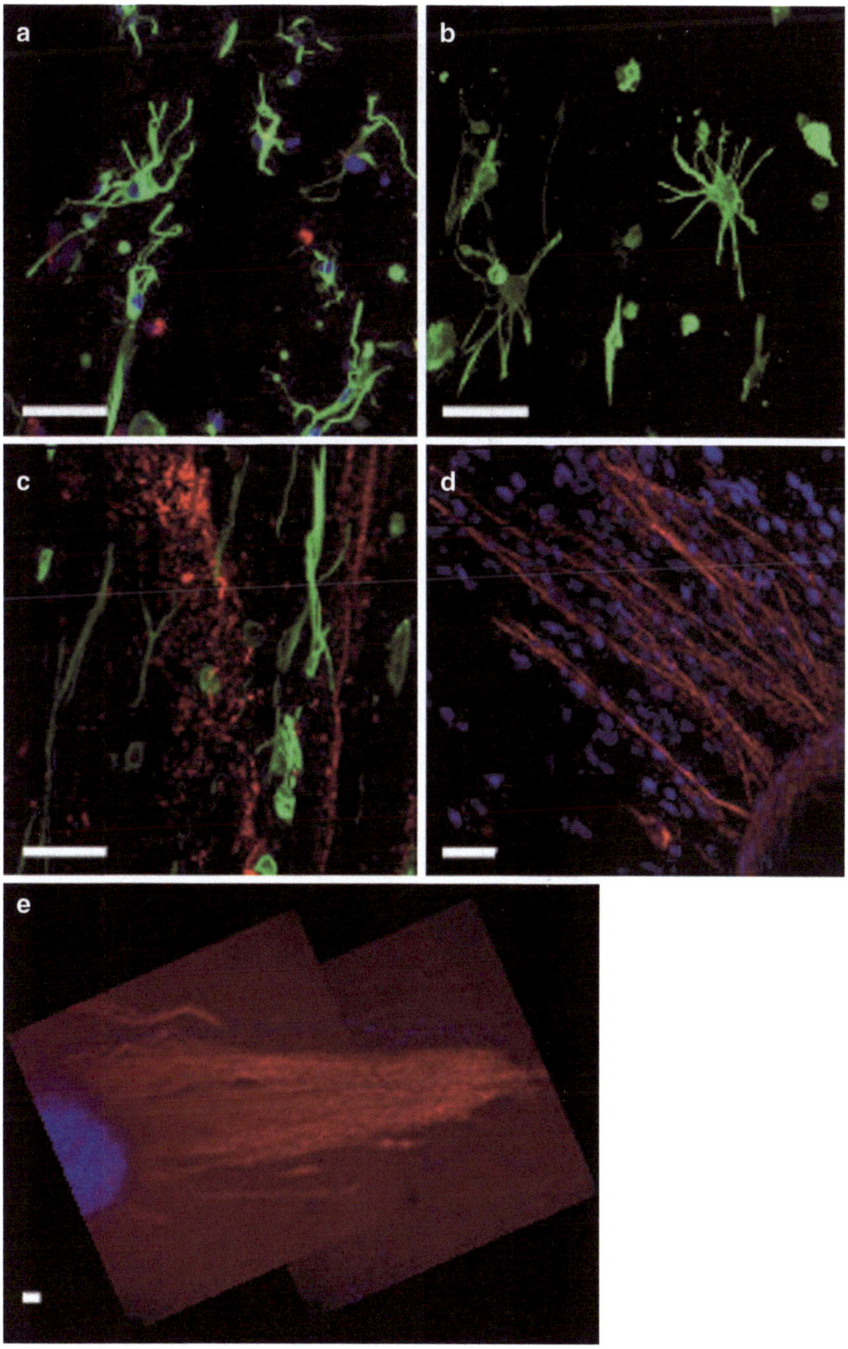

Fig. 2. Confocal micrographs showing GFAP expression (*green*) in astrocytes in 3D collagen gels (**a**, **b**) and patterns of immunoreactivity of deposited proteins (*red*) (**c**). Neurites grow out from the surface of explanted dorsal root ganglia embedded in the aligned region of a Schwann cell-populated tethered aligned gel (*red*; β-III tubulin) (**d**, **e**). All scale bars = 50 μm (thanks to Dr. Emma East and Mrs. Daniela Blum de Oliveira).

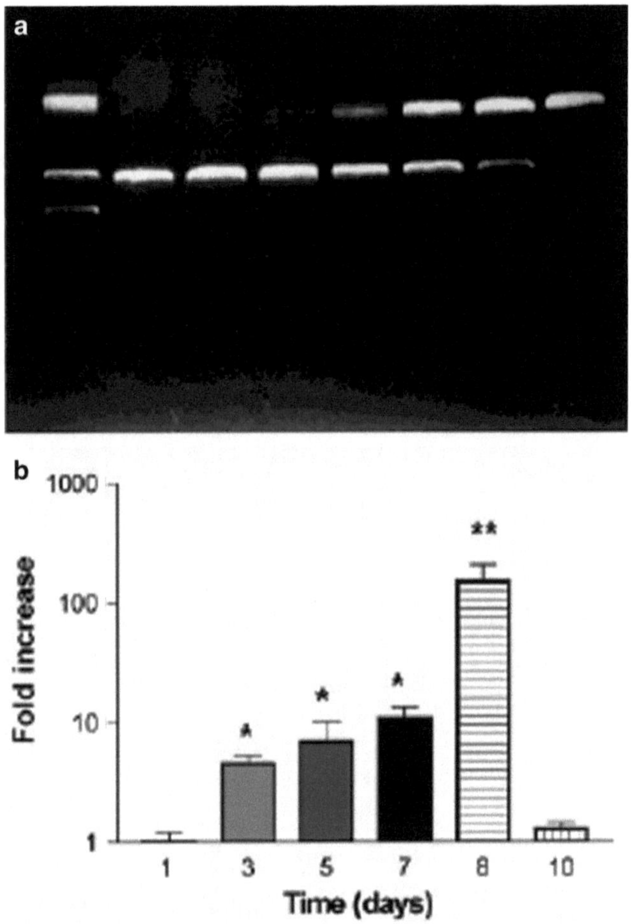

Fig. 3. (**a**) Photograph of RT-PCR result showing detection of MMP2 expression in a fibroblast-populated collagen gel (from Fig. 2 in Mudera et al. (15)). (**b**) Real-time quantitative PCR data showing the level of VEGF in plastic compressed constructs containing fibroblasts over a 10-day period (from Fig. 4 in Cheema et al. (16)).

4. RNA extraction can then be carried out using the standard protocols described by the supplier.

3.2. Towards Tissue Density Collagen Models

3.2.1. To Increase the Density of a Collagen Gel Using Plastic Compression

1. Follow instructions in Subheading 3.1.1, steps 1–4 to create a standard (~2 mg/mL) collagen gel (with or without cells as required).
2. Cast gels into moulds of any required dimensions or shape (typically rectangular 33 × 22 × 20 mm) and allow to set for 30 min at 37°C (to allow standardized gel maturation), trapping the cells within nano-fibrillar collagen mesh as it aggregates.
3. Once set, transfer the gel onto a prepared, standard blotting element, comprising a 165-μm thick stainless-steel support mesh (mesh size ~300 μm) covered by a layer of nylon mesh

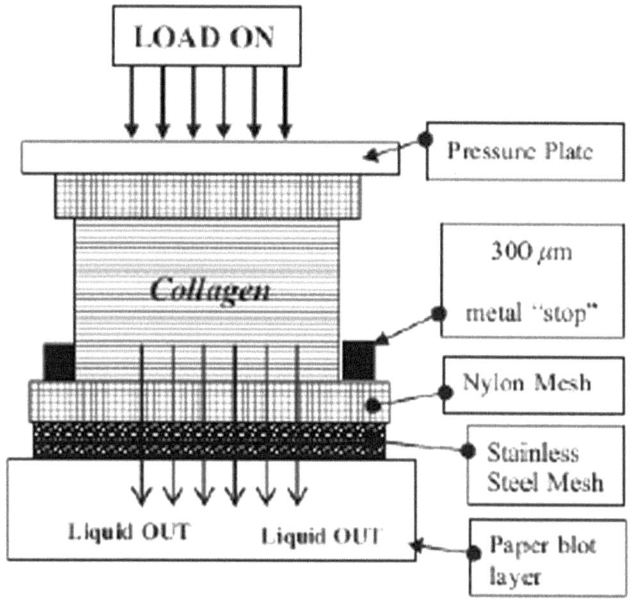

Fig. 4. Diagram showing the routine assembly for plastic compression of pre-formed collagen gels (from Fig. 1 in Brown et al. (4)).

(~50 μm mesh size), and placed onto three layers of circular Whatman (Grade 1; 240 mm diameter) filter paper.

4. Cover gel construct (now at room temperature) with a second nylon mesh and a glass sheet loaded with a 120-g weight (stress equivalent to approximately 1.8 kN/m^2) for 5 min to give a plastic compressed (PC) construct (see Fig. 4).
5. Thickness of such collagen constructs is typically 50–100 μm (from a 5-mm deep initial gel). However, this can be controlled by varying (1) the initial thickness and (2) degree of fluid removal. Alternatively, thicker (more conventionally 3D) constructs can be formed rapidly either by or rolling into tight spirals (or Swiss-rolls) or by layering with many other PC collagen sheets. Clearly, either of these latter techniques allows potential for building up complex, heterogeneous constructs by layering multiples of non-identical sheets to any required biomimetic design (4).
6. Micro–nano structuring depends on combinations of mechanical effects including fluid flow shear and compressive loads which shape the constructs and increase the density of the collagen fibrils and the cells trapped within the nano-porous mesh (see Note 10). Additional levels of compression (and with it, extension of the collagen density range) can be achieved by a second compression of the 3D-spiralled constructs, clearly adding to the anisotropic complexity which can be engineered (17).

3.2.2. Assembling Biomimetic Tissue Constructs from Plastic Compressed Collagen

The layering concept described above can be extended beyond the use of multiple collagen-cell sheets to include the addition/incorporation of completely distinct layers or zones which can contain other components. These can be designed to achieve a range of functional properties from controlled perfusion, or tensile/compressive strength to new micro-structures within the matrix. Some examples of approaches are detailed here, but many others are also feasible.

1. Incorporate new micro-structure or topographic features into each collagen sheet by embossing the surface using a support structure during compression with the appropriate embossing topography. Alternatively, internal micron-scale structure can be left permanently within each sheet by the "lost glass" technique (18) which involves compression of the collagen layer into which soluble (phosphate glass) fibres are pre-embedded. Dissolution and loss of such fibres leaves permanent open channels running through the 3D constructs in whatever density or orientation they were first laid down. Comparable micro-cavity structure generation is also possible using soluble glass (or other slowly soluble) particles of predetermined diameter in place of the fibres.
2. Incorporate desired mechanical properties using polymer meshes where a mechanically stronger layer of material is compressed between the collagen layers. A typical example here involves co-compression of a layer of poly-lactate–polyglycolate fibre mesh between two standard (cellular) collagen layers (19). The resultant composite is a highly mimetic asymmetric, cell-collagen tissue model but has enhanced overall tensile mechanical properties.
3. Generate new 3D structure by local osmotic swelling. An example of this technique is the insertion of an osmotically active material (e.g. hyaluronan) into a specific zone or layer of the construct (see Fig. 5a). The basis of the plastic compression is that the collagen itself has low inherent swelling capacity, but this can be locally disrupted by placing, for example, a small deposit of dry hyaluronan within the 3D collagen structure. The hyaluronan rapidly takes up water from the surroundings, swelling and expanding to form a cavity, channel, or tract in the surrounding collagen. The shape and positioning of such structures can be designed and controlled by the position of the hyaluronan depots. Once the hyaluronan or other agent dissolves and escapes, the expansion remains behind as a new local structural feature (again, with no detrimental effects on resident cells).
4. Incorporate mineral for biomimicry of hard tissues; controlled size hydroxyapatite can be added to mimic bone-like model tissues at the time of fabrication (21). Such incorporated par-

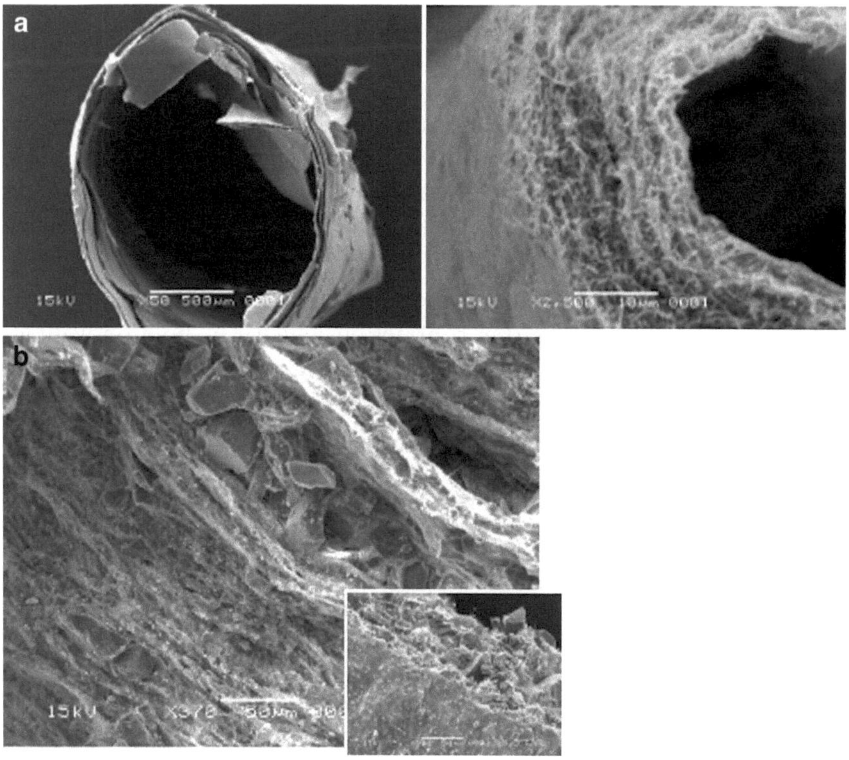

Fig. 5. (**a**) 3D plastic compressed, spiral collagen construct, the solid core of which was initially filled with a strip of hyaluronan. This swelled over the following minutes to generate a continuous full-length channel and so a tube (*left panel*). The *right hand panel* shows the fine detail of the compressed collagen mesh of the tube wall, with its multi-layers of dense packed collagen fibrils. (**b**) Plastic compressed collagen matrices as hard-tissue composites. In this example, PC collagen was seeded with soluble phosphate glass particles (<20 μm) at a density giving distinct zones of hard substrate between dense collagen layers. *Inset* shows the layer in cross section including its surface structure and embedded mineral particles (thanks to Ms. Tijna Alekseeva and Dr. Ensanya Abou-Neel).

ticles dramatically increase the compressive properties of the 3D constructs and alter subsequent resident cell (especially osteoblast-like) matrix remodelling. Comparable composite "hard-tissue" models can also be fabricated by embedding soluble glass into the collagen gels (see Fig. 5b).

5. Incorporate a cell population primed to produce local growth factors. For example allowing a high-density depot of cells to generate a localised hypoxia, enhancing the production of a range of angiogenic factors (16). Using the predictability of the 3D PC collagen model and real-time measurement of core oxygen gradients, it is possible to generate and time the appearance of such responses [in this example, vascular endothelial cell growth factor (VEGF)], so opening the way to a new form of dynamic cell therapy in which the cell component is essentially a disposable tool for controlled production of effector agents (e.g. growth factor, hormone, or drug) for engineering of timed, local biomimetic events.

4. Notes

1. Gentle mixing is required, but it is important to avoid fluid shearing in the viscous collagen solution or the introduction of bubbles. Use a 10-mL pipette to add the collagen slowly to the 10× MEM in a 20-mL plastic tube and swirl gently to mix.
2. Final cell density is critical in these models – higher densities cause faster contraction which can be an advantage or can result in the gel breaking during the experiment. Slow-contracting cell populations can be supplemented with other cells (e.g. fibroblasts) to ensure timely contraction. A useful exercise is to make small free-floating round gels with varying cell density in a multi-well plate to give an indication of the contraction properties of new cell populations.
3. Use a micropipette held above the solution to add NaOH one drop at a time. After each drop, swirl gently and watch carefully for colour changes (phenol red is the pH indicator in the MEM). As the solution changes from yellow towards orange, switch to the dilute NaOH. Stop as soon as the solution looks the same red colour as normal phenol red-containing culture medium.
4. In some instances a separate gel solution can be made for immersion of the tethering bars to ensure integration immediately prior to addition of the cellular gel. This is useful when cell numbers are limited as cells are all within the rectangular part of the gel and not around the tethering bars.
5. Initial setting should take about 5 min. A good indication of progress is to monitor the drop of gel that is left in the tube rather than disturb the experimental gel. A longer setting time is occasionally required, but for cellular gels more than about 10–15 min should be avoided because cells may die in the gel without medium.
6. The relative dimensions and aspect ratio of these gels (rectangular with distance between tethering bars 3× the width of the gel) is important since it influences the force vectors that lead to there being distinct and understandable regions of aligned and unaligned cells. Longer narrower gels lead to much larger aligned regions but relatively small delta zones (which may be desirable in some cases). Beware of the complex force vectors that are generated in squarer or round gels, however, as these are likely to have a profound influence on cell responses without being easily understandable, diminishing the usefulness of such gels as model systems.

7. The name of this protease is somewhat deceptive and deserves some caution. Being a bacterial enzyme, it is different from the well-known family of mammalian collagenases (e.g. MMP 1 and 9), particularly in its substrate specificity. Collagen degradation is rapid and complete (not to the characteristic large fragments of MMP1). Furthermore, most of the commercial sources are not pure and contain a rich selection of other bacterial proteinases and other ingredients, some of which are toxic to mammalian cells under extended treatments. As a result, it should be used with some caution, research, and preliminary testing in your cell system.
8. It can be useful to divide gels into smaller pieces after fixation, for example to enable immunostaining for a wider range of markers, or to use part of the gel for protein or mRNA extraction. Cutting fixed gels is best done within the mould using a sharp blade.
9. Handle the gel by lifting gently with a small spatula; do not try gripping it with forceps. Gels must be kept hydrated at all times to maintain their shape.
10. Importantly, seeded cells in the pre-compressed collagen gels are not significantly damaged despite the rapidity of the fluid flow, apparently due to the protective mechanical support from the mesh of (~30 nm diameter (22)) collagen fibrils surrounding them. Furthermore, since there is a strict direction to the fluid flow through the system (in this setup, it is from top to bottom) the collagen fibrils tend to be packed into a series of thin layers, perpendicular (i.e. in the x-y plane) to the fluid flow (which is in the z-plane). These are commonly 500 nm to approximately 2 μm thick. An especially thick/dense collagen layer forms on the fluid-leaving surface of the construct (i.e. the basal surface in this setup) where the initial deposition (filtration-caking) of collagen is greatest. This means that each PC collagen sheet made using this setup (uniaxial compression in the z-plane) is inherently and distinctly anisotropic in structure and properties at the nano–micro (meso-)-scale (4).

Acknowledgements

Work supported in part (RAB) by grants from BBSRC and TSB-EPSRC and (JBP) the Wellcome Trust. The authors would like to thank Emma East, Daniela Blum de Oliveira, Nelomi Anandagouda, and Tijna Alekseeva for contributing data used in figures.

References

1. Eastwood, M., Mudera, V. C., McGrouther, D. A., and Brown, R. A. (1998) Effect of precise mechanical loading on fibroblast populated collagen lattices: morphological changes. *Cell Motil. Cytoskeleton* **40,** 13–21.
2. Eastwood, M., McGrouther, D. A., and Brown, R. A. (1994) A culture force monitor for measurement of contraction forces generated in human dermal fibroblast cultures: evidence for cell-matrix mechanical signalling. *Biochim. Biophys. Acta* **1201,** 186–192.
3. Brown, R. A., Prajapati, R., McGrouther, D. A., Yannas, I. V., and Eastwood, M. (1998) Tensional homeostasis in dermal fibroblasts: mechanical responses to mechanical loading in three-dimensional substrates. *J. Cell. Physiol.* **175,** 323–332.
4. Brown, R. A., Wiseman, M., Chuo, C.-B., Cheema, U., Nazhat, S. N. (2005) Ultrarapid engineering of biomimetic materials and tissues: fabrication of nano- and microstructures by plastic compression. *Adv. Funct. Mater.* **15,** 1762–1770.
5. Brown, R. A., Sethi, K. K., Gwanmesia, I., Raemdonck, D., Eastwood, M., and Mudera, V. (2002) Enhanced fibroblast contraction of 3D collagen lattices and integrin expression by TGF-beta1 and -beta3: mechanoregulatory growth factors? *Exp. Cell Res.* **274,** 310–322.
6. Cheema, U., Yang, S. Y., Mudera, V., Goldspink, G. G., and Brown, R. A. (2003) 3-D in vitro model of early skeletal muscle development. *Cell Motil. Cytoskeleton* **54,** 226–236.
7. Marenzana, M., Pickard, D., MacRobert, A. J., and Brown, R. A. (2002) Optical measurement of three-dimensional collagen gel constructs by elastic scattering spectroscopy. *Tissue Eng.* **8,** 409–418.
8. Phillips, J. B., Bunting, S. C., Hall, S. M., and Brown, R. A. (2005) Neural tissue engineering: a self-organizing collagen guidance conduit. *Tissue Eng.* **11,** 1611–1617.
9. Porter, R. A., Brown, R. A., Eastwood, M., Occleston, N. L., and Khaw, P. T. (1998) Ultrastructural changes during contraction of collagen lattices by ocular fibroblasts. *Wound Repair Regen.* **6,** 157–166.
10. Phillips, J. B., Bunting, S. C. J., Hall, S. M., and Brown, R. A. (2005) Neural tissue engineering: a self-organizing collagen guidance conduit. *Tissue Eng.* **11,** 1612–1618.
11. East, E., Golding, J., and Phillips, J. B. (2009) A versatile 3D culture model facilitates monitoring of astrocytes undergoing reactive gliosis. *J. Tissue Eng. Regen. Med.* **3(8),** 634–646.
12. East, E., Golding, J., and Phillips, J. B. (2007) Increased GFAP immunoreactivity by astrocytes in response to contact with dorsal root ganglia cells in a 3D culture model. *Neuron Glia Biol.* **3,** S119.
13. East, E. and Phillips, J. B. (2008) Tissue Engineered Cell Culture Models for Nervous System Research. In: *Tissue Engineering Research Trends* (Greco, G. N., Ed.), Nova Science Publishers, New York.
14. Hadjipanayi, E., Mudera, V., and Brown, RA. (2009) Guiding cell migration in 3D: A collagen matrix with graded directional stiffness. *Cell Motil. Cytoskeleton* **66(3),** 121–128.
15. Mudera, V. C., Pleass, R., Eastwood, M., Tarnuzzer, R., Schultz, G., Khaw, P., McGrouther, D. A., and Brown, R. A. (2000) Molecular responses of human dermal fibroblasts to dual cues: contact guidance and mechanical load. *Cell Motil. Cytoskeleton* **45,** 1–9.
16. Cheema, U., Brown, R. A., Alp, B., and MacRobert, A. J. (2008) Spatially defined oxygen gradients and vascular endothelial growth factor expression in an engineered 3D cell model. *Cell. Mol. Life Sci.* **65,** 177–186.
17. Abou Neel, E. A., Cheema, U., Knowles, J. C., Brown, R. A., and Nazhat, S. N. (2006) Use of multiple unconfined compression for fine control of collagen gel scaffold density and mechanical properties. *Soft Matter* **2,** 986–992.
18. Nazhat, S. N., Neel, E. A., Kidane, A., Ahmed, I., Hope, C., Kershaw, M., Lee, P. D., Stride, E., Saffari, N., Knowles, J. C., and Brown, R. A. (2007) Controlled microchannelling in dense collagen scaffolds by soluble phosphate glass fibers. *Biomacromolecules* **8,** 543–551.
19. Ananta, M., Aulin, C. E., Hilborn, J., Aibibu, D., Houis, S., Brown, R. A., and Mudera, V. (2009) A poly(lactic acid-co-caprolactone)-collagen hybrid for tissue engineering applications. *Tissue Eng. Part A* **15,** 1667–1675.
20. East, E., Blum de Oliveira, D., Golding, J. P. and Philips, J. B. (2010) Alignment of Astrocytes Increases Neuronal Growth in Three-Dimensional Collagen Gels and Is Maintained Following Plastic Compression to Form a Spinal Cord Repair Conduit Tissue Engineering Part A. doi:10.1089/ten.tea.2010.0017
21. Buxton, P. G., Bitar, M., Gellynck, K., Parkar, M., Brown, R. A., Young, A. M., Knowles, J. C., and Nazhat, S. N. (2008) Dense collagen matrix accelerates osteogenic differentiation and rescues the apoptotic response to MMP inhibition. *Bone* **43,** 377–385.
22. Cheema, U., Chuo, C.-B., Sarathchandra, P., Nazhat, S. N., and Brown, R. A. (2007) Engineering functional collagen scaffolds: cyclical loading increases material strength and fibril aggregation. *Adv. Funct. Mat.* **17,** 2426–2431.

Chapter 13

Organotypic and 3D Reconstructed Cultures of the Human Bladder and Urinary Tract

Claire L. Varley and Jennifer Southgate

Abstract

Three-dimensional organotypic cultures of human urinary tract tissue have been established as intact and reconstituted tissues, with the latter generated by combining cultured normal human urothelial (NHU) cells with an appropriate stroma. Organoids may be maintained at an air–liquid interface in static culture for periods of up to 20 weeks, with analysis by immunohistology for expression of urothelial differentiation-associated markers providing a qualitative, but objective assessment criterion. Where reconstructed using bladder cancer cell lines, the resultant organoids recapitulate the invasive characteristics of the originating tumour, but the need to use authenticated cell line stocks is emphasised. The organoid approach represents an important tool for investigating urothelial–stromal cell interactions during homeostasis and disease, and for testing bladder tissue engineering and reconstructive strategies. Potential future developments of the technique are discussed and include genetic manipulation of the urothelial cells to generate disease models and incorporation of biomaterial scaffolds to support artificial stroma development.

Key words: Bladder, Urothelium, Organoids, Urothelial cell carcinoma, UCC, Cancer, Cell lines, Human, Organ culture, Tissue engineering

1. Introduction

The renal pelvis, ureters, bladder, and proximal urethra are lined by a transitional epithelium, known as urothelium. The urothelium is stratified into three zones: a single basal cell layer that is attached to a basement membrane and overlies a vascularised stroma, an intermediate zone of variable cell thickness, and a single layer of large overlying superficial or "umbrella" cells that abut onto the urinary space. The superficial cells are considered as terminally differentiated and are highly adapted, both morphologically and functionally, to their role as the primary urinary barrier.

John W. Haycock (ed.), *3D Cell Culture: Methods and Protocols*, Methods in Molecular Biology, vol. 695,
DOI 10.1007/978-1-60761-984-0_13,

The different urothelial zones can be distinguished by differential expression of various genes and their protein products, including cytokeratins (1, 2), claudins (3), and the urothelium-specific uroplakins (4, 5). These genes/antigens provide markers for objective assessment of differentiation status both in vivo and in vitro.

Urothelial cells isolated from surgical samples of normal human urinary tract tissues may be propagated as finite normal human urothelial (NHU) cell lines through multiple serial subculture (6, 7) (see Note 1). When maintained in a low calcium (0.09 mM) growth factor-supplemented keratinocyte serum-free medium, NHU cells proliferate rapidly as a monolayer, but do not spontaneously express genes or proteins associated with late/terminal urothelial cytodifferentiation. In medium containing 2 mM (near physiological) calcium, NHU cells interact with neighbouring cells to form a stratified epithelium bonded by intercellular junctions, but without differentiated characteristics (6, 7). Differentiation of NHU cell cultures may be achieved either (a) pharmacologically by activating peroxisome proliferator activated receptor gamma (PPARγ; e.g. with 1 μM troglitazone) when which autocrine epidermal growth factor receptor signalling is blocked (e.g. with 1 μM PD153035) (3, 8–10), or (b) transfer and sub-culture in medium containing 2 mM calcium and 5% bovine serum (11, 12). Both methods result in NHU cells acquiring a more differentiated phenotype; however, only cultures grown on permeable membranes and differentiated by the latter method develop as a "biomimetic" urothelium with functional urinary barrier properties (11, 12).

Urothelium may also be maintained in combination with a stroma as three-dimensional heterotypic or organotypic cultures. Such organoids may either be established from intact tissues, or recombined from reconstructed urothelial and stromal compartments prior to culture. The organotypic approach offers an important tool for investigating urothelial–stromal interactions, particularly in the context of (a) understanding basic paracrine signalling mechanisms that regulate tissue homeostasis, (b) developing approaches for bladder reconstruction and tissue engineering, and (c) generating models of benign and malignant disease.

The culture conditions for maintaining urothelial organoids were initially established and optimised for intact human urothelial tissues in order to ensure that the conditions were compatible with maintenance of a differentiated urothelial phenotype. These studies revealed that a differentiated urothelium could be maintained for some 20 weeks, although there was loss over time of some terminal differentiation-associated markers, such as cytokeratin 20 from superficial cells (13). The method was subsequently applied to reconstructed organoids, in which NHU cells propagated in monolayer cultures were harvested and combined with pieces of de-epithelialised, intact stroma. The results of this work

Table 1
Derivation and genotyping of UCC cell lines

Cell line	Grade and stage of original tumour	DNA profile
RT4	G1, T2 Recurrent papillary tumour	CSF1PO: 10,12; D13S317: 8; D16S539: 9; D5S818: 11,12; D7S820: 9,12; THO1: 9,9.3; TPOX: 8,11; vWA: 14,17
RT112	G2 Stage not reported	CSF1PO: 10,11; D13S317: 13,14; D16S539: 11,13; D5S818: 10,13; D7S820: 11,12; THO1: 7; TPOX: 8,11; vWA: 14,17
EJ	G3 T2 minimum	CSF1PO: 10,12; D13S317: 12; D16S539: 9; D5S818: 10,12; D7S820: 10,11; THO1: 6; TPOX: 8,11; vWA: 17

Genotyping was performed on UCC cell lines to verify the authenticity (see Note 2). DNA profiles for RT4 and EJ cell lines matched the ATCC® profiles for these cell lines, whereas the RT112 DNA profile has not been published by ATCC®

were important as they demonstrated that in vitro-propagated NHU cells retained the capacity to differentiate and reform a slow turnover, stratified transitional epithelium (13). In a further development, where organoids were constructed by combining established human bladder cancer-derived cell lines with de-epithelialised stromal tissues, the cancer cells recapitulated the grade (differentiation) and stage (invasive) characteristics of the originating tumours (see Table 1) (14) (see Note 2).

For organ culture, surgical specimens of normal human urothelium are collected aseptically in "Transport Medium" and a small representative sample is processed for histology to check the pathology and quality of the starting material. The remaining sample is dissected into two pieces, with one piece used for intact organ culture and the second piece used to isolate NHU cells for primary culture (Fig. 1). For intact organ culture, the tissue is dissected into 0.5 cm^2 pieces, which are orientated and placed on permeable membranes in tissue culture inserts, with medium introduced into the well underneath the insert, so that the tissue lies at the air–liquid interface. To establish primary NHU cell cultures, the urothelium is separated from the stroma by incubation in "Stripping Medium" for 16 h to release intact sheets of urothelium. The urothelial sheets are harvested, incubated with collagenase, then pipetted (without "frothing") to dissociate the sheets

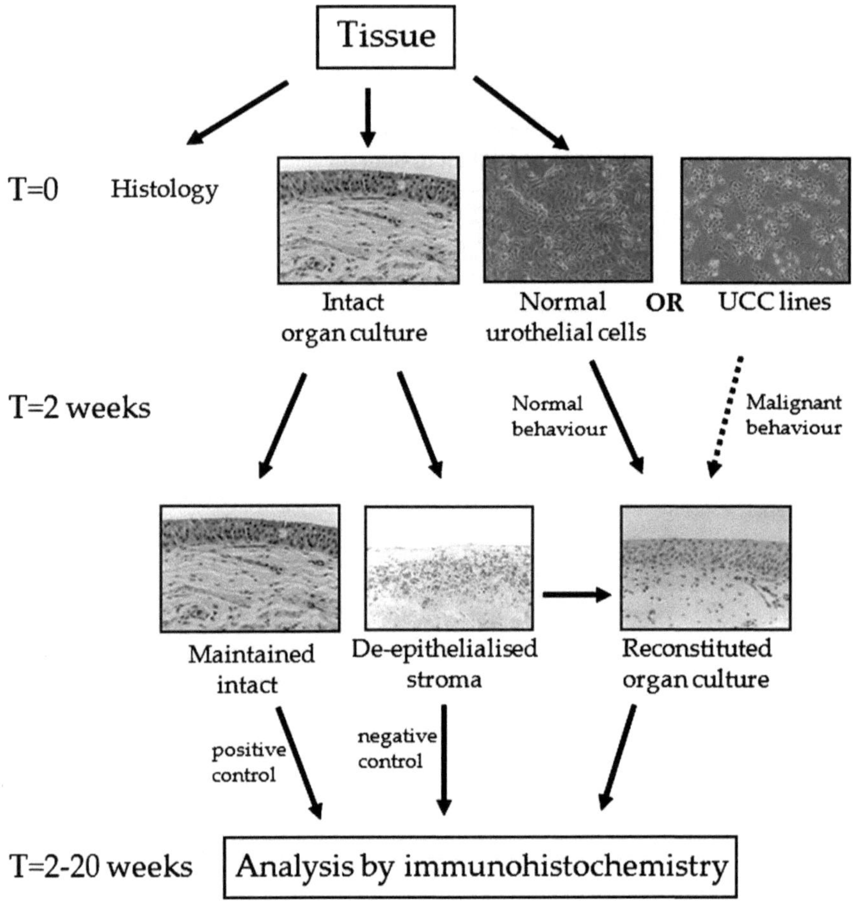

Fig. 1. Summary of the organ culture system. Three dimension reconstituted organ cultures seeded with normal or malignant urothelial cells are used to investigate normal or malignant cell behaviours, respectively.

and the cells seeded and maintained in monolayer culture in Primaria™ tissue culture flasks. The cultures are harvested at just confluence using a chelation and trypsinisation procedure (see Note 3) and either serially cultured by reseeding at a lower density, or cryopreserved for long-term storage.

To generate autologous reconstituted organoids, a sample of intact tissue is maintained in a viable status in organ culture until sufficient NHU cells have been generated in the parallel monocultures (approximately 2 weeks). The intact organoids are then subjected to the same protocol used to isolate the urothelium from fresh tissue. The de-epithelialised stromal portion is orientated, basement membrane side up, on a culture insert and seeded with a pellet of NHU cells harvested from monolayer culture. The organoids are maintained at an air–liquid interface (Fig. 2) and harvested at 2-week intervals for immunohistochemistry. As a control, de-epithelialised stromal tissue is maintained in parallel organoid culture to demonstrate the absence of regrowth from a

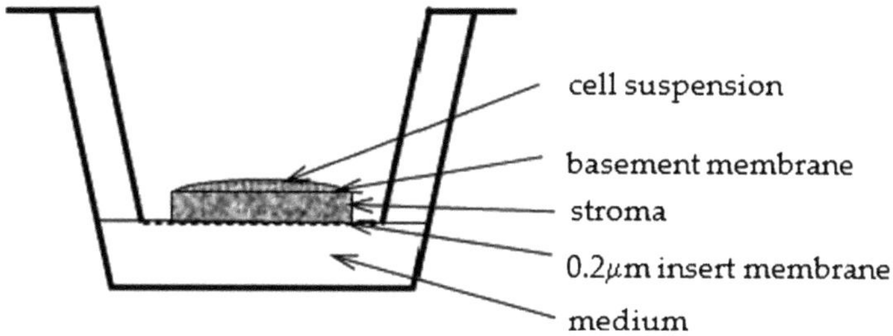

Fig. 2. Schematic diagram of composite organ culture.

residual epithelial cell population. Variation on these techniques includes seeding de-epithelialised stromal tissues with pellets of established human urothelial cell carcinoma (UCC) cell lines, such as RT4, RT112, and EJ (Fig. 1). These cell lines constitute a useful panel, as they are well-characterised and display a spectrum of differentiated and invasive phenotypes that epitomises UCC (14–18). The derivation and genotyping of the UCC cell lines is shown in Table 1 (see Note 2).

2. Materials

2.1. Tissue Collection and Processing Agents

1. Transport medium: Hanks' balanced salt solution (HBSS with Ca^{2+} and Mg^{2+}), 10 mM HEPES, pH 7.6, 20 Kallekrein inhibiting units (KIU)/mL aprotinin (Trasylol, Bayer plc, Newbury, UK).
2. Stripping medium: HBSS (without Ca^{2+} and Mg^{2+}), 10 mM HEPES, pH 7.6, 20 KIU/mL aprotinin, 0.1% (w/v) EDTA.
3. Collagenase type IV (100 U/mL; Sigma-Aldrich Ltd., Poole, UK) is dissolved in complete HBSS with 10 mM HEPES, pH 7.6 and stored in 2 mL aliquots at –20°C.

2.2. Culture Media

2.2.1. NHU Cell Cultures

Keratinocyte serum-free medium (KSFM) containing bovine pituitary extract and EGF at the manufacturer's recommended concentrations (Invitrogen) and supplemented with 30 ng/mL cholera toxin (Sigma-Aldrich) is referred to as KSFM complete (KSFMc) (see Note 4). Cholera toxin is dissolved in KSFM at 30 μg/mL and stored in aliquots at 4°C and is included as it aids initial cell attachment (19).

2.2.2. UCC Cell Lines

A 1:1 mixture of Dulbecco's Modified Eagle's Medium (DMEM) and Roswell Park Memorial Institute (RPMI) 1640 medium is supplemented with 5% (v/v) fetal bovine serum (FBS) (see Note 4). Due to its liability, L-glutamine (1%) is kept frozen and added to the medium before use.

2.2.3. Organ Culture

Organ culture medium: Waymouth's MB 752/1 (Invitrogen), 10% FBS (Harlan Sera-Lab Ltd.), 2 μg/mL hydrocortisone hemisuccinate (Sigma-Aldrich), 0.45 μg/mL ferrous sulphate, 300 μg/mL vitamin C (Sigma-Aldrich). The supplements are kept as 0.2 μm filter-sterilised aliquots at –20°C and added to the medium before use.

2.3. Cell Culture

1. EDTA (0.1% w/v) is prepared in PBS (without Ca^{2+} and Mg^{2+}), sterilised by autoclaving at 121°C and kept at ambient temperature.
2. Trypsin/versene: HBSS (without Ca^{2+} and Mg^{2+}), 0.25% (w/v) trypsin (Sigma-Aldrich), 0.02% (w/v) EDTA. Filter-sterilised and stored at –20°C, or for up to 1 week at 4°C.
3. Soybean trypsin inhibitor (Sigma-Aldrich) is required to inhibit the activity of trypsin in cultures maintained in medium without serum (which contains a natural trypsin inhibitor). Trypsin inhibitor is dissolved in PBS at 20 mg/mL, filter-sterilised (0.2 μm) and stored in 2 mg aliquots at –20°C. For use, one aliquot is thawed into 5 mL KSFMc and is adequate to inhibit up to 1 mL of trypsin/versene.

2.4. Immunohistochemistry

1. Specimen fixative: 10% (v/v) formalin in PBS containing 0.9 mM $CaCl_2$, 0.5 mM $MgCl_2$ (PBSc) (see Note 5).
2. Electrostatically charged Superfrost plus® microscope slides (VWR International, Poole, UK).
3. Haematoxylin: 850-mL polished water, 20 mL ethanol containing 15% (w/v) haematoxylin, 120 mL glycerol to which 0.3 g sodium iodide, 1 g citric acid, 50 g chloral hydrate, 50 g aluminium potassium sulphate are added sequentially.
4. Scott's tap water: 2% (w/v) $MgSO_4$ and 0.35% (w/v) $NaHCO_3$.
5. 1% Aqueous eosin (Raymond A Lamb Ltd., Eastbourne, East Sussex, UK).
6. Permanent mountant: DPX (Fluka, supplied by Sigma-Aldrich).
7. Antibody diluent: 50 mM Tris–HCl at pH 7.6, 150 mM NaCl, 0.1% (w/v) sodium azide, 0.1% (w/v) BSA.
8. Example antigen retrieval solutions: (a) 0.1% (w/v) trypsin (Sigma-Aldrich), 0.1% (w/v) $CaCl_2$ at pH 7.6; (b) 10 mM sodium citrate buffer, pH 6.0.
9. Avidin/biotin blocking kit and streptavidin-biotin-horseradish peroxidase from Vector Laboratories (Peterborough, UK).
10. Antibodies: Primary mouse monoclonal antibodies and suppliers are listed in Table 2. Secondary antibody is rabbit anti-mouse immunoglobulin [$F(ab')_2$ fragments] from Dako.

Table 2
Antibodies and antigen retrieval

Antibody specificity (clone[a])	Antigen retrieval	Supplier
CD44 v6 (2F10)	Trypsin	R&D systems, Abingdon, Oxfordshire, UK
E-cadherin (HECD1)	Microwave	Takara Biomedicals supplied by R&D systems
β1-integrin (BD-15)	Trypsin	Serotec Ltd., Kidlington, Oxford, UK

[a] The antibodies shown here are all mouse monoclonal antibodies

11. Immunoperoxidase substrate: 3,3′-Diaminobenzidine (DAB) solution is prepared fresh on the day by dissolving a Sigma Fast DAB tablet in 5 mL polished water.

2.5. Culture Plasticware

1. Primaria™ tissue culture plasticware (Falcon®, Becton-Dickinson supplied by Scientific Laboratories Supplies, Nottingham, UK) is used for all NHU monolayer cell cultures.
2. Organ culture: Falcon® tissue culture inserts (3.0 μm pore Cyclopore® membrane filters supplied by VWR International Ltd, Lutterworth, UK) to fit 6-well tissue culture plates (supplied by Scientific Laboratories Supplies).

3. Methods

3.1. Human Urinary Tract Tissue Collection and Processing

1. Surgical specimens of normal bladder, renal pelvis, and ureter, ranging in urothelial surface area from 1.5 to 3.5 cm^2, are obtained from consented patients with no history of UCC, undergoing open urological surgery. Tissues are placed immediately into Transport Medium for transfer to the laboratory and remain viable for several days.
2. In the laboratory, extraneous connective and haemorrhagic tissues are removed by dissection, a small representative piece of tissue is fixed and processed for histology and the remaining sample is cut into two. One piece is used for intact urothelial organ cultures and the remaining piece (minimum urothelial area 0.5 cm^2) is used to isolate and establish primary urothelial cell cultures (6, 7, 13) (Fig. 1).

3.2. Intact Organ Culture

1. Tissues are dissected into pieces of approximately 0.5 cm^2 urothelial area.
2. The specimens are orientated, urothelial surface up, onto Falcon® tissue culture inserts in a Falcon® 6-well plate (13).

3. Organ culture medium is introduced to the well underneath the insert, so that the underside of the filter is just in contact with the medium, positioning the tissue at the air–liquid interface (Fig. 2) (see Note 6).
4. Organ cultures are incubated at 37°C in a 95% humidified atmosphere of 5% (v/v) CO_2 in air (see Note 6).

3.3. Isolation and Culture of Primary Urothelial Cells

1. The urothelial tissue is incubated at 4°C for up to a maximum of 16 h in Stripping Medium. This releases the urothelium as a complete sheet of cells, leaving the stroma with basement membrane intact.
2. The urothelial sheets are gently teased from the stroma using Watchmaker's forceps, collected, and incubated for 20 min in collagenase type IV, before pipetting to disaggregate the urothelial sheets and centrifugation at 250 × *g* for 4 min. The medium is aspirated and the pellet is flicked to resuspend the cells in the residual volume before adding fresh KSFMc and seeding into Primaria™ tissue culture flasks at a minimum seeding density of 4×10^4 cells/cm^2 (6, 7). Cultures of NHU cells are maintained at 37°C in a humidified atmosphere of 5% CO_2 in air, with a medium change twice weekly.
3. Urothelial cell cultures are harvested at near or just confluence by incubation for 5 min in 0.1% EDTA in PBS, followed by incubation at 37°C in trypsin/versene for the minimum time needed for cells to detach when the flask is tapped sharply.
4. Cells are collected into 5 mL KSFMc containing 2 mg trypsin inhibitor, centrifuged at 250 × *g* for 4 min and either seeded into further Primaria™ flasks, cryopreserved, or used in tissue reconstitution experiments.

3.4. Bladder Cancer Cell Lines

1. RT4, RT112, and EJ cell lines are harvested by incubation in 0.1% EDTA in PBS for 5 min, followed by trypsin/versene for the minimum time required to detach the cells by sharp tapping of the flask.
2. Cells are collected into DMEM: RPMI medium containing 10% FBS, centrifuged at 250 × *g* for 4 min and reseeded into tissue culture flasks, or used for composite organ cultures.
3. EJ, RT112, and RT4 cells are routinely maintained using a 1:40, 1:20, and 1:10 split ratio, respectively (see Note 7). This relates to constitutive differences in rates of proliferation (EJ > RT112 > RT4) and reflects the characteristics of the originating tumours.

3.5. Composite Organ Culture

For reconstitution of autologous organoids, pieces of intact tissue are placed in organ culture at the time of initiation of urothelial cell cultures; this circumvents the problem of deterioration of the de-epithelialised stroma during the NHU cell culture expansion phase. Once adequate numbers of expanded NHU cell cultures are available (approximately 2 weeks), the intact tissue is de-epithelialised, washed to remove residual urothelial cells and either processed for histology to check for efficiency of urothelial removal, or seeded with the propagated autologous NHU cells (see below) (and see Note 8).

1. After removal of the urothelium as described in Subheading 3.3, the stroma is rinsed in Transport Medium and transferred, basement membrane uppermost, onto Falcon® tissue culture inserts placed inside a 6-well plate.
2. The de-epithelialised stroma is used to establish control organ cultures to check for regrowth of normal urothelial cells, or may be seeded with NHU or bladder cancer cell lines.
3. For seeding of de-epithelialised stromal fragments, NHU cells are harvested from one near-confluent 75 cm^2 flask (approximately 4×10^6 cells) and bladder cancer cells are harvested from one near-confluent 25 cm^2 flask (approximately 1.5×10^6 cells), as described in Subheadings 3.3 and 3.4, respectively.
4. After centrifugation and aspiration of the medium, the cell pellets are flicked to resuspend in the residual medium to give a thick slurry. The cell suspension is drawn into the fine bore of a sterile plugged Pasteur pipette and one drop, approximately 20 μL, is placed in the centre of the stroma, with care taken to avoid cells spilling over the sides.
5. The reseeded organ culture is placed in a humidified 5% CO_2 incubator for 1 h to allow cells to adhere, before the organ culture medium is added to the basal compartment (Fig. 2) (13, 14).

3.6. Maintenance and Harvesting of Organ Cultures

1. Both the intact and composite urothelial organoids are maintained in organ culture medium at 37°C in an incubator in 5% CO_2 in air, in a 95% humidified atmosphere.
2. Organ culture medium is applied below the insert, sufficient to just wet the underside of the membrane, so that the tissue is maintained at the air–liquid interface (see Note 6). The medium is replaced every 2–3 days.
3. The intact organ cultures and reconstituted organoids are maintained and harvested for immunohistochemistry at 2-week intervals, typically up to 20 weeks.
4. To harvest the intact organ cultures or organoids, the membrane of the insert is detached from the side walls using a scalpel, and the membrane plus culture is fixed in formalin.

3.7. Immunohistochemistry

1. Samples are fixed in 10% formalin overnight at ambient temperature (see Note 5), then transferred and stored in 70% ethanol until processed to prevent over-fixation.
2. The sample is dehydrated through 10 min changes in 70% and 2 × 100% ethanol, with final clearing through two changes of isopropanol, followed by three changes of xylene. Specimens are placed in four changes of paraffin wax at 60°C for 15 min each, embedded in a wax block and left to cool.
3. Sections (5 μm) are cut on a microtome and collected on serially numbered Superfrost plus® microscope slides. This facilitates using serially labelled sections to compare different antibodies, or to assess antigen expression at different tissue levels.
4. Paraffin wax-embedded sections are dewaxed in two changes of xylene at 37°C for 10 min, followed by two changes at ambient temperature and rehydration through 100 and 70% ethanol to polished water, as described above.
5. For haematoxylin and eosin staining, slides are immersed sequentially into haematoxylin for 2 min, Scott's tap water for 1 min and 1% eosin for 30 s, with washing under running tap water between each step. Finally, slides are dehydrated through the alcohol grades into xylene and mounted in DPX. Typical results are illustrated in Fig. 1.
6. For immunolabelling, some antibodies require that the immunoreactivity of antigens masked by the tissue processing is restored. Optimal retrieval protocols need to be determined empirically for each antigen, but typical protocols involve 10 min digestion in 0.1% (w/v) trypsin in 0.1% (w/v) $CaCl_2$ at pH 7.6, microwave heating in 10 mM citrate buffer at pH 6.0 for 13 min, or a combination of both, in which the incubation in trypsin is limited to 1 min (Table 2).
7. Immunodetection can be performed by a number of methods, including indirect immunofluorescence (see Note 9), but in all cases, antibodies concentrations must be determined empirically and appropriate background/specificity controls included. Our preference is for an indirect streptavidin ABC immunoperoxidase method, as it allows the tissue to be counterstained. Non-specific binding of the secondary antibody is blocked by incubation in 10% normal serum of the secondary antibody host (see Note 10). Endogenous peroxidase activity is blocked by treating the sections with 2% hydrogen peroxidase solution for 10 min followed by rinsing under running water for 10 min. Endogenous avidin-binding sites are blocked by treating the sections with an avidin/biotin blocking kit, using 100 μL avidin per section for 10 min, followed by two 5-min washes in Tris-buffered saline (TBS) and then 100 μL biotin for 10 min followed by two 5 min washes in TBS.

8. Sections are incubated sequentially in primary antibodies for 1 h, biotinylated secondary antibodies [e.g. rabbit anti-mouse immunoglobulin F(ab′)$_2$ fragments] for 30 min and streptavidin/horseradish peroxidase ABC complex for 30 min, with washing between each step. Sections of normal human ureter are used as positive control tissue and irrelevant primary antibodies serve as negative controls. 50 mM TBS (pH 7.6) is used throughout as wash buffer and all incubations are performed at ambient temperature.

9. Bound antibody is visualised by reaction with DAB substrate catalysed by H_2O_2 and stopped after 15 min by washing in running water. Sections are lightly counterstained with haematoxylin, dehydrated, cleared, and mounted in DPX for viewing by light microscopy (Fig. 3).

3.8. Future Perspectives

Composite organ culture has been important in demonstrating that in vitro-propagated NHU cells retain the capacity to form a stratified transitional epithelium when recombined with a stromal tissue, whereas using the same approach, UCC cell lines recapitulate the differentiation and invasion characteristics of the originating tumour (13, 14). This opens the potential of using composite organ cultures to investigate the molecular mechanisms of tumorigenesis, invasion, and metastasis. One approach would be to develop organoids from urothelial cells in which candidate cancer

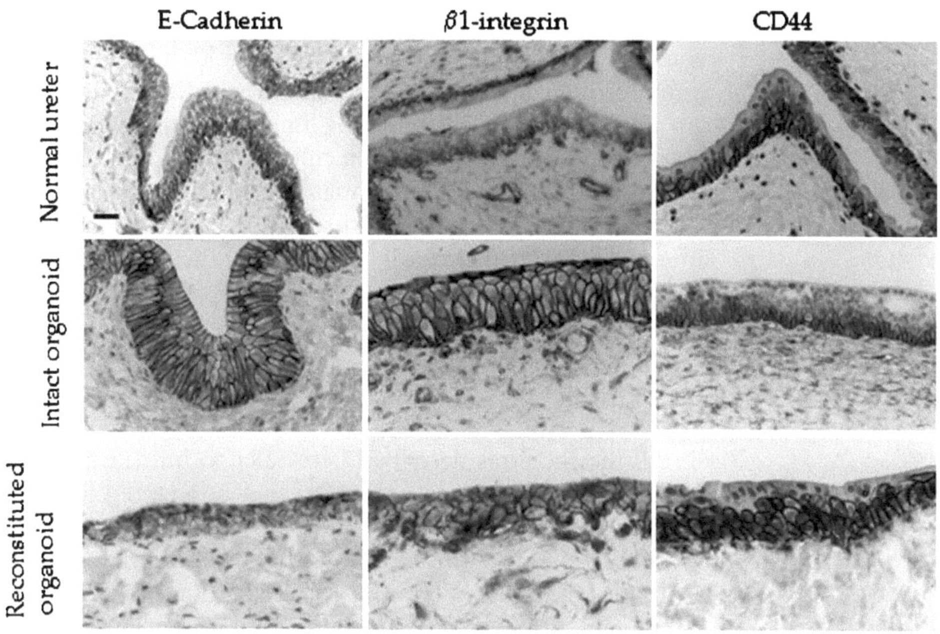

Fig. 3. Expression of cell-adhesion antigens in normal human urothelium and by intact and reconstituted organoids. E-cadherin and β1-integrin subunits were expressed at the intercellular junctions throughout all urothelial layers. CD44 expression was essentially limited to the basal and intermediate cell layers. Scale bar 20 μm for CD44 labelling of normal urothelium and E-cadherin and β1-integrin labelling of intact organ cultures; 40 μm for the rest.

genes have been modified. Thus, metastasis-associated oncogenes could be suppressed in UCC cell lines using a dominant-negative or small interfering RNA approach. Alternatively, "paramalignant" human urothelial cells have been developed following transduction with retroviruses to integrate key cancer-associated gene changes into NHU cells (20, 21). These strategies have the potential to give insight on the influence of specific genetic changes in bladder cancer development in a three-dimensional composite organ culture system.

Recent strategic interest in tissue engineering and systems biology has invigorated a change in focus from reductionist to synthetic biology. Tissue engineering involves combining in vitro-propagated cells with scaffolds of natural and synthetic biomaterials to form tissue constructs for incorporation in vivo, often via a period of nurture in vitro. Organ culture provides a method for maintaining intact heterotypic tissues in vitro for extended periods and by application, provides a test bed for producing differentiated tissues from the assorted tissue engineering parts (cells and biomaterials). Such an approach was used in one study to examine different strategies and biomaterials for transferring NHU cells onto a stroma (22). Earlier work demonstrated that rodent urothelium could be maintained on an artificial stroma formed from seeding fibroblasts within a collagen gel (23). More recent work has begun to examine the growth and differentiation potential of human urinary tract-derived stromal cells seeded into different natural and synthetic scaffolds (24). A natural extension would be to combine stromal-seeded scaffolds with in vitro-propagated urothelial cells in organ culture; with the ability to maintain intact urothelial tissues providing a critical baseline. In the long term, although static organ culture systems may be adequate to maintain an already differentiated urothelium, more dynamic systems may be necessary for tissue maturation by mechanically simulating the normal voiding function of the bladder.

4. Notes

1. All the methods described here refer to human urothelium, although most of the techniques may be adapted successfully, with minor modifications, to other large mammalian species such as the pig (25, 26). Note that we avoid the use of antibiotics in any of the buffers or media used for cell or organ culture, although we would add antibiotics to the Transport Medium if animal tissues are to be sourced from an abattoir.
2. Given the long-standing problems of cross-contamination of many established cell lines, as has recently again been highlighted (27), the importance of working with authenticated

cell stocks is emphasised as essential. We have found it useful to introduce genotyping as part of our quality assurance process to verify the authenticity of established cell lines used in the laboratory. In brief, DNA is isolated using the DNeasy™ Tissue kit according to the manufacturer's protocol (Qiagen Ltd., Crawley, West Sussex, UK) and a PCR performed using the PowerPlex® 1.2 system (Promega, Southampton, UK), which allows the co-amplification of eight STR loci. The resultant PCR products are processed using the ABI PRISM® Genetic Analyser and GeneMapper® software (Applied Biosystems, Warrington, UK) and the DNA profile for each cell line is compared to the allelic ladder.

3. If left too long at confluence, NHU cells become contact-inhibited and exit the cell cycle, which can compromise their growth potential. Sub-culture therefore tends to be most successful if cultures are harvested when in exponential growth and just as they reach confluence in order to ensure maximal cell numbers.
4. We routinely test FBS and KSFMc to select batches that provide optimum growth.
5. Instead of formalin, zinc salts fixative may be used. This has the advantage of preserving more antigens than formalin-fixed tissues, although it is not possible then to block endogenous peroxidise activity. To make zinc fixative, add 0.5% (w/v) zinc acetate and 0.05% (w/v) zinc chloride to 0.1 M Tris pH 7–7.4 containing 0.05% (w/v) calcium acetate.
6. The level of the medium is critical to ensure that the organ culture is neither submerged nor becomes dehydrated. For this reason too, it is essential that the humidity of the incubator was kept at near saturation.
7. In addition to checking the authenticity of cell stocks by genotyping, it is also important to check routinely that cell line stocks are not contaminated by *Mycoplasma* spp. (also known as PleuroPneumonia-Like Organisms, PPLO), which is an endemic problem in many cell culture laboratories. Various PCR-based screening methods are available, but for a simple, routine screening test, we find staining with the DNA intercalating dye, bis-benzimide to be simple, quick, cheap, and effective. Briefly, seed cells sparsely on a slide or coverslip and grow for 24–48 h. Fix in a 1:1 mixture of methanol and acetone, air-dry and stain with 0.1 μg/mL Hoechst 33258 in PBS for 5 min. Rinse in water, air-dry, and mount using a suitable mountant. Only nuclei should fluoresce blue when visualised by fluorescence microscopy, with any extranuclear fluorescence due potentially to contamination by *Mycoplasma* spp.

8. Although it should be possible to establish allogeneic combinations of cells and stroma, in our hands, autologous combinations using this strategy have been more successful than heterologous combinations.
9. Indirect immunofluorescence is particularly useful for dual labelling with two antibodies as it is possible to visualise the separate and combined patterns using different filter combinations. Immunoperoxidase or other substrate-based immunodetection methods are ideal for localising antigen expression on tissues/organoids as the tissue may be counterstained to reveal histological detail; however, such techniques are less appropriate for use in dual labelling as it can be difficult to distinguish individual substrates, particularly where they co-localise.
10. For example, 10% rabbit serum would be used as the blocking serum if a rabbit anti-mouse secondary antibody was used to detect a primary mouse monoclonal antibody.

Acknowledgements

The authors would like to thank York Against Cancer for funding and urology colleagues for providing tissues. We are grateful to Dr Sharon Scriven and Dr Catherine Booth who were originally involved in developing the organ culture methods described. Ms Lisa Clements and Mrs Rosalind Duke are thanked for providing the genotyping data and Ms Jenny Hinley is thanked for careful proof-reading of the manuscript.

References

1. Southgate, J., Harnden, P., and Trejdosiewicz, L. K. (1999) Cytokeratin expression patterns in normal and malignant urothelium: a review of the biological and diagnostic implications. *Histol. Histopathol.* **14**, 657–664.
2. Varley, C. L., Stahlschmidt, J., Smith, B., Stower, M., and Southgate, J. (2004) Activation of peroxisome proliferator-activated receptor-gamma reverses squamous metaplasia and induces transitional differentiation in normal human urothelial cells. *Am. J. Pathol.* **164**, 1789–1798.
3. Varley, C. L., Garthwaite, M. A., Cross, W., Hinley, J., Trejdosiewicz, L. K., and Southgate, J. (2006) PPARgamma-regulated tight junction development during human urothelial cytodifferentiation. *J. Cell Physiol.* **208**, 407–417.
4. Lobban, E. D., Smith, B. A., Hall, G. D., Harnden, P., Roberts, P., Selby, P. J., Trejdosiewicz, L. K., and Southgate, J. (1998) Uroplakin gene expression by normal and neoplastic human urothelium. *Am. J. Pathol.* **153**, 1957–1967.
5. Wu, X. R., Medina, J. J., and Sun, T. T. (1995) Selective interactions of UPIa and UPIb, two members of the transmembrane 4 superfamily, with distinct single transmembrane-domained proteins in differentiated urothelial cells. *J. Biol. Chem.* **270**, 29752–29759.
6. Southgate, J., Hutton, K. A., Thomas, D. F., and Trejdosiewicz, L. K. (1994) Normal human urothelial cells in vitro: proliferation and induction of stratification. *Lab. Invest.* **71**, 583–594.
7. Southgate, J., Masters, J. R., and Trejdosiewicz, L. K. (2002) Culture of Human Urothelium, in *Culture of Epithelial Cells* (Freshney, R. I. and Freshney, M. G., Eds.), 2nd ed., pp. 381–400, John Wiley and Sons, Inc., New York.

8. Varley, C. L., Stahlschmidt, J., Lee, W. C., Holder, J., Diggle, C., Selby, P. J., Trejdosiewicz, L. K., and Southgate, J. (2004) Role of PPARgamma and EGFR signalling in the urothelial terminal differentiation programme. *J. Cell Sci.* **117**, 2029–2036.
9. Varley, C., Hill, G., Pellegrin, S., Shaw, N. J., Selby, P. J., Trejdosiewicz, L. K., and Southgate, J. (2005) Autocrine regulation of human urothelial cell proliferation and migration during regenerative responses in vitro. *Exp. Cell. Res.* **306**, 216–229.
10. Varley, C. L., Bacon, E. J., Holder, J. C., and Southgate, J. (2009) FOXA1 and IRF-1 intermediary transcriptional regulators of PPARgamma-induced urothelial cytodifferentiation. *Cell Death Differ.* **16**, 103–114.
11. Cross, W. R., Eardley, I., Leese, H. J., and Southgate, J. (2005) A biomimetic tissue from cultured normal human urothelial cells: analysis of physiological function. *Am. J. Physiol. Renal Physiol.* **289**, F459–F468.
12. Cross, W. and Southgate, J. (2004) Biomimetic Urothelium, WO/2004/011630.
13. Scriven, S. D., Booth, C., Thomas, D. F., Trejdosiewicz, L. K., and Southgate, J. (1997) Reconstitution of human urothelium from monolayer cultures. *J. Urol.* **158**, 1147–1152.
14. Booth, C., Harnden, P., Trejdosiewicz, L. K., Scriven, S., Selby, P. J., and Southgate, J. (1997) Stromal and vascular invasion in an human in vitro bladder cancer model. *Lab. Invest.* **76**, 843–857.
15. Masters, J. R., Hepburn, P. J., Walker, L., Highman, W. J., Trejdosiewicz, L. K., Povey, S., Parkar, M., Hill, B. T., Riddle, P. R., and Franks, L. M. (1986) Tissue culture model of transitional cell carcinoma: characterization of twenty-two human urothelial cell lines. *Cancer Res.* **46**, 3630–3636.
16. Rieger, K. M., Little, A. F., Swart, J. M., Kastrinakis, W. V., Fitzgerald, J. M., Hess, D. T., Libertino, J. A., and Summerhayes, I. C. (1995) Human bladder carcinoma cell lines as indicators of oncogenic change relevant to urothelial neoplastic progression. *Br. J. Cancer* **72**, 683–690.
17. Southgate, J., Proffitt, J., Roberts, P., Smith, B., and Selby, P. (1995) Loss of cyclin-dependent kinase inhibitor genes and chromosome 9 karyotypic abnormalities in human bladder cancer cell lines. *Br. J. Cancer* **72**, 1214–1218.
18. Trejdosiewicz, L. K., Southgate, J., Donald, J. A., Masters, J. R., Hepburn, P. J., and Hodges, G. M. (1985) Monoclonal antibodies to human urothelial cell lines and hybrids: production and characterization. *J. Urol.* **133**, 533–538.
19. Hutton, K. A., Trejdosiewicz, L. K., Thomas, D. F., and Southgate, J. (1993) Urothelial tissue culture for bladder reconstruction: an experimental study. *J. Urol.* **150**, 721–725.
20. Crallan, R. A., Georgopoulos, N. T., and Southgate, J. (2006) Experimental models of human bladder carcinogenesis. *Carcinogenesis* **27**, 374–381.
21. Shaw, N. J., Georgopoulos, N. T., Southgate, J., and Trejdosiewicz, L. K. (2005) Effects of loss of p53 and p16 function on life span and survival of human urothelial cells. *Int. J. Cancer* **116**, 634–639.
22. Scriven, S. D., Trejdosiewicz, L. K., Thomas, D. F., and Southgate, J. (2001) Urothelial cell transplantation using biodegradable synthetic scaffolds. *J. Mater. Sci. Mater. Med.* **12**, 991–996.
23. Howlett, A. R., Hodges, G. M., and Rowlatt, C. (1986) Epithelial-stromal interactions in the adult bladder: urothelial growth, differentiation, and maturation on culture facsimiles of bladder stroma. *Dev. Biol.* **118**, 403–415.
24. Baker, S. C. and Southgate, J. (2008) Towards control of smooth muscle cell differentiation in synthetic 3D scaffolds. *Biomaterials* **29**, 3357–3366.
25. Fraser, M., Thomas, D. F., Pitt, E., Harnden, P., Trejdosiewicz, L. K., and Southgate, J. (2004) A surgical model of composite cystoplasty with cultured urothelial cells: a controlled study of gross outcome and urothelial phenotype. *BJU Int.* **93**, 609–616.
26. Turner, A. M., Subramaniam, R., Thomas, D. F., and Southgate, J. (2008) Generation of a functional, differentiated porcine urothelial tissue in vitro. *Eur. Urol.* **54**, 1423–1432.
27. Chatterjee, R. (2007) Cell biology. When 60 lines don't add up. *Science* **315**, 929.

Chapter 14

Ex Vivo Organ Culture of Human Hair Follicles: A Model Epithelial–Neuroectodermal–Mesenchymal Interaction System

Desmond J. Tobin

Abstract

The development of hair follicle organ culture techniques is a significant milestone in cutaneous biology research. The hair follicle, or more accurately the "pilo-sebaceous unit", encapsulates all the important physiologic processes found in the human body; controlled cell growth/death, interactions between cells of different histologic type, cell differentiation and migration, and hormone responsivity to name a few. Thus, the value of the hair follicle as a model for biological scientific research goes way beyond its scope for cutaneous biology or dermatology alone. Indeed, the recent and dramatic upturn in interest in hair follicle biology has focused principally on the pursuit of two of biology's holy grails; post-embryonic morphogenesis and control of cyclical tissue activity. The hair follicle organ culture model, pioneered by Philpott and colleagues, ushered in an exceptionally accessible way to assess how cells of epithelial (e.g., keratinocytes), mesenchymal (e.g., fibroblasts), and neuroectodermal (e.g., melanocytes) origin interact in a three-dimensional manner. Moreover, this assay system allows us to assess how various natural and pharmacologic agents affect complex tissues for growth modulation. In this article, I focus on the culture of the human hair follicle mini-organ, discussing both the practical issues involved and some possible research applications of this assay.

Key words: Keratinocytes, Fibroblasts, Melanocytes, Mesenchymal, Epithelial, Neuroectodermal, Neural crest, Stem cells

1. Introduction

After significant early success with the culture of fibroblasts from human skin dermis more than 70 years ago, strategies for the cultivation of keratinocyte from human skin epidermis emerged only in the 1970s (1). More resistant still to the development of long-term culture methodologies was the epidermal melanocyte (2) with their highly restricted requirements for growth factor combinations. However, hair biologists were quick to adapt these

John W. Haycock (ed.), *3D Cell Culture: Methods and Protocols*, Methods in Molecular Biology, vol. 695,
DOI 10.1007/978-1-60761-984-0_14,

culture protocols to the histologically similar cell populations in the hair follicle. This was first demonstrated for keratinocytes of the outer root sheath using plucked human anagen hair follicles (3). Protocols for the cultivation of the hair growth-inductive fibroblast-like cells of the follicular papilla (4), and the contractile fibroblast-like cells of the dermal sheath, which encases the entire hair follicle (5) were published a little later. The successful cultivation of matrix epithelial cells from the anagen bulb was not achieved till the early 1990s (6), followed by the first long-term cultivation of follicular melanocytes a few years later (7). Since then, there have been attempts to cultivate epithelial stem cells (8) and even the intact hair follicle itself by bringing together all the main sub-populations of hair follicle cells or via more primitive follicular spheroids. During this latter period, the maintenance of intact growing hair follicle isolated from human scalp in ex vivo culture was first reported by Philpott and colleagues (9) and subsequently developed to provide researchers access to a uniquely regenerative and cyclical mini-organ in the adult. The hair follicle mini-organ is formed from a bewilderingly complex set of interactions involving ectodermal, mesodermal, and neuroectodermal components, which go to elaborate five or six concentric cylinders of at least 15 distinct interacting cell sub-populations. Together these provide an exceptional tissue that rivals the vertebrate limb-bud as a model for studies of the genetic regulation of development (10).

2. Materials

2.1. Isolation of Human Scalp Hair Follicles

1. Normal human haired skin tissue, ideally obtained from scalp with terminal hair follicles in high anagen VI phase of the hair growth cycle. These can be routinely sourced from elective cosmetic surgery (e.g., post-auricular skin from face-lift procedures). Hair follicles should be harvested within 8 h of surgery; earlier if possible.
2. Transport medium to transport skin from clinic to laboratory: RPMI 1640 basal medium, 10% fetal bovine serum, 2 mM l-glutamine, antibiotics (2,500 U/mL penicillin and 2.5 mg/mL streptomycin), amphotericin B-containing anti-mycotic (60 μg/mL Fungizone). Fully supplemented medium is kept at 4°C until use.
3. Wash solution-1: To remove debris, clotted blood, and surface hair from skin specimens; 0.1 M phosphate buffered saline (PBS, pH 7.4), 5% fetal bovine serum, concentrated antibiotics – 1,000 U/mL penicillin and 1 mg/mL streptomycin, concentrated amphotericin B-containing anti-mycotic – 25 μg/mL Fungizone.

4. Wash solution-2: To rinse isolated hair follicles and as a holding solution before selection of best follicles for culture; 0.1 M PBS (pH 7.4) with 5% fetal bovine serum, concentrated antibiotics – 500 U/mL penicillin and 0.5 mg/mL streptomycin, concentrated amphotericin B-containing anti-mycotic – 12.5 μg/mL Fungizone.
5. Dissection and micro-dissection: Scalpel (No. 3) and blades (No. 11), extra-fine Dumont Biology Tweezers (No. 4, 5, 7), Serrated Dumont Tweezers (No. 24) (Agar Scientific Ltd, Stansted, UK, but several excellent alternative sources exist).
6. Stereomicroscope with wide zoom magnification control (up to at least 6×) and separate light source for maximum light contrasting of tissue components.
7. Cell culture plastic-ware: 24- or 48-well culture-treated dishes.
8. CO_2 incubator: A Heracell incubator (Thermo Fisher Scientific Ltd, Loughborough, UK) is used here, although several excellent alternative sources exist.

2.2. Ex Vivo Culture of Isolated Human Scalp Hair Follicles

1. William's E Medium (Invitrogen Ltd, Paisley, Scotland). Store at 4°C.
2. Medium supplements (store at 4°C): 2 mM glutamine (store frozen at –20°C), 10 ng/mL hydrocortisone (store at 4°C), 10 μg/mL insulin (store at 4°C). Several excellent alternative sources exist for these reagents.

2.3. Immunohistochemistry of Cultured Human Scalp Hair Follicles (Cryosections)

1. Optimal cutting temperature (OCT) compound (Tissue-Tek, Sakura Finetek, USA).
2. Aluminum foil (domestic grade).
3. Liquid nitrogen.
4. Liquid nitrogen Dewar flask.
5. Cryostat (Leica Microsystems UK Ltd., Milton Keynes, Buckinghamshire, UK).
6. Microscope slides: Ground edges, twin frosted glass.
7. 10% Poly-L-lysine solution in distilled water.
8. Bench-top oven.
9. Primary and secondary antibodies with chromogen or fluorochrome detection systems.
10. Histological counterstain, e.g., Mayer's haematoxylin.

2.4. High-Resolution Examination of Cultured Human Scalp Hair Follicles (Plastic Sections)

1. Fixative: Karnovsky's half-strength fixative (Agar Scientific Limited, Stansted, Essex, UK), 0.2 M sodium cacodylate buffer (pH 7.4), 0.1 M sodium cacodylate buffer (pH 7.4), 2 g paraformaldehyde powder (EM Grade), 20 mL H_2O distilled, 1 M sodium hydroxide (NaOH) (three drops), 10 mL, 25%

glutaraldehyde solution (EM Grade), 25 mg calcium chloride ($CaCl_2$), 2% osmium tetraoxide (in water).

2. Refrigerator at 4°C.
3. Dehydration in graded alcohols: 50, 70, 80, 90, 95, and 100% Ethanol.
4. Propylene oxide.
5. Epoxy resin (Agar Scientific Limited, Stansted, Essex, UK): 10 mL Araldite CY212 resin, 10 mL dodecenyl succinic anhydride DDSA, 0.4 mL benzyl dimethylamine BDMA.
6. Embedding moulds: Silicone rubber (TAAB Laboratories Equipment Ltd, Aldermaston, Berkshire, UK).
7. Embedding oven (TAAB Laboratories Equipment Ltd, Aldermaston, Berkshire, UK).
8. Ultra-microtome (Reichert-Jung, Leica, Germany).
9. Glass knifes (406 mm × 25 mm × 6 mm; Agar Scientific Limited, Stansted, Essex, UK).
10. Microscope slides: Ground edges, twin frosted glass.
11. Histological stains: We have successfully used (a) toluidine blue (1%) in 1% Borax; and (b) 1% toluidine blue in 1% Borax in a 2:1 ratio of 1% pyronin (Agar Scientific Ltd., Stansted, Essex, UK).
12. Permanent mountant: Histomount or DPX (Agar Scientific Ltd., Stansted, Essex, UK).

2.5. Transmission Electron Microscopy Examination of Cultured Human Scalp Hair Follicles

Identical to items 1–9 above.

1. Section collection grids: Copper or nickel, 175 or 200 mesh grids and can be sourced from Agar Scientific Ltd.
2. Anti-magnetic ultra-fine tweezers, Dumont Biology Tweezers (No. 4, 5, 7) (Agar Scientific Ltd, Stansted, Essex, UK).
3. Heavy metal stains (Agar Scientific Ltd. Stansted, Essex, UK): (a) Reynold's lead citrate; 1.33 g lead nitrate, and 1.76 g sodium citrate boiled and cooled in 30 mL double-distilled water. (b) Uranyl acetate: 2% in distilled water or saturated in 50% methanol. Uranyl acetate may be difficult to obtain in some jurisdictions (dependent on levels of application of the *Treaty on the Non-Proliferation of Nuclear Weapons*). The author understands that Nissin EM Co. Ltd, Tokyo, Japan is currently trialing an alternative platinum staining method. Sodium hydroxide (NaOH) pellets, double-distilled water, plastic staining dishes with black-out capability.
4. Transmission electron microscope. A Joel 100CX was used in this case, but several excellent alternatives exist.

3. Methods

3.1. Isolation of Human Scalp Hair Follicles

Several important issues need to be borne in mind when attempting to culture normal human hair follicles. The most critical of these include: (a) the source of haired skin should be as fresh as possible, ideally within 8 h of surgery and (b) hair follicles should only be isolated if they exhibit full anagen (anagen VI) morphology (see Note 1).

1. Scalp specimens are obtained in toto from an appropriate clinic with informed consent and with local ethics committee approval, and collected into labelled vessels containing transport medium at 4°C.
2. Scalp pieces are washed with wash solution-1 until all traces of blood and debris are removed. The skin is transected into pieces of approximately 3 cm^2, from which the hair is removed by shaving in the growth direction using a fresh scalpel blade. The skin is washed several times in wash solution-1 to remove any remaining debris.
3. The skin is cut into 1 cm^2 pieces and the epidermis and most of the dermis is carefully removed using scalpel (with No. 11 blade) (Fig. 1) and discarded (or used separately for cell isolation) (see Note 2).
4. Individual hair follicles are isolated under the dissecting microscope using two pairs of fine tweezers. A larger blunt pair is used to squeeze the soft subcutaneous tissue thereby exposing the distal region of the hair follicles from the surrounding tissue, while a fine tweezers is used to draw out the hair follicles via clasping the protruding upper hair follicle (Fig. 1).
5. The isolated hair follicles are collected in quick succession into fully supplemented William's E medium. Only hair follicles depicting the morphology of full anagen VI hair follicles were collected (Fig. 2). Any damaged follicles or follicles with morphologic features of non-growing catagen or telogen hair follicles (e.g., loss of pigmentation from a sub-optimally expansive bulb) were discarded.
6. Hair follicles of optimal quality were washed carefully twice in fully supplemented William's E medium.
7. Hair follicles were prepared for culture by placing them carefully and singly (and always by holding only the upper distal part of the hair follicle i.e., away from the growth area of the hair bulb) into 48-well tissue culture plastic-ware (see Note 3).
8. Control hair follicles can be collected immediately for analysis by: (a) immunohistochemistry (by placing into a droplet of

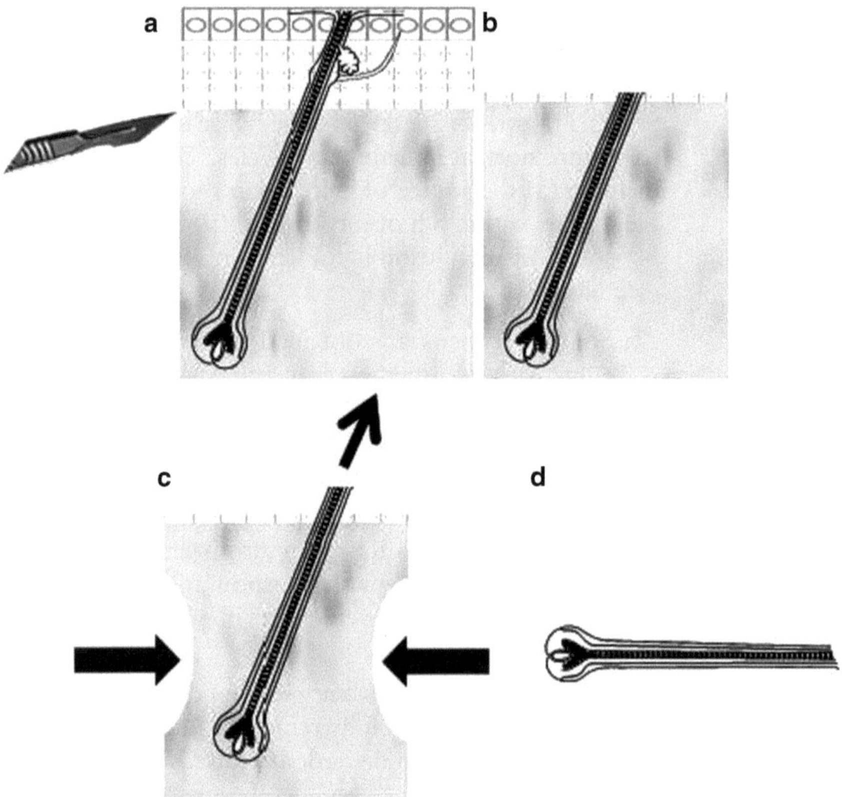

Fig. 1. Schematic showing steps in isolation of intact hair follicles. (**a**) Removal of the epidermis and most of dermis. This will also remove the upper hair follicle containing sebaceous gland. (**b**) Remaining lower dermis and subcutaneous tissue containing lower 75% of the hair follicle. (**c**) Depression or squeezing of the lower dermis and subcutaneous tissue causes the hair follicles to move up in the tissue with resulting extrusion of the distal parts of the follicle. This can then be specifically grasped and removed using fine tweezers. (**d**) Isolated hair follicle, ready for ex vivo culture.

OCT compound) and stored at –80°C until used for immunostaining (see *below*), or (b) by high-resolution light microscopy and transmission electron microscopy (TEM) by fixation in half-strength Karnovsky's fixative (see *below*).

9. The culture dishes containing the isolated hair follicles are then placed in an incubator at 37°C and 95% air/5% CO_2.

3.2. Culture of Isolated Human Scalp Hair Follicles

1. The culture dishes containing the hair follicles are examined after 12 h to assess their growth potential. This will be evident, if present, by some elongation of the hair fiber at the distal end of the reflected follicle (Fig. 3) and the appearance of a "shoulder" demarcating the new growth from the original length of the follicle upon isolation (see Note 4).
2. Fully supplemented medium (or medium with test reagents) is best changed every 48 h. A further benefit of frequent, even daily, checking of the follicles is to ensure that they remain free-floating. If follicles begin to adhere to the surface of the

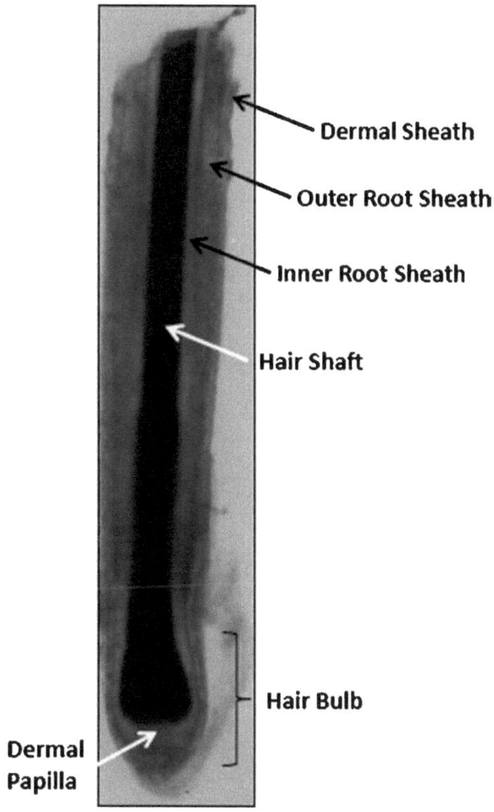

Fig. 2. Freshly isolated human scalp anagen hair follicle revealing dermal sheath, outer root sheath, inner root sheath, hair shaft, hair bulb, and dermal papilla.

plate, their growth may be impaired, but if caught before full adherence, they can be re-floated using a fine sterile needle. Much information of interactions between different cell types can be obtained during the maximum 2-week duration of the culture (Fig. 4). Alternatively, if single cell cultures of hair follicle-derived keratinocytes or fibroblasts are sought, the adhering follicles can be left to fully attach to the plastic surface (Fig. 5) (see Note 5).

3. Hair follicles can be photo-documented using low power objectives (2.5×) to capture the full growing hair follicle in a single image (Fig. 4a) or higher power objectives (4×, 10×, 20×) to capture more morphologic information (Fig. 4c).

3.3. Immunohistological Examination of Cultured Human Scalp Hair Follicles (Cryosections)

1. Microscope slides are coated with 10% poly-l-lysine solution to facilitate the secure bonding between glass slide and tissue section. Slides are pre-coated and dried overnight in an oven.
2. Hair follicles selected for immunohistochemical analysis are placed flat onto a smooth aluminum foil-covered card.

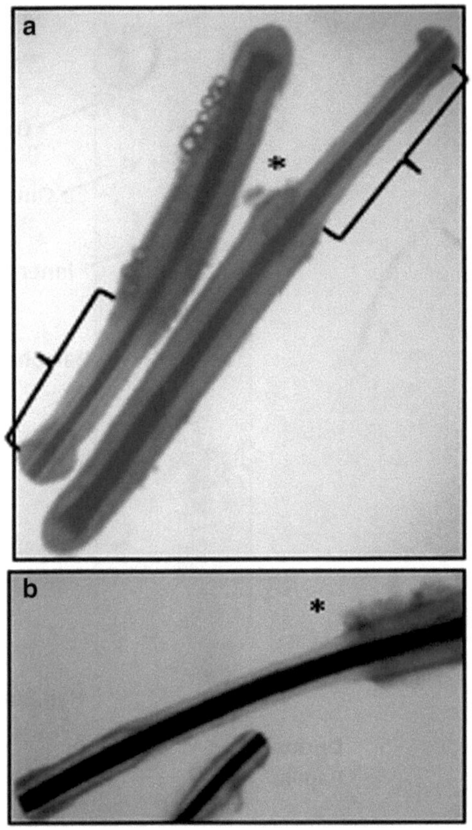

Fig. 3. (**a**) Two isolated human scalp hair follicles after 3 days in culture. Note extent of hair shaft elongation (*brackets*) as indicated by the shoulder at the asterisk. (**b**) Further elongation of the hair fiber at culture day 6 showing marked interface (*asterisk*) between original follicle length and new growth.

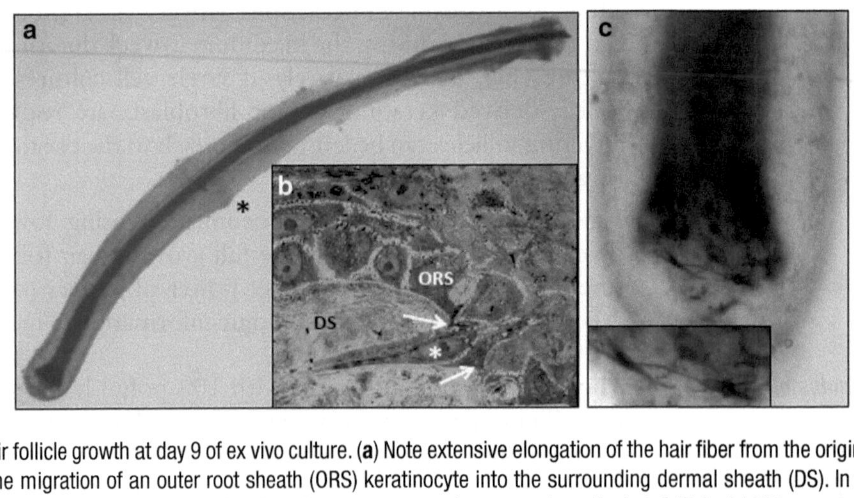

Fig. 4. Hair follicle growth at day 9 of ex vivo culture. (**a**) Note extensive elongation of the hair fiber from the origin (*asterisk*) and (**b**) the migration of an outer root sheath (ORS) keratinocyte into the surrounding dermal sheath (DS). In this way it can be appreciated why wound-healing keratinocytes commonly emerge from the hair follicle. (**c**) This assay can also be used to assess the nature of the keratinocytes–melanocyte interaction. Note the alteration in the position of hair bulb melanocytes after 5 days in culture. *Inset*: higher power view of migrating melanocytes.

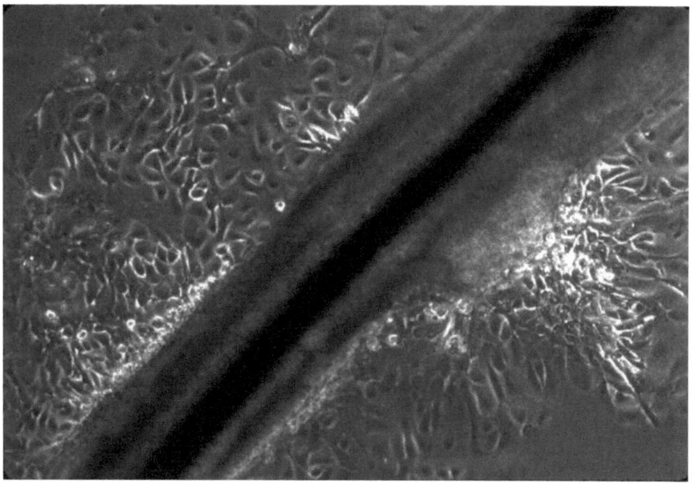

Fig. 5. Some hair follicles adhere to the surface of the culture dish and begin to exhibit explantation of both dermal sheath fibroblasts and outer root sheath keratinocytes as seen in this image. In this way, cultures of outer root sheath keratinocytes can be established using this assay, though this can more efficiently be achieved from single-cell suspensions after hair follicle trypsinization.

3. A small drop of OCT medium is placed on top of the hair follicle and is immediately plunged into liquid nitrogen, thereby freezing the follicle within the OCT and ensuring it remains with a flat aspect to facilitate fully longitudinal sectioning.
4. The button of frozen OCT containing the hair follicle is removed from the foil, carefully and squarely placed onto a frozen cryostat chuck with some freshly poured molten OCT and returned to freeze hard in the cryostat.
5. 5–7 μm sections are cut from the follicle and collected onto room temperature microscope slides (sections will not adhere securely to cold slides) checking under a light microscope (with or without staining) for the desired region of the hair follicle – typically through the centre of the follicle (see Note 6).
6. Sections on slides can be stored at −20°C until ready for analysis, but should be air-dried for 1 h before staining with or without additional fixation – typically in acetone for 10 mins.
7. The mode of immunohistochemical analysis will be variable and specific for the question asked by the researcher (Fig. 6) (see Note 7).
8. Typically, for immunostaining using light microscopy, a counter-stain with a color distinct from the chromogen can also provide optimal discrimination between the antibody-targeting epitope/antigen and the surrounding tissue architecture. Including DAPI (4′,6-diamidino-2-phenylindole) into the mountant can provide a similar counter-stain (albeit of nuclei only) for immunofluorescence views of the hair follicle.

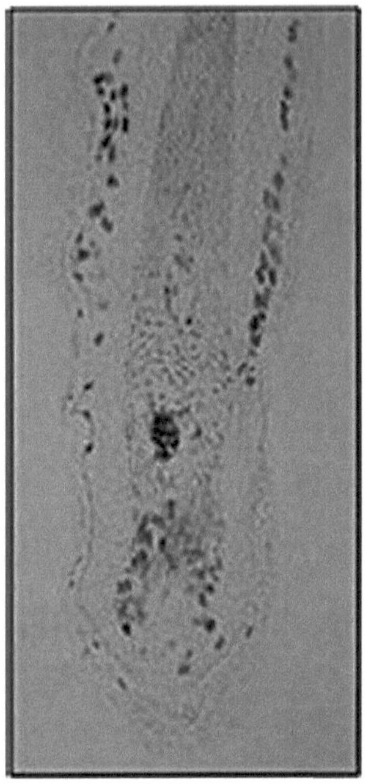

Fig. 6. Immunohistochemical analysis of cell proliferation (using Ki67 antibody) on cryosections of hair follicles after 5 days in ex vivo culture. Positive-staining nuclei (*brackets*) can be seen against a non-counterstained background, particularly in the outer root sheath and hair bulb.

3.4. High-Resolution Examination of Cultured Human Scalp Hair Follicles (Plastic Sections)

1. Hair follicles selected for high-resolution light microscopy are placed into a small volume of half-strength Karnovsky's fixative, prepared as follows:
   (a) Add 2 g paraformaldehyde powder (EM Grade) to 20 mL distilled H_2O and stir for 40 min at 70°C (do not boil).
   (b) Add approximately three drops of 1 M NaOH to clear the solution (may take a few minutes).
   (c) Allow solution to cool.
   (d) Separately add 10 mL of 25% glutaraldehyde solution (EM Grade) to 20 mL of sodium cacodylate buffer (0.2 M, pH 7.4). Add 25 mg $CaCl_2$.
   (e) Add paraformaldehyde solution (20 mL) to glutaraldehyde solution (30 mL) and check if pH is between 7.2 and 7.4 (see Note 8).
   (f) Add 50 mL of sodium cacodylate buffer (0.1 M, pH 7.4) to step (e) above for 1:1 dilution (i.e., this gives "*half-strength*" Karnovsky's fixative). Check if pH is still between 7.2 and 7.4. Refrigerate for a maximum of 2 weeks to ensure consistent good quality.

2. Fix the hair follicles for 2–3 h at 4°C (see Note 9).
3. Remove fixative and dispose of carefully. Wash hair follicles in three changes of sodium cacodylate buffer (0.1 M).
4. Remove buffer and add 2% osmium tetraoxide (in water) for 1 h at 4°C (see Note 10).
5. Remove osmium tetraoxide and rinse hair follicles three times in sodium cacodylate buffer (0.1 M).
6. Begin dehydration in 50% ethanol (in distilled water) for 5 min at 4°C with gentle agitation.
7. Transfer hair follicles to four 5-min changes in 70% ethanol at 4°C with gentle agitation (see Note 11).
8. Transfer hair follicles into the following graded ethanol; 95% ethanol for two 5-min changes at 4°C with gentle agitation, then 100% (absolute ethanol) for three 20-min changes at 4°C with gentle agitation.
9. Make epoxy resin as follows: Mix araldite CY212 Resin (10 mL) with Dodecenyl succinic anhydride DDSA (10 mL) and Benzyl dimethylamine BDMA (0.4 mL) by stirring and gentle heating.
10. Transfer hair follicles to the following mixtures of araldite resin and propylene oxide:
    (a) Propylene oxide: resin mixture (2:1) for 1 h at room temperature.
    (b) Propylene oxide: resin mixture (1:1) for 3 h at room temperature.
    (c) Propylene oxide: resin mixture (1:2) for 18 h at room temperature.
    (d) Full (freshly prepared) resin for 24 h at room temperature.
11. Transfer the resin-impregnated hair follicles (removing much of the adherent old resin) to a silicone rubber embedding mould, placing the follicles longitudinally (if this is the preferred aspect for subsequent analysis) across either flat end of the mould. Fill each well of the mould with fresh complete resin and add laser-printed label to surface of resin in each well for hair follicle identification.
12. Polymerize the resin for 3 days at 60°C. Cool and remove resin blocks containing the hair follicles from the silicone rubber mould. Store until ready for section cutting.
13. Add resin blocks to ultra-microtome and cut longitudinally at 1 μm using freshly cut glass knifes. Stain sections periodically with a histological stain (e.g., toluidine blue in borax) to orient to desired part of the hair follicle – typically through the centre of the hair follicle.

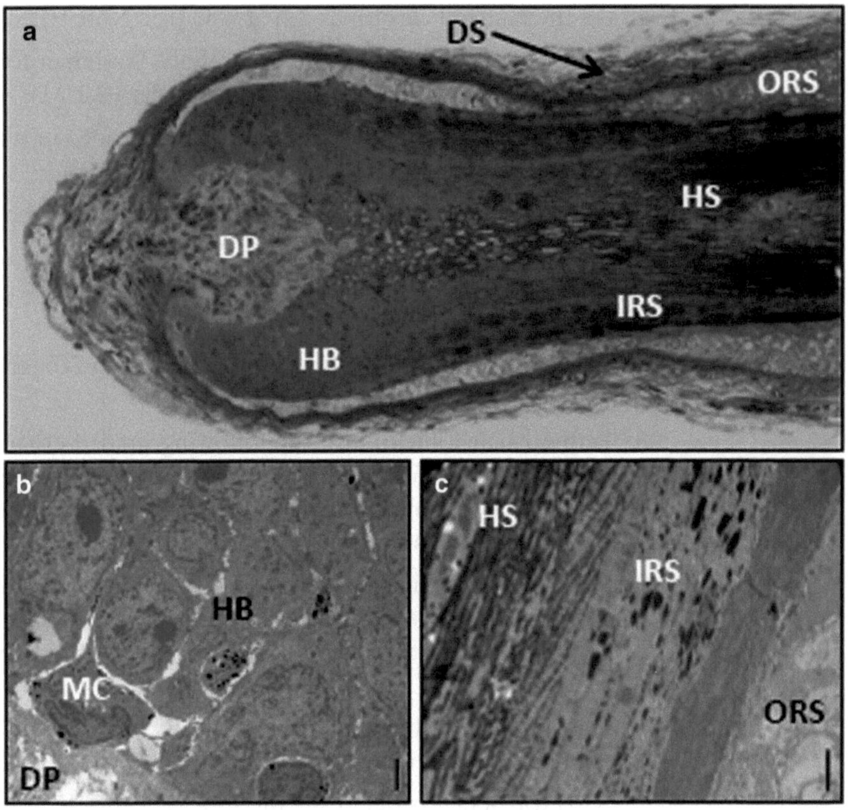

Fig. 7. High resolution and transmission electron microscopic views of the lower hair follicle in ex vivo culture. (**a**) Note the retention of the typical anagen VI morphology in the hair bulb (HB), including its expansive bulbous shape, its deep pigmentation, and close engagement with the dermal papilla (DP). Also visible in this view are the loose connective tissue of the dermal sheath (DS), the outer root sheath (ORS), the inner root sheath (IRS), and hardening hair shaft (HS). (**b**) Transmission electron microscopic view of part of the lower hair bulb showing a hair melanocyte (MC) at the hair matrix: DP interface. (**c**) Transmission electron microscopic view of part of the keratogeneous zone where rapid differentiation of matrix keratinocytes into all the layers of the hair shaft occur, including three layers of the HS, the three layers of the IRS and the ORS. In this way this assay can assess the regulation of differentiation in the hair follicle under the influence of test agents.

14. Rinse off the histological dye after the desired staining intensity (one may need to stain on hot plate) and coverslip with mountant DPX or Histomount.
15. Assess by light microscopy and photo-document as appropriate (Fig. 7a).

3.5. Transmission Electron Microscopy Examination of Cultured Human Hair Follicles

Same as in Subheading 3.2 above until step 12.

1. After determining the preferred hair follicle location for subsequent analysis by TEM, the hair follicles are cut to yield sections of approximately 70 nm using the automatic motorized control of the ultra-microtome. Ideally, the preferred section thickness corresponds to a pale golden color upon

reflection of light off the section as it floats on the water bath constructed at the knife edge (see Note 12).

2. Optimally cut sections (i.e., without surface defects including micro-tears, scratches, ridging, etc.) are collected onto the center of copper or nickel grids (see Note 13). The grids are drained of their surface water by placing their undersides onto Whatman® filter papers and storing them until they are ready for heavy metal staining.
3. Hair follicle ultrathin sections can be stained for electron contrast as follows:
   (a) Place 10 μL droplets of centrifuged Reynold's Lead Citrate (5 min at 12,000 × *g*) onto a sheet of Parafilm® on the base of a small plastic box. Add some sodium hydroxide pellets around the sides of the box to draw carbon dioxide from the air (see Note 14). Cover box and leave for 2 min.
   (b) Place grids by floating, section-side down, onto the small droplets of lead citrate for 10 min. The sections are rinsed by rapid and repetitive plunging and removal (20–30 times) from a series of three small beakers of distilled water before blotting dry (section-side up) on Whatman® filter paper.
   (c) Place 10 μL droplets of centrifuged uranyl acetate (5 min at 12,000 × *g*) onto a sheet of Parafilm® within a small plastic box. The grids are stained by floating, section-side down, onto the small droplets of light-sensitive uranyl acetate for 10 min in the dark. The sections were rinsed by rapid plunging and removal (20–30 times) from a series of three small beakers of distilled water before blotting dry (section-side up) on Whatman® filter paper. Leave to dry for 10 min.
4. The hair follicle ultrathin sections are now ready for examination in the transmission electron microscope (Figs. 4b and 7b, c).

4. Notes

1. As anagen VI typically extends for 3–5 years on the average adult human scalp, it is not possible, based on morphologic assessment alone, to determine whether the hair follicle is at the beginning, middle, or end of this growth IV phase. To the impact of obviate this mosaicism a minimum of ten morphological intact anagen VI hair follicles (i.e., without isolation-associated physical damage) need to be included in any single test group.

2. The subcutaneous fat layer, which contains the majority of the hair follicle mass below the level of their sebaceous gland, is left intact in order to provide maximum protection for the deep-seated lower growing hair bulbs. Damage to the hair bulb upon isolation can significantly reduce viability of the hair follicles.
3. For specific purposes 96-well plates may be used, though it is difficult to photo-document hair follicles in this well format due to light aberrations from the wall of these small wells.
4. The "shoulder" effect is created in part because no further growth of the dermal sheath is possible ex vivo, hence the abrupt narrowing of the elongating fiber with its associated outer and inner root sheath. If no elongation is apparent check the status of the hair bulb, if this exhibits significant change from the original, this follicle is then likely to be precipitating into a regressive "catagen-like" state incompatible with further growth.
5. Some manipulations may need to be conducted outside of the laminar flow hood. This is usually not a significant problem if conducted carefully with minimal time exposure to non-aseptic environment.
6. The hair shaft is very tough and can easily be dislodged from the follicle upon sectioning. This can be obviated somewhat by cutting in a proximal (i.e., bulbar) to distal direction.
7. The hair fiber can exhibit significant auto-fluorescence of a greenish-yellowish type and this should not be confused with FITC-sourced fluorescence.
8. If the pH is outside this range, do not alter. This indicates sub-optimal preparation of the solutions. Check age of glutaraldehyde as it can deteriorate on standing, even in fridge.
9. If larger pieces of skin tissue are required, these should be no bigger than 2 mm^2 and will need 3–4 h fixation time.
10. Hair follicles should go black in color. If they do not, check quality of the osmium tetraoxide.
11. The process can be stopped at this point for up to a week without significant loss in quality.
12. Given the particular hardness of the hair shaft within the much "softer" hair follicles, some workers may find it useful to use a diamond knife for ultrathin sections of hair follicles.
13. Choose a particular grid surface consistently (i.e., either matt or shiny side up) to avoid accidental loss of sections during subsequent staining, rinsing, and dry stages.
14. This reduces the tendency for electron-dense lead carbonate to form and precipitate onto the sections.

References

1. Rheinwald, J.G. and Green, H. (1975) Serial cultivation of strains of human epidermal keratinocytes: the formation of keratinizing colonies from single cells. *Cell* **6,** 331–343.
2. Eisinger, M. and Marko, O. (1982) Selective proliferation of normal human melanocytes in vitro in the presence of phorbol ester and cholera toxin. *Proc. Natl. Acad. Sci. USA* **79,** 2018–2022.
3. Weterings, P.J., Roelofs, H.M., Jansen, B.A., and Vermorken, A.J. (1983) Serially cultured keratinocytes from human scalp hair follicles: a tool for cytogenetic studies. *Anticancer Res.* **3,** 185–186.
4. Messenger, A.G. (1984) The culture of dermal papilla cells from human hair follicles. *Br. J. Dermatol.* **110,** 685–689.
5. Katsuoka, K., Schell, H., Hornstein, O.P., Deinlein, E., and Wessel, B. (1986) Comparative morphological and growth kinetics studies of human hair bulb papilla cells and root sheath fibroblasts in vitro. *Arch. Dermatol. Res.* **279,** 20–25.
6. Reynolds, A.J., Lawrence, C.M., and Jahoda, C.A. (1993) Human hair follicle germinative epidermal cell culture. *J. Invest. Dermatol.* **101,** 634–638.
7. Tobin, D.J., Colen, S.R., and Bystryn, J.C. (1995) Isolation and long-term culture of human hair-follicle melanocytes. *J. Invest. Dermatol.* **104(1),** 86–89.
8. Michel, M., Torok, N., Godbout, M.J., Lussier, M., Gaudreau, P., et al. (1996) Keratin 19 as a biochemical marker of skin stem cells in vivo and in vitro: keratin 19 expressing cells are differentially localized in function of anatomic sites, and their number varies with donor age and culture stage. *J. Cell. Sci.* **109,** 1017–1028.
9. Philpott, M.P., Green, M.R., and Kealey, T. (1990) Human hair growth in vitro. *J. Cell Sci.* **97,** 463–471.
10. Tobin, D.J. (2005) The biogenesis and growth of human hair. In: Hair in toxicology: An important biomonitor, Tobin, D.J. (ed.). The Royal Society of Chemistry, Cambridge, pp. 3–29.

Chapter 15

Human Endothelial and Osteoblast Co-cultures on 3D Biomaterials

Ronald E. Unger, Sven Halstenberg, Anne Sartoris, and C. James Kirkpatrick

Abstract

Increasingly, in vitro experiments are being used to evaluate the cell compatibility of novel biomaterials. Single cell cultures have been used to determine how well cells attach, grow, and exhibit characteristic functions on these materials and the outcome of such tests is generally accepted as an indicator of biocompatibility. However, organs and tissues are not made up of one cell type and the interaction of cells is known to be an essential factor for physiological cell function. To more accurately examine biomaterials for bone regeneration, we have developed methods to coculture osteoblasts, which are the primary cell type making up bone, and endothelial cells, which form the vasculature supplying cells in the bone with oxygen and nutrients to survive on 2- and 3-D biomaterials.

Key words: Human, Endothelial cell, Osteoblast, Coculture, Biomaterials, Tissue regeneration, Bone

1. Introduction

Many different types of biomaterials are being considered for use in tissue engineering. The ultimate goal of tissue engineering is to replace diseased or damaged tissue with a similar-shaped scaffold structure regenerated in vitro or induced in vivo. This approach uses the patients' or donors' specific cells (stem cells and progenitor cells) grown on a scaffold material in vitro to give a three-dimensional structure which is then implanted into the patient. Once in vivo, cells differentiate or migrate into the biomaterial, blood vessels attach and spread into the "new tissue", and the cells growing on the scaffold recreate the tissue or organ.

In vitro studies using human cell lines or primary human cell isolates are used to examine the cell compatibility of biomaterials (1).

John W. Haycock (ed.), *3D Cell Culture: Methods and Protocols*, Methods in Molecular Biology, vol. 695, DOI 10.1007/978-1-60761-984-0_15, © Springer Science+Business Media, LLC 2011

Generally, these cells in vitro exhibit many characteristics in culture flasks as they do in vivo. In cell compatibility studies, cells are added to 2- or 3-D biomaterials and are compared to those growing on plastic. These single cell cultures have been used to determine how well cells attach, grow, and exhibit characteristic functions on these materials, and the outcome of such tests is generally accepted as an indicator of biocompatibility (2–10). However, organs and tissues are not made up of one cell type and interaction of cells is known to be essential for normal tissue function. To examine biomaterials more accurately for bone regeneration, we have developed methods to coculture osteoblasts, which are the primary cell type making up bone, and endothelial cells, which make up the vasculature, supplying cells in the bone with oxygen and nutrients to survive (11–15).

The successful coculture of cells on 2- and 3-D substrates for extended periods of time with normal growth and functional characteristics requires knowledge of the amount of cells to add, the medium to culture the cells in, and methods to distinguish and follow cells with time. We have developed straightforward methods for the in vitro coculture of primary human osteoblasts and endothelial cells on 3-D biomaterials for extended periods of time. Under these conditions cells exhibit characteristics similar to those observed in vivo and assays for the detection of the cells involve commercially available reagents that can be adapted to a great variety of applications.

2. Materials

2.1. Isolation of Cells

1. HDMEC growth medium: Endothelial Basal Medium MV (PromoCell), 15% fetal bovine serum (FBS), and 100 U/100 μg/mL penicillin/streptomycin (see Note 1).
2. Sodium heparin is dissolved in PBS at 10 μg/mL. Store in aliquots at −20°C.
3. Basic fibroblast growth factor is reconstituted in 20 mM Tris–HCl, pH 7.0, to prepare a stock solution of 25 μg/mL.
4. HDMEC growth medium: Endothelial Basal Medium MV, 10 μg/mL heparin, and 2.5 ng/mL bFGF.
5. Osteoblast growth medium: DMEM 1,000 mg/L glucose, 10% FBS, 2 mM Glutamax I, 100 U/100 μg/mL penicillin/streptomycin, and 75 mg/L ascorbic acid.
6. Trypsin/EDTA is a solution of trypsin (0.25%) and ethylenediamine tetraacetic acid (EDTA, 1 mM).
7. Prepare a 0.2% collagenase working stock solution by dissolving collagenase in 50 mM TES buffer. Aliquot and store at −20°C.

8. Trypsin/versene solution is prepared by diluting 2.5% trypsin into versene to obtain 0.04% trypsin in versene.
9. Phosphate buffered saline (PBS).
10. Dynabeads® CD31 beads are used to select endothelial cells following the manufacturer's protocol.
11. A 0.2% gelatin solution is prepared by dilution of 2% gelatin in water.
12. 0.2% gelatin-coated flasks are prepared by pipeting a 0.2% gelatin solution into a flask, incubating at 37°C for 1 h and then carefully removing excess solution. These flasks can be used immediately or stored at 4°C.
13. 100-μm nylon cell strainer.

2.2. Biomaterial Preparation

1. Sterile porous 3-D biomaterials can be obtained from a number of sources such as Berkeley Advanced Biomaterials (Berkeley, CA, USA), DOT GmbH (Rostock, Germany), or Curasan AG (Kleinostheim, Germany). They are best obtained round or square-shaped so that they fit into 35-mm Petri dishes or into 12- or 24-well plates (8–10 mm in diameter or 5–10 mm^2).
2. Fibronectin 100 μg/mL is prepared by dissolving powder to yield a final concentration of 1 mg/mL. Incubate 30–60 min at 37°C to dissolve. Store at −20°C.
3. Sterile water (endotoxin free) (see Note 2).
4. 70% EtOH is prepared by diluting 100% EtOH with sterile water.
5. 0.1% w/v trypan blue in PBS solution.

2.3. Visualization of Cell Growth

1. Calcein AM (4 mM in anhydrous DMSO) is stored in single use aliquots at −20°C.
2. Cell-culture medium containing 0.1 μM Calcein-AM prepared fresh prior to use (see Note 3).
3. Qtracker® 655 (red) cell-labeling kit.

2.4. Confocal and Fluorescence Microscopy

1. Primary antibody: mouse anti-PECAM-1 (DAKO).
2. Secondary antibody: anti-mouse Alexa Fluor 488.
3. Antibody dilution buffer: PBS supplemented with 3% bovine serum albumin (BSA).
4. Nuclear stain: DAPI (4′,6-diamidino-2-phenylindole, dihydrochloride (DAPI) *FluoroPure™ grade*). To make a 5 mg/mL DAPI stock solution, dissolve 10 mg DAPI in 2 mL of deionized water. Aliquot and store at 2–6°C.
5. Permeabilization solution: 0.5% Triton-X 100 in PBS.
6. Paraformaldehyde: Prepare a 3.7% (w/v) solution of paraformaldehyde in CS buffer (0.1 M PIPES, 1 mM EGTA, 4%

polyethyleneglycol, and 0.1 M NaOH, pH 6.9). The solution may need to be carefully heated (use a stirring hot-plate in a fume hood) to dissolve and then be cooled to room temperature for use. Aliquots may be stored at −20°C and thawed prior to use.

7. Microscope cover slips.
8. Mounting solution: Gelmount.

3. Methods

3.1. Isolation of Cells

1. Osteoblasts are easily isolated from normal human hip-bone tissue obtained from surgical operations. Chop/crush bone into as-small-as-possible pieces using sterile equipment. Keep bone material moist by adding sterile water or medium. Wash 3× with sterile PBS. Resuspend in 0.2% collagenase at 37°C for 30 min then centrifuge and repeat with fresh collagenase. Wash 1× with PBS and then centrifuge suspension at 200 × *g* for 5 min. Resuspend pellet in osteoblast medium and place into a sterile cell-culture flask. After about 2 weeks, osteoblasts will have grown out from bone fragments. When confluent, passage 1:3 with Trypsin/EDTA to remove cells from bone fragments. Figure 1 shows osteoblast cells in culture at various time points after isolation.
2. Human dermal microvascular endothelial cells (HDMEC) are isolated from human foreskin tissue removed at surgery. Foreskin should be worked up as soon as possible. Macroscopic vessels and connective tissue should be removed carefully. The remaining tissue is cut into small (3–5 mm^2) squares and is then placed into 4 mL of 0.4% dispase and incubated overnight at 4°C. The epidermis (darker tissue) is then carefully

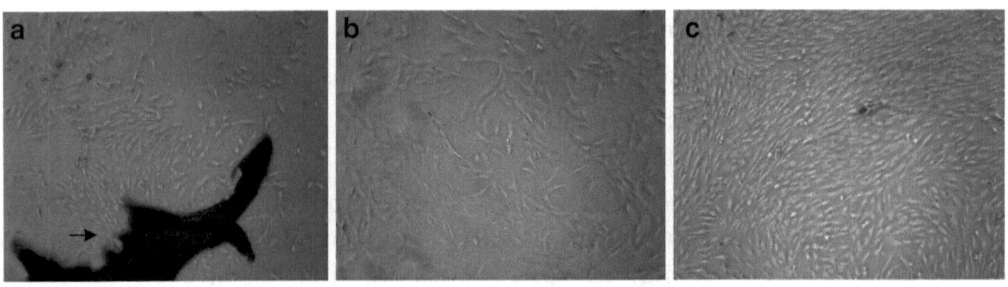

Fig. 1. Images of human osteoblasts in culture. (**a**) Is an image of osteoblasts growing out from a piece of bone fragment (*arrow*) about 5 days after placing into culture. (**b**) Is an image of cells 2 days after trypsinization of cells in (**a**) to remove bone fragments and replating in a new flask. (**c**) Is an image of confluent osteoblast cells after 5–7 additional days in culture.

removed after washing 3× with PBS and discarded. The dermis is placed into 5 mL 0.04% trypsin/versene solution and incubated for 2 h at 37°C. After this, 2 mL of FCS is added to stop the reaction and tissue pieces are placed into 50 mL PBS. Using the side of a scalpel, the tissue is compressed to press out cells. The cell/tissue suspension is then added to a 100-μm cell strainer and centrifuged at 300 × *g* for 5 min. The cell pellet is resuspended in HDMEC growth medium and placed into a 75-cm^2 cell-culture flask that was previously coated with 0.2% gelatin. Cells are passaged 1:3 when confluent, generally after about 7 days, with Trypsin/EDTA.

3.2. Purification of Endothelial Cells

1. To purify endothelial cells from potentially contaminating fibroblasts, endothelial cells are subjected to a Dynabeads® CD31 selection protocol and then plated into new flasks.
2. Add 12 μL of washed Dynabeads per mL of medium and incubate for 20 min at 37°C with gentle shaking.
3. Wash cells 1× with PBS and then add 0.5 mL trypsin/EDTA solution, tilt the flask back and forth two times and then remove the trypsin/EDTA solution. Cells should begin to detach and after 15–60 s, hit the flask firmly with the palm of the hand to dislodge remaining cells and then add PBS with 5% FCS. Wash once by centrifugation and resuspend cells at approximately 2 × 10^6 cells/mL.
4. Increase the solution twofold with PBS and place the tube in a magnet for 2 min.
5. Discard the supernatant and wash the cells 3× with PBS.
6. Finally, after removal of the third supernatant wash, add HDMEC medium, remove tube from magnet and gently pipet the mixture up and down a number of times to distribute cells and place into a new flask previously coated with 0.2% gelatin. At this point, remaining cells are generally greater than 99% endothelial cells. After 5–7 days cells are ready for placing onto biomaterials. Figure 2 shows images of HDMEC at various time points after isolation and purification with Dynabeads (see Note 4).

3.3. Labeling Endothelial Cells with Red Quantum Dots (Optional)

1. At this point, endothelial cells can be specifically labeled with red quantum dots using the Qtracker® 655 cell labeling kit. To prepare a 1 nM labeling solution, mix 1 μL of Qtracker® component A and 1 μL Component B in a 1.5-μL microcentrifuge tube.
2. Incubate for 5 min at room temperature.
3. Add 0.2 mL of fresh endothelial cell-growth medium to the tube and vortex for 30 s.

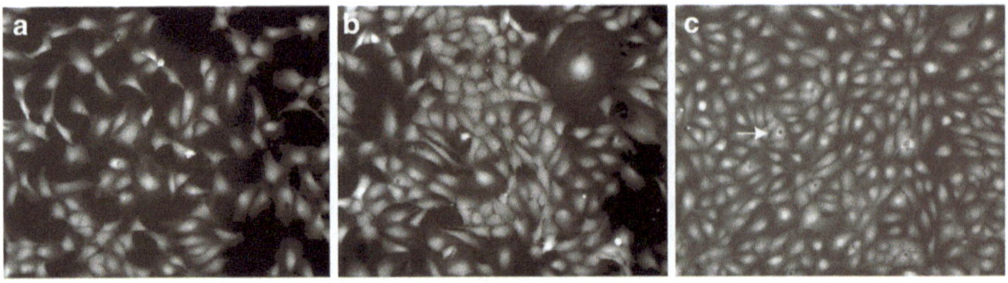

Fig. 2. Images of HDMEC at various timepoints of culture after selection with Dynabeads. (**a**) Is an image of cells 1 day after selection with Dynabeads. After an additional 3 days in culture (**b**), cell replication is evident and cells are beginning to spread out onto the available surface area. (**c**) Shows a confluent monolayer of HDMEC and the *arrow* shows a residual Dynabeads located within a cell.

Fig. 3. Microscopic images of quantum *dot*-labeled endothelial cells growing in a flask. HDMEC were labeled with quantum *dots* and cultured for 1 day and then examined by phase contrast microscopy (**a**), by fluorescent microscopy (**b**) and by phase/fluorescent microscopy. Only cells are visible in (**a**), in (**b**) only quantum *dots* are visible and in (**c**) quantum *dots* can be seen within endothelial cells.

4. Add 1×10^6 cells to this tube (approx. 1 mL, expand all volumes proportionally for labeling of higher numbers of cells).
5. Incubate at 37°C for 60 min and then wash cells 2× with HDMEC medium. Cells are now labeled with red quantum dots. A drop of cells can be examined by fluorescent microscopy. Also, prior to using in coculture experiments it is best to culture endothelial cells for an additional 1–2 days to remove all attached, but not internalized quantum dots. Figure 3 shows images of quantum dots in HDMEC in culture (see Note 5). A more specific way of detecting and specifically identifying HDMEC in coculture is through PECAM-1 staining (see below).

3.4. Preparation of Biomaterials

1. Sterile biomaterials are placed into 35-mm Petri dishes. In certain cases, biomaterials may contain substantial amounts of dust which must be removed by extensive washing of the biomaterials with sterile water or PBS. Non-sterile biomaterials may be sterilized with 70% EtOH for 15–30 min, followed by rinsing with water.
2. Biomaterials may require treatment with extracellular matrix proteins for endothelial cells to attach. To determine whether

this is necessary, HDMEC are added to the biomaterial, incubated for 4 h, and then the sample is removed, stained with calcein-AM and examined by confocal microscopy or fluorescent microscopy. If cells have not attached and appear rounded-up, then biomaterials must be pre-treated with an extracellular matrix protein (for example, fibronectin, below).

3. Fibronectin (100 μg/mL) is added to cover the biomaterial and incubated for 1 h at 37°C. After this, the biomaterial is removed from the solution and placed into the 35-mm Petri dish and cells are added.

3.5. Counting Cells

1. HDMEC (quantum dot-labeled or unlabeled) in the third or fourth passage, or osteoblasts are removed from the plastic culture flasks with trypsin/EDTA, washed once, resuspended in cell-culture medium, and counted with a cell counting chamber/hemocytometer (improved Neubauer).
2. Dilute 0.2 mL of cell suspension in 0.2 mL trypan blue. Viable cells remain unstained and non-viable cells take up the dye and are stained blue.
3. Add the coverslip to the hemocytometer and fill both counting chambers of the hemocytometer chamber with the diluted, stained cell suspension.
4. View cells under the microscope with a 40× objective and count viable cells in each of the four corner 1 mm^2 squares and then repeat for the second side of the chamber (total 8×1 mm^2). Cell suspensions should be dilute enough so that the total count is approximately 100 cells.
5. The calculation of cell density is based on the volume underneath the cover glass. To get the final count in cells/mL, first divide the total count by 0.1 (chamber depth in mm) then divide the result by the total surface area (mm^2) counted. For example, if a total of 100 cells were counted in the four large corner squares altogether, divide 100 by 0.1, then divide the result by 4 mm^2 squares, which is the total area counted (each large square is 1 mm^2), i.e., $100/0.1 = 1{,}000$ and $1{,}000/4 = 250$ cells/mm^3. There are 1,000 mm^3/mL, therefore, the solution contains 250,000 cells/mL. If a 1:2 dilution was initially prepared, then the final concentration of cells is 500,000 cells/mL.

3.6. Adding Cells to Biomaterials

1. Endothelial cells can either be added alone or together with osteoblasts. The endothelial cells (maximum passage number 4) are mixed with primary osteoblasts (passage 4) in HDMEC medium (see Note 6).
2. The number of endothelial cells added to a biomaterial depends on the shape of the biomaterials: on a 1×1 cm biomaterial in a 24-well plate, add 150,000 HDMEC; on 1-cm

discs in a 48-well plate, add 100,000 HDMEC; and in 35-mm cell-culture plates, add 300,000 HDMEC.

3. The optimal ratios for mixing cells are between 5:1 and 10:1 HDMEC:osteoblast. Depending on biomaterial shape, mixtures of cells are added to biomaterials in cell numbers as follows: on 1 × 1 cm in a 24-well plate (130,000 HDMEC and 20,000 osteoblast cells), on 1-cm discs in a 48-well plate (85,000 HDMEC and 15,000 osteoblast cells), and in 35-mm cell-culture plates (260,000 HDMEC and 40,000 osteoblast cells).
4. After 4 h, biomaterials are gently removed from the wells and placed into six-well plates with 3 mL fresh medium (see Note 7).
5. Medium is changed regularly at 3–4 day intervals. All incubations are carried out in a humidified atmosphere at 37°C (5% CO_2).

3.7. Visualization of Cells on Biomaterials

1. In cocultures, cell growth can be followed by phase contrast microscopy and some distinction between osteoblasts and endothelial cell locations (red quantum-dot-labeled cells, if labeled) can be observed directly by a combination of phase contrast and fluorescent microscopy or by confocal microscopy (Fig. 4a, b).
2. Biomaterials with cells are placed into medium containing Calcein-AM so that both HDMEC and osteoblasts will be stained. Calcein-AM becomes fluorescent when taken up by viable cells and the fluorescence is spread throughout the cell.
3. The stained samples are then placed onto a microscope slide and examined by confocal laser scanning microscopy (CLSM: Leica TCS NT). In this case all viable cells stain green and

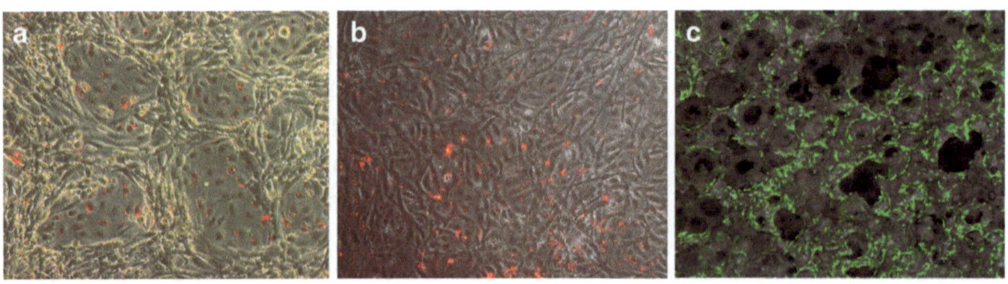

Fig. 4. Images of cocultures on cell culture plastic and bone biomaterial. (**a**) and (**b**) Are a combination of fluorescent and phase contrast images of HDMEC labeled with quantum *dots* and cocultured with osteoblasts. Although some information on cell identity can be gained, little detail can be observed. (**c**) Is a confocal image of a calcein-AM stained coculture of HDMEC and osteoblasts imposed on a reflected image of the porous hydroxyapatite biomaterial (see also Fig. 6b). No differentiation between the two cell types can be made, but it can be seen that cells are spreading over the entire surface and into the pores of the biomaterial. Thus this method can be used to follow cell spread on a biomaterial with time but not to identify individual cell types.

endothelial cells can be identified due to the presence of the red quantum dots in the cytoplasm (Fig. 4c).

3.8. Visualization of Endothelial Cells on Biomaterials

1. To identify endothelial cells and microcapillary-like structure formation HDMEC can be specifically stained with anti-PECAM-1 (Fig. 5a, b).
2. Cells on biomaterials are rinsed briefly with PBS and then fixed with paraformaldehyde for 15 min at room temperature, washed 4× with PBS and then permeabilized with 0.5% Triton-X 100

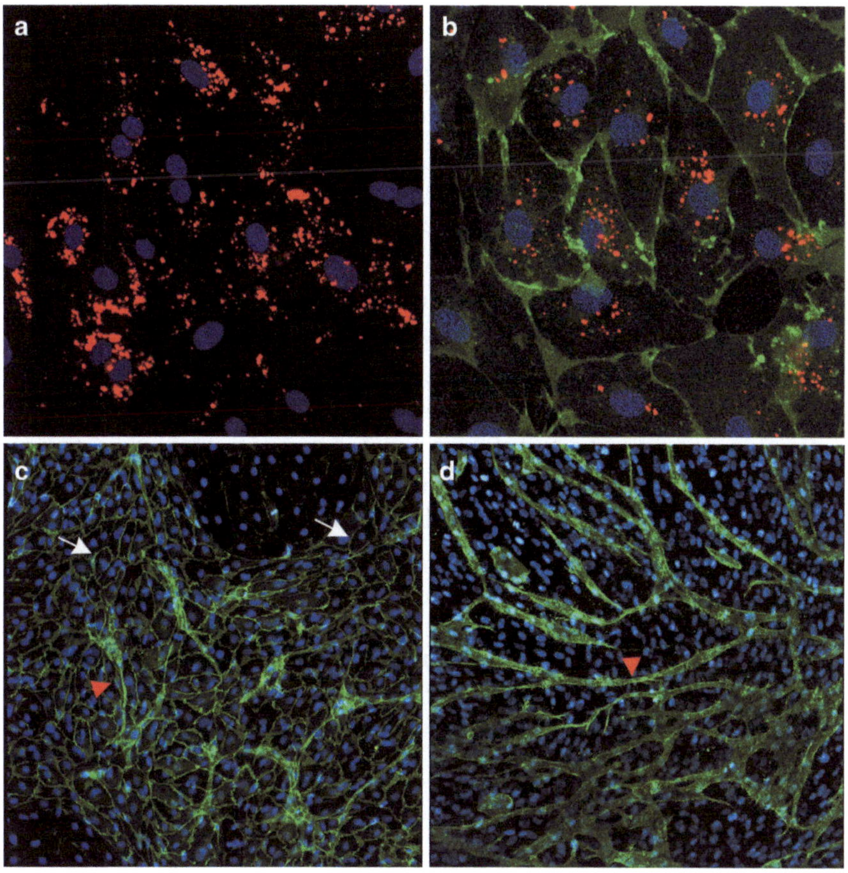

Fig. 5. Confocal images of HDEMC in single and coculture on cell culture plastic. (**a**) Is an image of HDMEC containing quantum *dots* (*red*) grown on plastic, fixed in paraformaldehyde, and stained with DAPI in order to visualize nuclei (*blue*). (**b**) Is similar to (**a**) with an additional staining for endothelial cell PECAM-1 (*green*). PECAM-1 is localized at endothelial cell–cell contacts and this staining method is useful to visualize endothelial cell shape. (**c**) and (**d**) Are images of cocultures of HDMEC and HOS on day 14 and 28 after begin of coculture. Staining is with PECAM-1 (*green*, endothelial cells) and DAPI (*blue*, nuclei). No HOS-specific stain is available and therefore these cells were not counterstained. As can be seen in (**c**) many endothelial cells are present, some still as a monolayer with a typical PECAM-1 staining pattern (*white arrows*) as seen in (**b**) and some beginning to form microcapillary-like structures (*red arrowhead*). In (**d**) most endothelial cells have migrated to form microcapillary-like structures containing nuclei. Remaining nuclei without PECAM-1 staining are HOS cells.

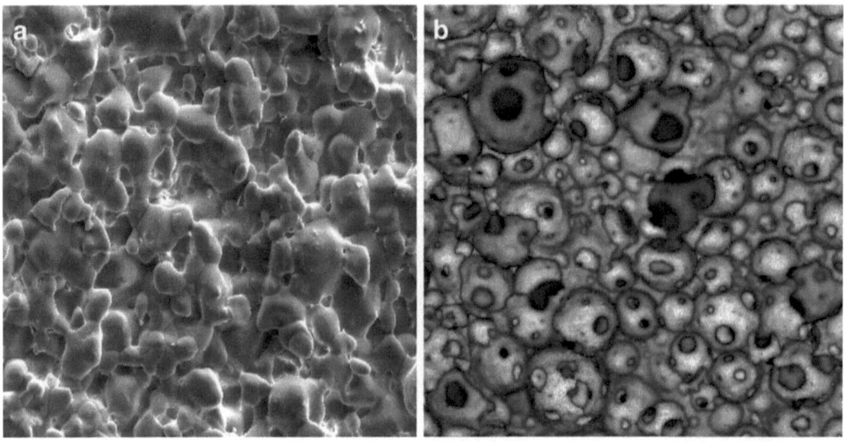

Fig. 6. Scanning electron micrographs of 3-D biomaterials. (**a**) Is an image of porous tri-calcium phosphate (Curasan, Germany) and (**b**) is an image of porous hydroxyapatite (HiPor, England).

for 10 min. Figure 6 shows SEM images of two commonly available biomaterials used in bone tissue engineering.

3. After washing four times with PBS, anti-PECAM-1 antibody is added and incubated overnight at 4°C.
4. The biomaterials are then washed four times with PBS and then anti-mouse Alexa Fluor 488 is added and incubated for 1 h at room temperature (see Note 8).
5. This is followed by a brief rinse in PBS after which the nuclei are stained with Hoechst 33342 solution (1 μg/mL) for 5 min followed by washing with PBS.
6. Biomaterials are then placed in a 35-mm plastic Petri plate, Gelmount is added and a coverslip is placed on top of the biomaterial. The biomaterials can then be examined by CLSM (Fig. 5c, d, 2-D cultures on cell-culture plastic and Fig. 7, 3-D cocultures on biomaterials).

4. Notes

1. We have examined a number of different culture media for the growth of HDMEC and have found the Endothelial Basal Medium MV to be the best. However, we have also found that good growth and expression of phenotypes of HDMEC in this medium is highly dependent on the source of FBS. Therefore, we recommend getting a number of different lots and sources of FBS, testing one or two isolations of HDMEC in the various media for growth, morphology, and phenotypes and then selecting the FBS for further studies, which gives the best results. This will ensure reproducible isolations

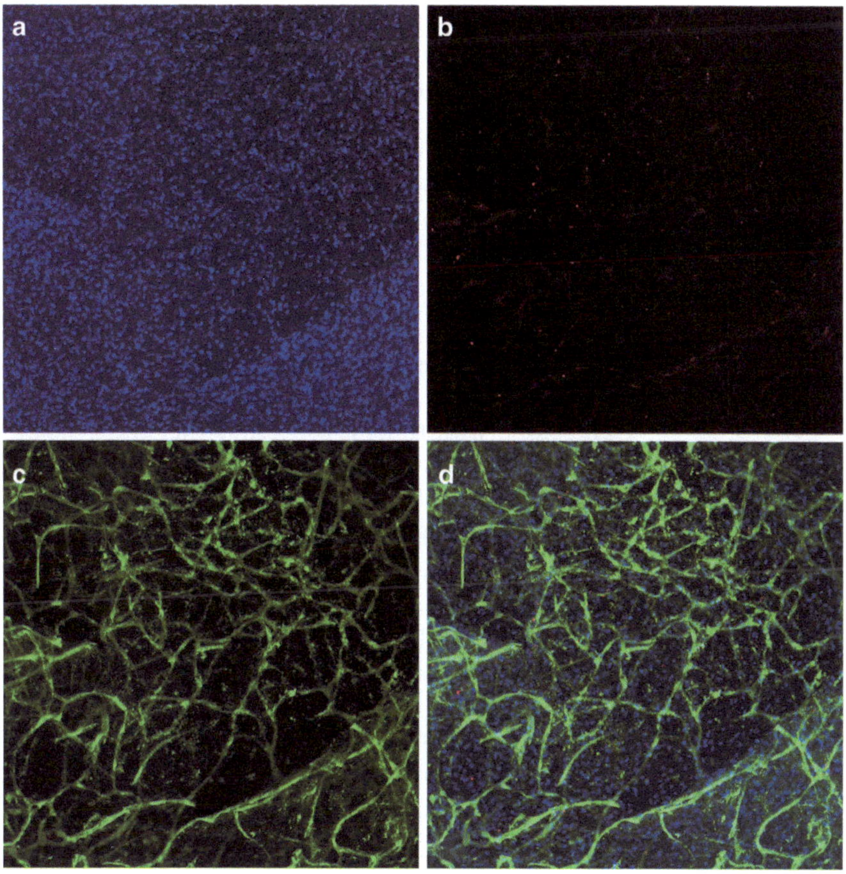

Fig. 7. Confocal images of one view of a coculture of HDMEC labeled with quantum *dots* and HOS on tri-calcium phosphate shown in Fig. 6a. Cells were cultured for 28 days on the biomaterial and then fixed and stained. (**a**) Is a DAPI stain showing only nuclei, (**b**) is an image showing quantum *dots*, (**c**) is an image of cells stained with PECAM-1 and (**d**) is a superimposed image of (**a**), (**b**) and (**c**). All images show an identical plane of focus. As can be seen in (**a**) many cells are present. In (**b**) few quantum *dots* are observed and this may be due to the length of culture time, i.e., the continuous division of endothelial cells with less and less quantum *dots*, resulting finally in numbers of quantum *dots* in cells below the level of detection. In (**c**), endothelial cells were stained with PECAM-1 and have formed capillary-like structures, and image (**d**) shows these capillary-like structures with respect to the total number of cells indicated by the *blue* nuclei. It can be seen that many nuclei lie between the capillary-like structures, indicating the presence of osteoblasts.

and populations of cells with similar characteristics from different donors.

2. It is very important to use endotoxin-free water and other reagents for the culture of endothelial cells. Even very small amounts of endotoxin can result in the induction of a cascade of proteins in endothelial cells. These cells will then not be exhibiting normal endothelial cell phenotype in culture and endotoxin may disrupt structure and functions of the cells.
3. Unused medium should not be stored longer than 1 day and then be used again. Calcein-AM becomes inactive with time. It is best to prepare fresh and use directly.

4. We have found that the Dynabeads® CD31 Beads when used following the manufacturers description result in highly pure populations of endothelial cells free from any potentially contaminating cells. It is best to test cells after the selection, for example, for the presence of vWF, to determine the purity of the culture. If contaminating cells are still present, it is possible to carry out a second CD31 bead selection. However, this should be done before contaminating cells are given too much time to grow, preferably no longer than 24–48 h after plating from the first CD31 bead selection.
5. Quantum dots are very intense, but with each division of a cell, the degree of label within each cell becomes less. Therefore, with time, it is possible that either quantum dots are diluted so much by division that detection is not possible or that some cells do not distribute quantum dots to daughter cells, resulting in cells that are not labeled.
6. HDMEC are primary cells. This means that they are not a cell line and have a limited life-span. We have found that characteristics of HDMEC begin to change after the fifth passage. In addition, prior to two passages, cells are not as homogenous as they are in passage 3 and 4. Therefore, for the most consistent results we use cells in the third or fourth passage.
7. Biomaterials are removed at this point to separate cells attached to the biomaterials from cells that have not attached and fallen to the side and have attached to the cell-culture plastic. These cells would also continue to grow and both the small size of the well and the growth of cells on biomaterial and plastic would require frequent changing of medium. Therefore, the biomaterials are removed after cell attachment and are placed into a larger well with excess medium.
8. We have found that the above sequence of incubation times of the primary and secondary antibody works best in most cases. However, depending on the biomaterial and cell densities, some adjustments in time may be necessary to get the best penetration and staining with the antibodies. In this case, a number of samples may need to be prepared and different concentrations of antibodies and/or incubation times need to be tested.

Acknowledgments

This work was supported by the German Federal Ministry of Education and Research (Ref. Nr. 0313405C) and the NoE EXPERTISSUES (Contract No. 500283-2) from the EU. We also wish to thank DOT GmbH and Curasan AG for generously supplying us with the biomaterials used in these studies.

References

1. Kirkpatrick, C. J., Fuchs, S., Iris Hermanns, M., Peters, K., and Unger, R. E. (2007) Cell culture models of higher complexity in tissue engineering and regenerative medicine. *Biomaterials* **28**, 5193–5198.
2. Campillo-Fernandez, A. J., Unger, R. E., Peters, K., Halstenberg, S., Santos, M., Sanchez, M. S., Duenas, J. M., Pradas, M. M., Ribelles, J. L., and Kirkpatrick, C. J. (2008) Analysis of the biological response of endothelial and fibroblast cells cultured on synthetic scaffolds with various hydrophilic/hydrophobic ratios: Influence of fibronectin adsorption and conformation. *Tissue Eng. Part A* **5**, 1331–1341.
3. Dubruel, P., Unger, R., Vlierberghe, S. V., Cnudde, V., Jacobs, P. J., Schacht, E., and Kirkpatrick, C. J. (2007) Porous gelatin hydrogels: 2. In vitro cell interaction study. *Biomacromolecules* **8**, 338–344.
4. Santos, M. I., Fuchs, S., Gomes, M. E., Unger, R. E., Reis, R. L., and Kirkpatrick, C. J. (2007) Response of micro- and macrovascular endothelial cells to starch-based fiber meshes for bone tissue engineering. *Biomaterials* **28**, 240–248.
5. Santos, M. I., Tuzlakoglu, K., Fuchs, S., Gomes, M. E., Peters, K., Unger, R. E., Piskin, E., Reis, R. L., and Kirkpatrick, C. J. (2008) Endothelial cell colonization and angiogenic potential of combined nano- and micro-fibrous scaffolds for bone tissue engineering. *Biomaterials* **29**, 4306–4313.
6. Thimm, B. W., Unger, R. E., Neumann, H. G., and Kirkpatrick, C. J. (2008) Biocompatibility studies of endothelial cells on a novel calcium phosphate/SiO2-xerogel composite for bone tissue engineering. *Biomed. Mater.* **3**, 15007.
7. Unger, R. E., Huang, Q., Peters, K., Protzer, D., Paul, D., and Kirkpatrick, C. J. (2005) Growth of human cells on polyethersulfone (PES) hollow fiber membranes. *Biomaterials* **26**, 1877–1884.
8. Unger, R. E., Krump-Konvalinkova, V., Peters, K., and Kirkpatrick, C. J. (2002) In vitro expression of the endothelial phenotype: comparative study of primary isolated cells and cell lines, including the novel cell line HPMEC-ST1.6R. *Microvasc. Res.* **64**, 384–397.
9. Unger, R. E., Peters, K., Huang, Q., Funk, A., Paul, D., and Kirkpatrick, C. J. (2005) Vascularization and gene regulation of human endothelial cells growing on porous polyethersulfone (PES) hollow fiber membranes. *Biomaterials* **26**, 3461–3469.
10. Unger, R. E., Wolf, M., Peters, K., Motta, A., Migliaresi, C., and James Kirkpatrick, C. (2004) Growth of human cells on a non-woven silk fibroin net: a potential for use in tissue engineering. *Biomaterials* **25**, 1069–1075.
11. Bondar, B., Fuchs, S., Motta, A., Migliaresi, C., and Kirkpatrick, C. J. (2008) Functionality of endothelial cells on silk fibroin nets: comparative study of micro- and nanometric fibre size. *Biomaterials* **29**, 561–572.
12. Fuchs, S., Ghanaati, S., Orth, C., Barbeck, M., Kolbe, M., Hofmann, A., Eblenkamp, M., Gomes, M., Reis, R. L., and Kirkpatrick, C. J. (2009) Contribution of outgrowth endothelial cells from human peripheral blood on in vivo vascularization of bone tissue engineered constructs based on starch polycaprolactone scaffolds. *Biomaterials* **30**, 526–534.
13. Fuchs, S., Hofmann, A., and Kirkpatrick, C. J. (2007) Microvessel-like structures from outgrowth endothelial cells from human peripheral blood in 2-dimensional and 3-dimensional co-cultures with osteoblastic lineage cells. *Tissue Eng.* **13**, 2577–2588.
14. Fuchs, S., Jiang, X., Schmidt, H., Dohle, E., Ghanaati, S., Orth, C., Hofmann, A., Motta, A., Migliaresi, C., and Kirkpatrick, C. J. (2009) Dynamic processes involved in the pre-vascularization of silk fibroin constructs for bone regeneration using outgrowth endothelial cells. *Biomaterials* **30**, 1329–1338.
15. Unger, R. E., Sartoris, A., Peters, K., Motta, A., Migliaresi, C., Kunkel, M., Bulnheim, U., Rychly, J., and Kirkpatrick, C. J. (2007) Tissue-like self-assembly in cocultures of endothelial cells and osteoblasts and the formation of microcapillary-like structures on three-dimensional porous biomaterials. *Biomaterials* **28**, 3965–3976.

Chapter 16

Assessment of Nanomaterials Cytotoxicity and Internalization

Noha M. Zaki and Nicola Tirelli

Abstract

The impact that nanotechnology may have on life and medical sciences is immense and includes novel therapies as much as novel diagnostic and imaging tools, often offering the possibility to combine the two. It is, therefore, of the essence to understand and control the interactions that nanomaterials can have with cells, first at an individual level, focusing on, e.g., binding and internalization events, and then at a tissue level, where diffusion and long-range transport add further complications. Here, we present experimental methods based on selective labeling techniques and the use of effectors for a qualitative and quantitative evaluation of endocytic phenomena involving nanoparticles. The understanding of the cell–material interactions arising from these tests can then form the basis for a model-based evaluation of nanoparticles behavior in 3D tissues.

Key words: Nanotechnology, Cytotoxicity, Endocytosis, Microfluorimetry, Internalization, Endocytic-effectors

1. Introduction

Nanoparticles or other forms of nanocarriers can be used as means to deliver active payloads, such as low MW drugs, genetic material or other complex biopharmaceutical agents, performing their action in the extracellular environment or intracellularly. The latter mode of action is modulated and sometimes hampered by the presence of a barrier to nanoparticles by passive diffusion, i.e. the cell membrane: the phospholipid bilayer is naturally impermeable to molecules of large molecular weight. The possibility for a substance to permeate through a membrane depends both on its diffusion coefficient and on its solubility and the latter depends mostly on the hydrophilic/hydrophobic balance for low molecular weight

John W. Haycock (ed.), *3D Cell Culture: Methods and Protocols*, Methods in Molecular Biology, vol. 695, DOI 10.1007/978-1-60761-984-0_16, © Springer Science+Business Media, LLC 2011

compounds; on the other hand, the molar mass becomes the essential factor for larger molecules (approximately anything larger than 1 kDa), the reason being the drop in the entropic contribution to solubility with increasing molecular weight. However, cells possess a variety of active (energy dependent) internalization mechanisms to accommodate the entry of large objects.

The surface properties of nanocarriers play a crucial role in determining their interactions with cells and, as a result, the possibility and the mechanism of intracellular transport, which significantly affects the efficacy of encapsulated drugs (1). A nanocarrier can be internalized simply because of its physical proximity to the membrane or because of interactions with specific receptors, originating via a number of different internalization paths (2). In both cases, the cell membrane will invaginate to engulf these colloidal objects and a variable amount of extracellular fluid in an intracellular vesicle, which takes the generic name of an (early) endosome, The formation and subsequent trafficking of endosomes through the cell is a process generically known as endocytosis, although a number of different endocytic processes can be recognized (Fig. 1). Once inside the cell, the intracellular fate of the endosomal contents is an important determinant of successful delivery of the cargo. Depending on the membrane interactions and on the nature of the components involved in vesicle formation, endosomes will mature into acidic vesicles

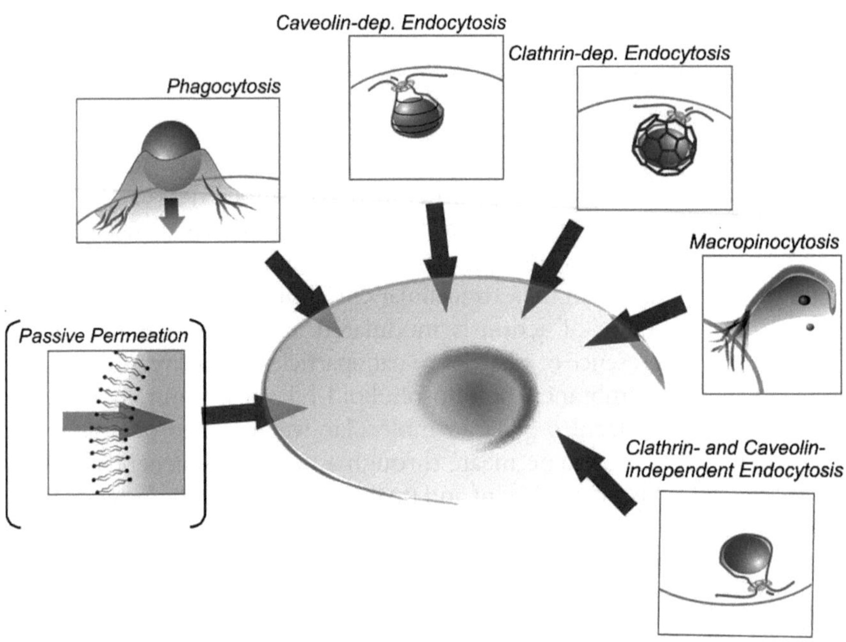

Fig. 1. Different portals of entry into the cell. Passive permeation is substantially excused for macromolecules or colloidal (nano-) carriers. An essential review is provided in (2).

which may or may not fuse with lysosomes, where hydrolytic and enzymatic reactions may lead to the complete destruction of the macromolecular or nanocarrier material (3, 4).

The methods described herein provide essential information about the interactions of nanoparticles with cells grown in culture; specifically, we focus on the evaluation of cytotoxic effects, the quantification of cellular uptake and the assessment of its mechanism.

Two points to note: (a) The formazan assay is most widely used for studying the cytotoxicity of nanosystems (5–9) and we have developed a modification for testing high concentrations of nanoparticles, overcoming problems that arise from possible osmotic shock. (b) Cellular uptake can be conveniently observed and quantified employing fluorescently labeled nanoparticles (10–15); however, unequivocal data of intracellular fluorescence emitted by the engulfed particles can be obtained only after an effective quenching of the extracellular fluorescence using a nonmembrane-permeable probe, in our case trypan blue.

2. Materials

2.1. Cell Culture

1. Semiadherent murine macrophage J774.2 cells (ECACC no. 85011428, Rockville, UK) cultured in Dulbecco's modified Eagle's medium (DMEM) (Gibco/BRL, Bethesda, MD), 10% fetal calf serum (FCS), 10% glutamine, 450 U/mL penicillin, and 420 μg/mL streptomycin. Other cell lines, e.g., mouse fibroblasts L929 may be used. We use only about ten passages (i.e., cells within a defined interval of passages). We use macrophages (J774.2 cells) between passage numbers 5 and 15 and mouse fibroblasts L929 between passage numbers 10 and 20 (see Note 1).
2. Teflon cell scrapers (Fisher, UK).
3. Phosphate-buffered saline (PBS), Dulbecco's, without Ca, Mg, 1× (Gibco/Invitrogen, 500 mL).
4. FCS; not heat inactivated, store at −70°C (long term)/−20°C (short term). Thaw in fridge, aliquot after thawing and store aliquots at −20°C.
5. Powder DMEM (Sigma-Aldrich, Gillingham, UK), 10% FCS, and 2 mM glutamine. Dissolve one vial containing powder DMEM in 50 mL FCS, add 50 mL water until complete dissolution using a magnet stirrer and sterilize by filtration through sterile Millipore filter (0.22 μm).
6. Trypsin solution 2.5% (10×) (Gibco/Invitrogen, 100 mL).

7. EDTA 2% (sodium salt) solution (ICN Biomedicals).
8. Hemocytometer (cell-number counter bright-line/dark-line counting chambers 0.1 μL volume, Hausser Scientific, UK).
9. Nanoparticles (NPs): Nanoparticles of any composition may be used. Our description specifically refers to chitosan-triphosphate nanoparticles with positive Zeta potential (+50 mV) and sizes in the range 100–300 nm. It is essential that the nanoparticle size and size distribution is checked under cell culture conditions (i.e., in serum or other culture medium), in order to highlight possible agglomeration phenomena.

2.2. Confocal Microscopy

1. Microscope cover slips (22 mm × 22 mm × 0.15 mm) from Fisher, Pittsburgh, PA.
2. Fixation solution: Methanol-free formaldehyde, prepare a 4% (w/v) solution in DMEM fresh for each experiment (Fisher, Pittsburgh, PA).
3. Postuptake quench solution: Trypan blue, final concentration = 0.25 mg/mL in PBS pH 7.4, by dilution of stock solution 0.4% (Sigma, UK).
4. Permeabilization solution: 0.5% (v/v) Triton X-100 in PBS.
5. Nuclear stain: 300 nM DAPI (4,6-diamidino-2-phenylindole) dilactate in PBS (Sigma, UK).
6. Lysosomal stain: 50 nM Lysotracker red-DND99 in PBS; excitation 577 nm, emission 590 nm (Molecular Probes, Eugene, OR, USA).
7. Actin stain: Texas red-phalloidin (Molecular Probes, Eugene, OR, USA), dissolve vial in 1.5 mL methanol to give a stock solution, store at −20°C.
8. Mounting medium: Antifade (Molecular Probes, Eugene, OR, USA).
9. More generally, a list of most common stains is provided in Table 1.

2.3. Cellular Uptake by Microfluorimetry

1. Fluorescently labeled NPs (e.g., isothiocyanate-labeled chitosan NPs) dispersed in water (see Note 2) (concentrations at least one order of magnitude lower than the recorded IC50 for any given kind of NP). Store in the dark, in the fridge.
2. 96-well microplate (black with transparent bottom) with cover and fluorescence microplate reader capable of measuring fluorescence emission at ~520 nm with an excitation at ~480 nm.
3. Cell-viability assay reagents.

Table 1
Different fluorescent markers/probes commonly used for studying internalization processes

Marker	Application/relevance	Excitation λ_{max} (nm)	Emission λ_{max} (nm)	References
Dextran 70,000, Texas red labeled	Macropinocytosis	595	615	(14)
Lucifer yellow	Internalized via noncoated vesicles/fluid phase marker	428	535	(16)
Chlolera Toxin, subunit B, Alexa 594 labeled	Caveolae-mediated endocytosis	589	616	(17)
Transferrin, Alexa 594 labeled	Clathrin-mediated endocytosis	589	617	(18)
Lysotracker red-DND99	Late endosomes and lysosomes	577	590	(19, 20)
Phalloidin, Texas red labeled	Actin filaments	595	615	(21)
DAPI	Nuclear stain	358	461	(22, 23)
Concanavalin A, Alexa Fluor® 594	Cell surface glycoprotein	590	617	(24)
FITC (or others)	Fluorescent label for NPs	490	518	(25–28)
Trypan blue	Extracellular probe (quench noninternalized FITC-labeled nanoparticles fluorescent signals)	525	–	(29–32)

4. DMSO >99.9% (ACS reagent, Sigma), sterile filtrated across hydrophobic fluorophore (PTFE) membrane for fine particle removal from organic solvents 0.2 μm Millex FG (Millipore), stored at −20°C in sterile vials.
5. Endocytosis effectors (see Table 2): For example, nocodazole, bafilomycin A1, cytochalasin D, and filipin are dissolved in DMSO to give a stock solution of 1 μg/mL; store at −20°C in the dark. Appropriate dilution in culture medium is done at the time of the experiment.
6. Potassium-free buffer containing 140 mM NaCl, 20 mM HEPES, 1 mM $CaCl_2$, 1 mM, $MgCl_2$, 1 mg/mL D-glucose calibrated to pH 7.4, store in the fridge.

Table 2
Effectors/pharmacological treatments for studying different endocytic pathways

Endocytosis effector	Modus operandi	Dose/preincubation time	References
Chlorpromazine Hypertonic challenge Intracellular K+ depletion	Inhibits clathrin-mediated endocytosis	65 µg/mL (0.45 M sucrose) (K+ free buffer) (30 min – 37°C)	(33) (34) (35)
Filipin Nystatin	Inhibits caveolae-mediated and caveolae-like endocytosis by binding to 3β-hydroxysterols (distinguish between clathrin and caveolae)	5 µg/mL – 30 min –37°C	(29, 34, 36)
LY294002 Dimethylamiloride Ethylisopropylamiloride Wortmannin	Inhibits macropinocytosis [selective inhibitors of all mammalian isoforms of phosphoinositide 3-kinase (PI 3-kinase)). Inhibits the Na+/H+ exchange protein in the plasma membrane causing cytosolic acidification	20 µM 100 µM 100 µM 10 nM (60 min – 37°C)	(37) (38) (39) (40)
4°C Sodium azide	Inhibits all energy-dependent processes	100 mM (30 min – 4°C)	(41)
Chlororquine Bafilomycin A1 Brefeldin A	Inhibits acidification of endosomes (affects the clathrin pathway in its very late stage). Useful to highlight phenomena of acidity-induced endosomal escape	200 µM 200 nM 100 nM (30 min – 37°C)	(42) (43) (44)
Cytochalasin D (actin filament barbed-end capping) Latrunculin B (actin dimer sequestering and severing) Swinholide A (actin monomer sequestering)	Depolymerization of actin; inhibits membrane ruffling	1 µM/30 min – 37°C 1 µM/30 min – 37°C 10 nM/1 h – 37°C	(45)
Nocodazole Colchicine	Disruption of microtubules	0.1 µg/mL; 30 min – 37°C 1 µM/2 h – 37°C	(18, 46, 47) (48–50)

3. Methods

3.1. Cytotoxicity of Nanoparticles

For cell culture studies, unpurified (e.g., commercially available) NPs should not be used owing to the nonphysiological medium in which NPs are dispersed/prepared and the possible residual presence of reactants, catalysts, initiators, or other deleterious reagents which could be cytotoxic. To obtain reliable and reproducible results, we generally follow a three-step protocol. NP dispersions should be purified, e.g., membrane or hollow fiber dialysis or ultrafiltration, optimizing the conditions (pH and temperature) for different kinds of NPs. Subsequently, they are freeze dried in the presence of a cryoprotectant [generally a sugar, such as sucrose, or poly(vinyl alcohol)] to avoid aggregation upon redispersion. Ultimately, the colloids are dispersed in the least volume of water and dialyzed against water to remove sucrose then redispersed in culture medium with the pH and osmotic pressure of the dispersion adjusted before incubation with cells. High concentrations of the colloidal dispersion are prepared to serve as stock from which appropriate dilutions are made at the time of the experiment.

1. Freeze dry the NPs in presence of 1% sucrose solution as a cryoprotectant (see Note 3).
2. Redisperse the NPs in the least amount of MilliQ water (pH 6), dialyze against water to remove the sucrose (membranes with molecular weight cut-off = 50 kDa) and then sterilize by UV irradiation overnight. This would serve as a stock NPs dispersion.
3. Split the cells by trypsinization (fibroblasts L929) or mechanical scraping (macrophages J774.2), centrifuge; disperse the pellet by addition of complete culture medium to give an exact count of 1×10^5 cells/mL.
4. Seed cells in 96-well microtiter plate at density of 10,000 cells/well (i.e., 100 μL of cell suspension in culture medium) and incubate cells for 24 h to allow attachment to plate.
5. In order to test high concentrations of NPs, prepare five times concentrated culture medium using powder DMEM.
6. Prepare final dispersions of NPs to be tested on the cells under LAF bench by mixing one part 5× culture medium and four parts of the different NPs dispersion on a vortex mixer. The isotonicity should be checked with an osmometer (see Note 4). Prepare the control by mixing one part 5× culture medium with four parts sterile MilliQ water.
7. Carefully remove the culture medium on top of the cells in the 96-well plate and replace it with 100 μL of the different NP dispersions and control. Incubate for 24 h at 37°C in a humid atmosphere with 5% CO_2.

8. Carefully remove the colloidal dispersion by gentle aspiration (vacuum suction can detach macrophages J774.2 cells) and wash the cells three times with prewarmed PBS pH 7.4.
9. Add 100 μL of MTT solution (0.5 mg/mL in DMEM) to each well and incubate the cells for 4 h.
10. Remove the stain by very gentle aspiration (to avoid suction of formazan crystals).
11. Add 100 μL of sterile DMSO to each well and shake the plate for 5 min to solubilize the formazan crystals.
12. Measure the absorbance (A) at 550 nm in a microtiter plate reader.
13. Cell viability is calculated by the following equation:

$$\text{Cell viability } (\%) = A_{\text{test}} / A_{\text{control}} \times 100$$

where A_{test} is the absorbance of the cells incubated with the different NP dispersions and $A_{control}$ is the absorbance of the cells incubated with the culture medium only (positive control). IC_{50}, the drug concentration at which inhibition of 50% cell growth occurs in comparison with that of the control sample, is calculated by the curve fitting of the cell viability data.

3.2. Qualitative Assessment of Nanoparticle Internalization by Confocal Microscopy

These studies should be performed at concentrations of at least one order of magnitude lower than the recorded IC_{50} for any given kind of NP, in order to provide results that are applicable to fully viable cells. Even lower concentrations should be used if the effect of cell activation on the internalization efficiency is to be avoided.

1. Preheat culture medium at 37°C for at least 30 min before use.
2. Scrape macrophage J774.2/trypsinized fibroblast L929 cells on the day before uptake experiment, count them using a hemocytometer and dilute the cells to exactly 10^5 cells/mL in complete culture medium.
3. Seed cells on sterile cover slips (22 mm × 22 mm × 1.5 mm) placed in six-well plate by adding 1-mL cell suspension to each well under laminar flow bench (LAF bench). Coverslips are sterilized in a hot air oven at 200°C for 30 min. It might be necessary to coat the coverslips to improve cell attachment to the glass (see Note 5).
4. Cover the six-well plate and transfer to CO_2 incubator in a humid atmosphere at 37°C for 24 h to allow cell attachment to coverslips.
5. Carefully remove the culture medium on top of the cells and replace it with 750 μL solution of different effectors (see Notes 6 and 7) of appropriate concentration, incubate for the appropriate time (Table 2).

6. Add 250 μL of NP dispersion (such that the final concentration in each well is at least one order of magnitude lower than the recorded IC_{50} for any given kind of NPs). Incubate for 30 min in a humid atmosphere at 37°C in a CO_2 incubator.
7. Carefully remove the colloidal dispersion by gentle aspiration. Immediately add 1 mL trypan blue solution to each well, leave in contact with cells for 1 min.
8. Wash cells at least four times with pre-warmed phosphate-buffered saline, pH 7.4 (PBS).
9. Fix the sample in 4% formaldehyde solution in complete DMEM for 20–30 min at room temperature (see Notes 8 and 9).
10. Rinse three or more times with PBS.
11. Permeabilize the sample in 0.05% Triton X-100 in PBS for 5 min at room temperature.
12. Wash three or more times with PBS.
13. For actin staining with Texas red-phalloidin, dilute 5 μL methanolic stock solution into 200 μL PBS for each coverslip to be stained (see Note 10). The staining solution should be placed on the coverslip for 20 min at room temperature (if a temperature of 37°C is used it is useful to keep the coverslips inside a covered container during the incubation to avoid evaporation).
14. Wash at least three times with PBS.
15. For nuclear staining, add 500 μL DAPI solution in PBS on the coverslips, make sure that the cells are completely covered. Leave for 3 min at room temperature.
16. Wash at least three times with PBS.
17. The samples are then ready to be mounted. The coverslip is carefully inverted into a drop of mounting medium on a microscope slide.
18. Samples can be stored in the dark at 4°C for up to 1 month.
19. The slides are viewed under phase contrast microscopy (to locate the cells and identify the focal plane) and under confocal microscopy. Excitation at 485 nm induces the FITC fluorescence (green emission) for the labeled NPs, while excitation at 595 and 364 nm induces Texas red (red emission) and DAPI fluorescence (blue emission), respectively. ImageJ® software can be used to overlay the phase contrast and fluorescence images. Each condition is performed in triplicate.
20. Examples of the confocal images showing internalization of NPs are shown in Fig. 2.
21. For lysosomal staining, dilute 1 mM LysoTracker Red DND-99 stock solution to the final working concentration 50 nM in the culture medium. Add 200 μL to each well and incubate

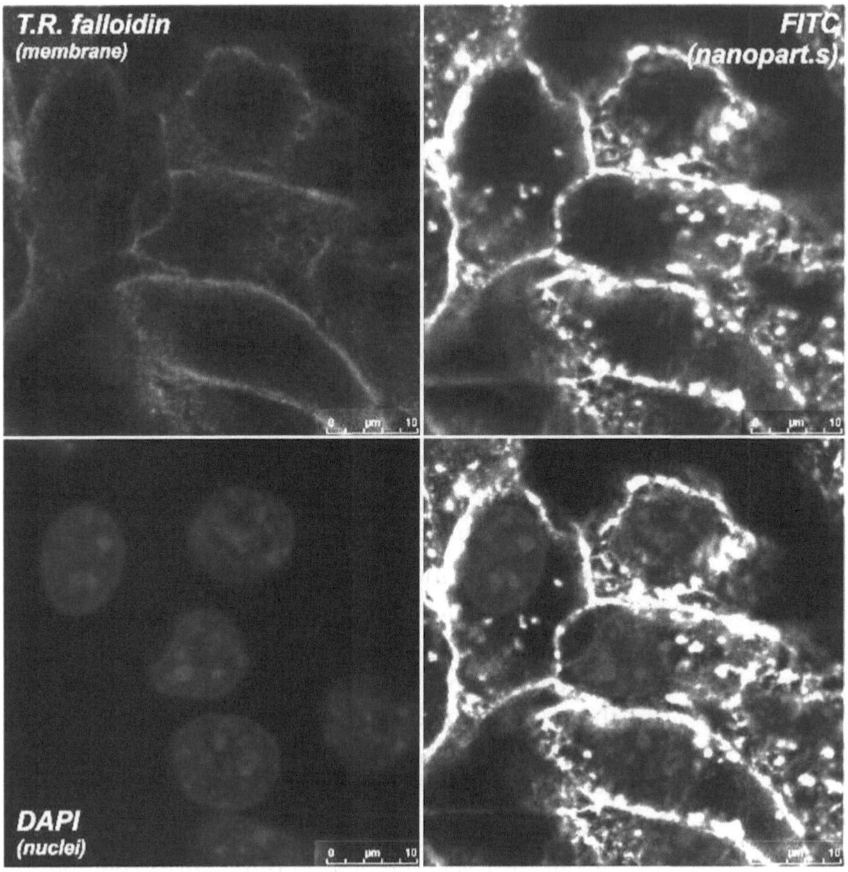

Fig. 2. Internalization of FITC-labeled chitosan nanoparticles by fibroblast L929 cells stained with Texas red-phalloidin for actin to reveal cell borders and DAPI for nuclear staining, showing localization of nanoparticles in the cytoplasm.

with cells in presence of NPs dispersion for 30 min after step 6 then bypass steps 9–16 and continue as before. Examples of the confocal images showing endolysosomal escape of chitosan-tripolyphosphate NPs are shown in Fig. 3.

3.3. Quantitative Assessment of Nanoparticle Internalization by Microfluorimetry

After a qualitative assessment and localization, the internalization can be quantified by measuring the emission intensity of the nanoparticles in the cell body. The main assumption of this part of the study is that the internalization has not dramatically altered the emission properties of the fluorescently labeled NPs. This may indeed happen either because of chemical degradation in endosomal compartments or because of agglomeration that causes a local increase of the NP concentration, and resulting partial auto-absorption of the emitted radiation.

The first point should be checked by running parallel experiments in the presence of bafilomycin or other effectors capable to

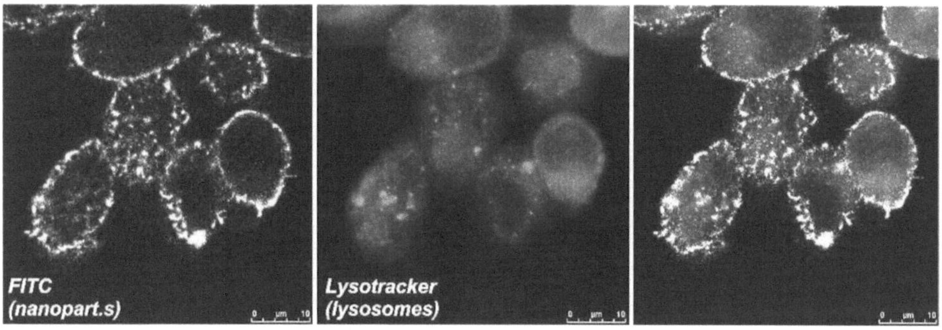

Fig. 3. Confocal images of L929 fibroblasts stained with Lysotracker red-DND99 to reveal late endosomes and lysosomes. FITC-labeled nanoparticles are localized intracellularly because they are not quenched by the action of membrane-impermeable trypan blue but are not present in lysosomal compartments; hence, an endosomal escape path should be supposed.

inhibit the proton pump in early endosomes (see Table 2), which therefore can minimize possible acidity-induced damage to fluorophores. Possible agglomeration should be checked by monitoring the size of the nanoparticles in the presence of cell lysates prior to running the experiments, and how the NPs fluorescence may be affected by it.

1. Subculture macrophages J774.2/fibroblast L929 cells (or other preferred cell types) in complete DMEM.
2. Harvest the cells prior to use by scraping/trypsinizing them from the surface of the tissue culture dish.
3. Centrifuge the cell suspension and resuspend the cell pellet with the DMEM.
4. Determine the cell viability using cell-viability assay. The cell suspension for use should have >90% viability.
5. The final cell concentration is adjusted to 10^5/mL by adding DMEM to the suspension.
6. The negative-control wells are prepared by adding 150 μL of DMEM to six wells on the microplate.
7. The positive-control and experimental wells are prepared by first pipetting 100 μL of the adjusted cell suspension into 45 wells on the microplate.
8. The loaded microplate is covered and transferred to the incubator for 24 h to allow the cells to adhere to the plate surface.
9. The positive-control wells are further prepared by adding 50 μL of the DMEM to six of the cell-containing wells.
10. Finally the experimental wells are completed by adding 50 μL of the endocytosis effector at desired concentrations in

DMEM to the remaining cell-containing wells. In order to minimize effects of experimental errors, at least six replicate tests for each experimental condition are made.

11. Add 50 μL of the prepared fluorescent NP suspension to all the negative-control, positive-control, and experimental wells.
12. The microplate is covered and transferred to the CO_2 incubator for 30 min (in case of quickly internalizable NPs, e.g., positively charged chitosan NPs) or for 2 h (in case of other, more slowly internalizable NPs). In order to further optimize the time of incubation, a separate time course experiment is needed.
13. The NP suspension is then removed from all of the microplate wells by gentle aspiration.
14. The wells are then rinsed three times with cold PBS (to terminate the experiment) followed by 20 min incubation at 4°C with 100 μL of 0.5% Triton X-100 in 0.2 N NaOH to lyse the cells. Caution: avoid cold PBS with macrophages J774.2 (see Note 11).
15. The amount of fluorescence of the cell lysate in each well is then measured using a fluorescence plate reader (TECAN Safire, Tecan Austria GmbH, Grödig, Austria) with excitation wavelength at 485 nm and emission wavelength at 520 nm [assuming the fluorophore used for NP labeling is fluorescein (i.e., FITC)].
16. Cell-bound and internalized NPs are quantified by making use of a calibration curve obtained with fluorescently labeled NPs in a cell lysate solution (10^6 untreated fibroblast L929 cells dissolved in 1 mL of the Triton X-100 solution).
17. The protein content of the cell lysate in each well was determined using the QuantiPro micro BCA protein assay kit (Sigma, MO, USA).
18. Uptake is expressed as the amount (μg) of NPs associated by unit weight (mg) of cellular protein (see Notes 12 and 13).
19. The nanoparticle uptake by cells should be done also at different colloid concentrations, in step 11.
20. To determine the kinetics of NP uptake by cells, different incubation times 0, 10, 15, 30, 45, 60, 90, and 120 min should be used, in step 12. For slowly internalizable nanoparticles times of up to 1 day can be used.
21. To determine whether nanoparticle uptake by cells is energy dependent, repeat the experiment but incubate at 4°C, in step 12.

4. Notes

1. After thawing a vial of macrophage J774.2 or fibroblast L929 cells from the stock, cultivate the cells for two passages before seeding cells for the uptake experiments to stabilize the cell phenotype.
2. Unless stated otherwise, all solutions should be prepared in water with a resistivity of 18.2 MΩ cm and total organic content of less than five parts per billion. This standard is referred to as "water" in this text.
3. Freeze drying is used to concentrate the NPs after purification by appropriate techniques viz. membrane or hollow fiber dialysis, ultrafiltration, or gel filtration chromatography.
4. The isotonicity of nanoparticles dispersion should be checked with a calibrated osmometer and, if necessary, adjusted with mannitol to 290–310 mOsm. The pH should be almost physiological (pH 7–7.6) unless this adversely affects the nanoparticles stability.
5. For some cell lines it might be necessary to coat the cover slips with poly-L-lysine to maximize the cell attachment. Coating is done by dipping cover slips in sterile filtered 0.1% poly-L-lysine solution in PBS and leaving to dry under a LAF bench.
6. Endocytosis effectors that are sparingly water-soluble compounds are often dissolved in DMSO. In this case, it is important to ensure that the compounds do not precipitate after dilution in culture medium. DMSO may affect the cell monolayer viability and/or integrity. It is therefore recommended that the maximum final concentration of DMSO should not exceed 1%.
7. It is advisable to ensure that the cell monolayers tolerate the co-solvent at the concentration applied. This is done by conducting an uptake experiment in the absence and in the presence of the co-solvent. Note that DMSO might interact with membrane proteins.
8. Methanol can disrupt actin during the fixation process. Therefore, it is best to avoid any methanol-containing fixatives. The preferred fixative is methanol-free formaldehyde.
9. The use of formaldehyde (4%) diluted in DMEM (instead of buffer) can help preserve cytoskeletal ultrastructure.
10. Nonspecific background staining with phalloidin can be reduced by addition of 1% bovine serum albumin (BSA) to the staining solution. Hence, preincubate fixed cells with PBS containing 1% BSA or with the Image-iT™ FX signal enhancer

(Molecular probes) for 20–30 min prior to adding the phalloidin staining solution.

11. To stop nanoparticle uptake by fibroblast L929 cells, ice-cold PBS should be used for rinsing after removal of the nanoparticle dispersion. If macrophages J774.2 are used ice-cold PBS should be avoided since it causes the detachment of this semi-adherent cell line. Pre-warmed PBS can be used instead.
12. The cell uptake efficiency, which is the ratio between the amount of particles taken up in cells and the amount of those in the control, can be calculated from the following equation:

$$\text{Uptake efficiency }(\%) = \frac{I_{\text{sample}} - I_{\text{negative}}}{I_{\text{postive}} - I_{\text{negative}}} \times 100$$

where I_{sample}, I_{positive}, and I_{negative} are the fluorescence intensities of the sample, positive control, and negative control, respectively.

13. Cellular uptake has also been expressed as the cellular uptake efficiency (51, 52). The cellular uptake is assessed quantitatively using flow cytometry (14, 53–56), microfluorometry (28, 51, 57, 58), or by quantitative extraction of the markers from the cells (12, 45, 59).

Acknowledgments

This work was supported by a Science and Technology Development Fund (STDF) grant to Noha M. Zaki.

References

1. Labhasetwar, V. (2005) Nanotechnology for drug and gene therapy: the importance of understanding molecular mechanisms of delivery. *Curr. Opin. Biotechnol.* **16**, 674–680.
2. Conner, S. D. and Schmid, S. L. (2003) Regulated portals of entry into the cell. *Nature* **422**, 37–44.
3. Surti, N., Naik, S., Bagchi, T., Garcion, E., and Misra, A. (2008) Intracellular delivery of nanoparticles with biological systems. *Biomaterials* **9**, 217–223.
4. Basarkar, A. and Misra, A. (2008) Intracellular delivery of nanoparticles of an antiasthmatic drug. *AAPS PharmSciTech* **9**, 217–223.
5. Lemarchand, C., Gref, R., Passirani, C., Garcion, E., Petri, B., Muller, R., Costantini, D., and Couvreur, P. (2006) Influence of polysaccharide coating on the interactions of nanoparticles with biological systems. *Biomaterials* **27**, 108–118.
6. Basarkar, A., Devineni, D., Palaniappan, R., and Singh, J. (2007) Preparation, characterization, cytotoxicity and transfection efficiency of poly(D,L-lactide-co-glycolide) and poly(dl-lactic acid) cationic nanoparticles for controlled delivery of plasmid DNA. *Int. J. Pharm.* **343**, 247–254.
7. Lee, M.-K., Lim, S.-J., and Kim, C.-K. (2007) Preparation, characterization and in vitro cytotoxicity of paclitaxel-loaded sterically stabilized solid lipid nanoparticles. *Biomaterials* **28**, 2137–2146.
8. Lin, W., Huang, Y.-W., Zhou, X.-D., and Ma, Y. (2006) In vitro toxicity of silica nanoparticles in human lung cancer cells. *Toxicol. Appl. Pharmacol.* **217**, 252–259.

9. Azarmi, S., Tao, X., Chen, H., Wang, Z., Finlay, W. H., Lobenberg, R., and Roa, W. H. (2006) Formulation and cytotoxicity of doxorubicin nanoparticles carried by dry powder aerosol particles. *Int. J. Pharm.* **319**, 155–161.
10. Panyam, J., Sahoo, S. K., Prabha, S., Bargar, T., and Labhasetwar, V. (2003) Fluorescence and electron microscopy probes for cellular and tissue uptake of poly(D,L-lactide-co-glycolide) nanoparticles. *Int. J. Pharm.* **262**, 1–11.
11. Panyam, J., Zhou, W. Z., Prabha, S., Sahoo, S. K., and Labhasetwar, V. (2002) Rapid endo-lysosomal escape of poly(D,L-lactide-co-glycolide) nanoparticles: implications for drug and gene delivery. *FASEB J.* **16**, 1217–1226.
12. Davda, J. and Labhasetwar, V. (2002) Characterization of nanoparticle uptake by endothelial cells. *Int. J. Pharm.* **233**, 51–59.
13. Kawaura, C., Noguchi, A., Furuno, T., and Nakanishi, M. (1998) Atomic force microscopy for studying gene transfection mediated by cationic liposomes with a cationic cholesterol derivative. *FEBS Lett.* **421**, 69–72.
14. Huth, U. S., Schubert, R., and Peschka-Suss, R. (2006) Investigating the uptake and intracellular fate of pH-sensitive liposomes by flow cytometry and spectral bio-imaging, *J. Control Release* **110**, 490–504.
15. Simoes, S., Slepushkin, V., Duzgunes, N., and Pedroso de Lima, M. C. (2001) On the mechanisms of internalization and intracellular delivery mediated by pH-sensitive liposomes. *Biochim. Biophys. Acta* **1515**, 23–37.
16. Panyam, J. and Labhasetwar, V. (2003) Dynamics of endocytosis and exocytosis of poly(D,L-lactide-co-glycolide) nanoparticles in vascular smooth muscle cells. *Pharm. Res.* **20**, 212–220.
17. Huth, U. S., Schubert, R., and Peschka-Süss, R. (2006) Investigating the uptake and intracellular fate of pH-sensitive liposomes by flow cytometry and spectral bio-imaging. *J. Control Release* **110**, 490–504.
18. van der Aa, M. A., Huth, U. S., Hafele, S. Y., Schubert, R., Oosting, R. S., Mastrobattista, E., Hennink, W. E., Peschka-Suss, R., Koning, G. A., and Crommelin, D. J. (2007) Cellular uptake of cationic polymer-DNA complexes via caveolae plays a pivotal role in gene transfection in COS-7 cells. *Pharm. Res.* **24**, 1590–1598.
19. Nayak, S., Lee, H., Chmielewski, J., and Lyon, L. A. (2004) Folate-mediated cell targeting and cytotoxicity using thermoresponsive microgels. *J. Am. Chem. Soc.* **126**, 10258–10259.
20. de Diesbach, P., N'Kuli, F., Berens, C., Sonveaux, E., Monsigny, M., Roche, A. C., and Courtoy, P. J. (2002) Receptor-mediated endocytosis of phosphodiester oligonucleotides in the HepG2 cell line: evidence for non-conventional intracellular trafficking. *Nucleic Acids Res.* **30**, 1512–1521.
21. des Rieux, A., Ragnarsson, E. G., Gullberg, E., Preat, V., Schneider, Y. J., and Artursson, P. (2005) Transport of nanoparticles across an in vitro model of the human intestinal follicle associated epithelium. *Eur. J. Pharm. Sci.* **25**, 455–465.
22. Kim, S. H., Choi, H. J., Lee, K. W., Hong, N. H., Sung, B. H., Choi, K. Y., Kim, S. M., Chang, S., Eom, S. H., and Song, W. K. (2006) Interaction of SPIN90 with syndapin is implicated in clathrin-mediated endocytic pathway in fibroblasts. *Genes Cells* **11**, 1197–1211.
23. Kahn, E., Menetrier, F., Vejux, A., Montange, T., Dumas, D., Riedinger, J. M., Frouin, F., Tourneur, Y., Brau, F., Stoltz, J. F., and Lizard, G. (2006) Flow cytometry and spectral imaging multiphoton microscopy analysis of CD36 expression with quantum dots 605 of untreated and 7-ketocholesterol-treated human monocytic cells. *Anal. Quant. Cytol. Histol.* **28**, 316–330.
24. Mo, Y. and Lim, L. Y. (2005) Preparation and in vitro anticancer activity of wheat germ agglutinin (WGA)-conjugated PLGA nanoparticles loaded with paclitaxel and isopropyl myristate. *J. Control Release* **107**, 30–42.
25. Yumoto, R., Nishikawa, H., Okamoto, M., Katayama, H., Nagai, J., and Takano, M. (2006) Clathrin-mediated endocytosis of FITC-albumin in alveolar type II epithelial cell line RLE-6TN. *Am. J. Physiol. Lung Cell Mol. Physiol.* **290**, L946–L955.
26. Amidi, M., Romeijn, S. G., Borchard, G., Junginger, H. E., Hennink, W. E., and Jiskoot, W. (2006) Preparation and characterization of protein-loaded N-trimethyl chitosan nanoparticles as nasal delivery system. *J. Control Release* **111**, 107–116.
27. Cherukuri, A., Frye, J., French, T., Durack, G., and Voss, E. W. Jr. (1998) FITC-poly-D-lysine conjugates as fluorescent probes to quantify hapten-specific macrophage receptor binding and uptake kinetics. *Cytometry* **31**, 110–124.
28. Huang, M., Ma, Z., Khor, E., and Lim, L. Y. (2002) Uptake of FITC-chitosan nanoparticles by A549 cells. *Pharm. Res.* **19**, 1488–1494.

29. Rejman, J., Bragonzi, A., and Conese, M. (2005) Role of clathrin- and caveolae-mediated endocytosis in gene transfer mediated by lipo- and polyplexes. *Mol. Ther.* **12**, 468–474.
30. Huang, M., Khor, E., and Lim, L. Y. (2004) Uptake and cytotoxicity of chitosan molecules and nanoparticles: effects of molecular weight and degree of deacetylation. *Pharm. Res.* **21**, 344–353.
31. Ma, Z. and Lim, L. Y. (2003) Uptake of chitosan and associated insulin in Caco-2 cell monolayers: a comparison between chitosan molecules and chitosan nanoparticles. *Pharm. Res.* **20**, 1812–1819.
32. Serpe, L., Guido, M., Canaparo, R., Muntoni, E., Cavalli, R., Panzanelli, P., Della Pepal, C., Bargoni, A., Mauro, A., Gasco, M. R., Eandi, M., and Zara, G. P. (2006) Intracellular accumulation and cytotoxicity of doxorubicin with different pharmaceutical formulations in human cancer cell lines. *J. Nanosci. Nanotechnol.* **6**, 3062–3069.
33. Gotte, M., Sofeu Feugaing, D. D., and Kresse, H. (2004) Biglycan is internalized via a chlorpromazine-sensitive route. *Cell Mol. Biol. Lett.* **9**, 475–481.
34. Qaddoumi, M. G., Gukasyan, H. J., Davda, J., Labhasetwar, V., Kim, K. J., and Lee, V. H. (2003) Clathrin and caveolin-1 expression in primary pigmented rabbit conjunctival epithelial cells: role in PLGA nanoparticle endocytosis. *Mol. Vis.* **9**, 559–568.
35. Bauer, I. W., Li, S. P., Han, Y. C., Yuan, L., and Yin, M. Z. (2008) Internalization of hydroxyapatite nanoparticles in liver cancer cells. *J. Mater. Sci. Mater. Med.* **19**, 1091–1095.
36. Manunta, M., Nichols, B. J., Tan, P. H., Sagoo, P., Harper, J., and George, A. J. (2006) Gene delivery by dendrimers operates via different pathways in different cells, but is enhanced by the presence of caveolin. *J. Immunol. Methods* **314**, 134–146.
37. Corvera, S. and Czech, M. P. (1998) Direct targets of phosphoinositide 3-kinase products in membrane traffic and signal transduction. *Trends Cell Biol.* **8**, 442–446.
38. Shepherd, P. R., Reaves, B. J., and Davidson, H. W. (1996) Phosphoinositide 3-kinases and membrane traffic. *Trends Cell Biol.* **6**, 92–97.
39. Nakase, I., Niwa, M., Takeuchi, T., Sonomura, K., Kawabata, N., Koike, Y., Takehashi, M., Tanaka, S., Ueda, K., Simpson, J. C., Jones, A. T., Sugiura, Y., and Futaki, S. (2004) Cellular uptake of arginine-rich peptides: roles for macropinocytosis and actin rearrangement. *Mol. Ther.* **10**, 1011–1022.
40. Ui, M., Okada, T., Hazeki, K., and Hazeki, O. (1995) Wortmannin as a unique probe for an intracellular signalling protein, phosphoinositide 3-kinase. *Trends Biochem. Sci.* **20**, 303–307.
41. Enriquez de Salamanca, A., Diebold, Y., Calonge, M., Garcia-Vazquez, C., Callejo, S., Vila, A., and Alonso, M. J. (2006) Chitosan nanoparticles as a potential drug delivery system for the ocular surface: toxicity, uptake mechanism and *in vivo* tolerance. *Invest. Ophthalmol. Vis. Sci.* **47**, 1416–1425.
42. Dijkstra, J., Van Galen, M., and Scherphof, G. L. (1984) Effects of ammonium chloride and chloroquine on endocytic uptake of liposomes by Kupffer cells *in vitro*. *Biochim. Biophys. Acta. Mol. Cell Res.* **804**, 58–67.
43. Issa, M. M., Koping-Hoggard, M., Tommeraas, K., Varum, K. M., Christensen, B. E., Strand, S. P., and Artursson, P. (2006) Targeted gene delivery with trisaccharide-substituted chitosan oligomers in vitro and after lung administration *in vivo*. *J. Control Release* **115**, 103–112.
44. Hunziker, W., Whitney, J. A., and Mellman, I. (1992) Brefeldin A and the endocytic pathway. Possible implications for membrane traffic and sorting. *FEBS Lett.* **307**, 93–96.
45. Qaddoumi, M. G., Ueda, H., Yang, J., Davda, J., Labhasetwar, V., and Lee, V. H. (2004) The characteristics and mechanisms of uptake of PLGA nanoparticles in rabbit conjunctival epithelial cell layers. *Pharm. Res.* **21**, 641–648.
46. Goncalves, C., Mennesson, E., Fuchs, R., Gorvel, J. P., Midoux, P., and Pichon, C. (2004) Macropinocytosis of polyplexes and recycling of plasmid via the clathrin-dependent pathway impair the transfection efficiency of human hepatocarcinoma cells. *Mol. Ther.* **10**, 373–385.
47. Hashimoto, M., Morimoto, M., Saimoto, H., Shigemasa, Y., and Sato, T. (2006) Lactosylated chitosan for DNA delivery into hepatocytes: the effect of lactosylation on the physicochemical properties and intracellular trafficking of pDNA/chitosan complexes. *Bioconj. Chem.* **17**, 309–316.
48. Dijkstra, J., van Galen, M., and Scherphof, G. (1985) Effects of (dihydro)cytochalasin B, colchicine, monensin and trifluoperazine on uptake and processing of liposomes by Kupffer cells in culture. *Biochim. Biophys. Acta* **845**, 34–42.
49. Ramge, P., Unger, R. E., Oltrogge, J. B., Zenker, D., Begley, D., Kreuter, J., and Von Briesen, H. (2000) Polysorbate-80 coating enhances uptake of polybutylcyanoacrylate

(PBCA)-nanoparticles by human and bovine primary brain capillary endothelial cells. *Eur. J. Neurosci.* **12**, 1931–1940.

50. Boyd, A. E. III, Bolton, W. E., and Brinkley, B. R. (1982) Microtubules and beta cell function: effect of colchicine on microtubules and insulin secretion in vitro by mouse beta cells. *J. Cell Biol.* **92**, 425–434.
51. Hu, Y., Xie, J., Tong, Y. W., and Wang, C.-H. (2007) Effect of PEG conformation and particle size on the cellular uptake efficiency of nanoparticles with the HepG2 cells. *J. Control Release* **118**, 7–17.
52. Zhang, Z., Huey Lee, S., and Feng, S. S. (2007) Folate-decorated poly(lactide-co-glycolide)-vitamin E TPGS nanoparticles for targeted drug delivery. *Biomaterials* **28**, 1889–1899.
53. Park, J. S., Han, T. H., Lee, K. Y., Han, S. S., Hwang, J. J., Moon, D. H., Kim, S. Y., and Cho, Y. W. (2006) N-acetyl histidine-conjugated glycol chitosan self-assembled nanoparticles for intracytoplasmic delivery of drugs: endocytosis, exocytosis and drug release. *J. Control Release* **115**, 37–45.
54. Coester, C., Nayyar, P., and Samuel, J. (2006) In vitro uptake of gelatin nanoparticles by murine dendritic cells and their intracellular localisation. *Eur. J. Pharm. Biopharm.* **62**, 306–314.
55. Richard, J. P., Melikov, K., Brooks, H., Prevot, P., Lebleu, B., and Chernomordik, L. V. (2005) Cellular uptake of unconjugated TAT peptide involves clathrin-dependent endocytosis and heparan sulfate receptors. *J. Biol. Chem.* **280**, 15300–15306.
56. Perumal, O. P., Inapagolla, R., Kannan, S., and Kannan, R. M. (2008) The effect of surface functionality on cellular trafficking of dendrimers. *Biomaterials* **29**, 3469–3476.
57. Win, K. Y. and Feng, S. S. (2005) Effects of particle size and surface coating on cellular uptake of polymeric nanoparticles for oral delivery of anticancer drugs. *Biomaterials* **26**, 2713–2722.
58. Russell-Jones, G. J., Arthur, L., and Walker, H. (1999) Vitamin B12-mediated transport of nanoparticles across Caco-2 cells. *Int. J. Pharm.* **179**, 247–255.
59. Chavanpatil, M. D., Khdair, A., and Panyam, J. (2007) Surfactant-polymer nanoparticles: a novel platform for sustained and enhanced cellular delivery of water-soluble molecules. *Pharm. Res.* **24**, 803–810.

Chapter 17

Practical Aspects of OCT Imaging in Tissue Engineering

Stephen J. Matcher

Abstract

Optical coherence tomography (OCT) is a non-destructive, non-invasive imaging modality conceptually similar to ultrasound imaging but uses near-infrared radiation rather than sound. It is attracting interest throughout the medical community as a tool for ophthalmic scanning (especially of the retina) and potentially for the diagnosis of many other illnesses such as epithelial cancer, connective tissue disorders, and atherosclerosis, as well as for surgical guidance. More recently, it has begun to be explored as a tool for the real-time monitoring of the growth and development of tissue-engineered products. OCT has certain unique advantages over traditional confocal microscopy; in particular, it can image to depths measured in hundreds of microns rather than tens of microns in intact biological tissues and with working distances in excess of 1 cm. Also it possesses label-free contrast for imaging ordered collagen (via birefringence), flow velocity and local shear-rate (via Doppler shifts), and sub-cellular structure (via coherent speckle contrast). The purpose of this short review is to introduce OCT technology and also give guidelines on its practical implementation to the interested researcher.

Key words: Optical coherence tomography, Optical imaging, Tissue engineering, Fluid shear stress, Extracellular matrix

1. Introduction

OCT is a rapidly emerging optical imaging modality that has already established itself as the method of choice for retinal scanning in vivo and which has the potential to make a similar impact in many other areas (1). OCT is attracting interest because, as Fig. 1 shows, it fills a void in a parameter map of imaging resolution vs. imaging depth that lies between established tools such as confocal microscopy and ultrasound/MRI. OCT is a 3D optical imaging tool similar to confocal microscopy in certain regard. Confocal microscopy achieves resolution in the third dimension (depth) by using "confocal" gating. A high numerical aperture

John W. Haycock (ed.), *3D Cell Culture: Methods and Protocols*, Methods in Molecular Biology, vol. 695,
DOI 10.1007/978-1-60761-984-0_17, © Springer Science+Business Media, LLC 2011

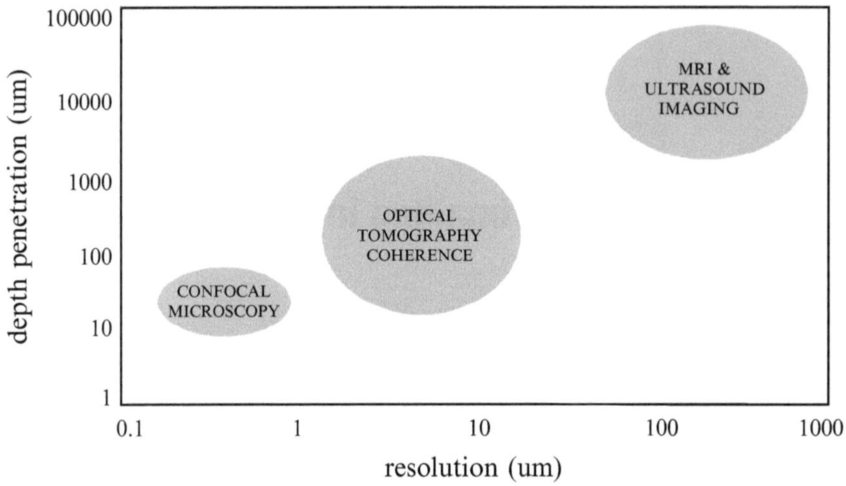

Fig. 1. Comparison of depth penetration vs. depth resolution for OCT and other common bioimaging modalities.

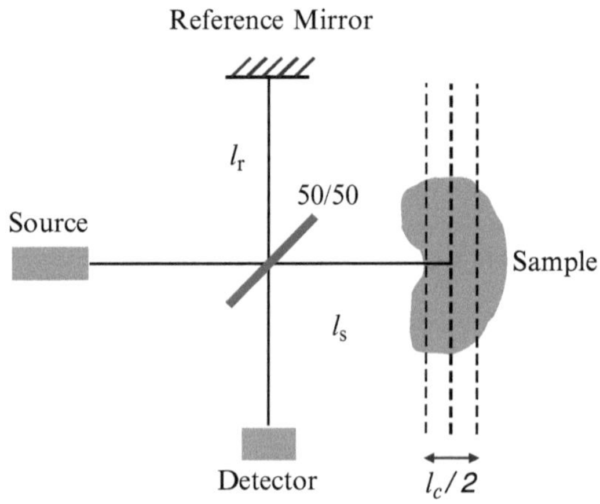

Fig. 2. Schematic diagram of a time-domain OCT system.

(NA) microscope objective and pinhole arrangement is used to ensure that only light back-scattered from a specific depth in the sample is efficiently detected. OCT achieves its depth discrimination by a completely different means: "coherence gating".

Figure 2 shows a schematic of a typical OCT system, which is basically a Michelson Interferometer. An incident light wave from the source is split into two identical copies by a beam splitter. One beam travels to the sample whilst the other travels to a "reference mirror." Waves back-scattered from structures at different depths in the sample are combined with the beam reflected from the reference mirror onto a detector. The crucial feature is that the light from the source has a low "coherence length", which means that the beam effectively consists of a train of independent "wave packets." It turns out that a "coherent" signal is only formed on the detector when the same wave packet from the sample and reference beams overlap temporally on the detector.

Hence, only structures in the sample that are located exactly the same distance from the beam splitter as the reference mirror (to a precision given by the wave-packet length) are detected: the signal from shallower and deeper structures is rejected by the "coherence gate." By analogy with US imaging, OCT scans can be 3D volume images (C-scans), 2D slices through a volume (B-scans), or 1D depth-resolved scans at a point (A-scans). An OCT A-scan is the most fundamental image and is acquired with the illuminating light beam held at one position on the tissue surface and the depth information obtained by scanning the coherence gate down into the tissue. In first-generation OCT systems designed in the mid-1990s, this was accomplished by mechanically moving the reference mirror. B scans are assembled from A scans, and C scans from B scans, by raster-scanning the light beam over the tissue surface. The depth resolution of coherence gating is unaffected by the NA of the focusing optics but instead is determined by the wave-packet length, which in turn is determined by the bandwidth (i.e., the spread in emission wavelengths) of the illuminating light source, with the resolution increasing as the bandwidth increases. For typical first-generation OCT systems, the depth resolution was typically 10–20 μm, whereas with state-of-the-art ultra-high resolution (UHR) systems, this can be 1–2 μm typically.

A number of important differences between OCT and confocal microscopy should be noted:

(a) OCT is fundamentally insensitive to fluorescence emission because it requires the signal from the sample to be mixed with a coherent reference wave. Thus, fluorescence labelling is of limited value in OCT imaging. OCT can be made indirectly sensitive to the presence of fluorophores, however, using an extension known as "pump-probe" OCT. However, this technique has not yet found widespread use, and the fundamental limits of sensitivity require careful evaluation.

(b) OCT generally uses much lower NA optics than confocal microscopy (typically less than 0.1). Hence, lateral resolution is much lower than confocal microscopy.

(c) OCT images display a form of noise called "speckle," which do not appear in confocal images.

(d) The fundamental image format is a depth-resolved 2D slice in OCT – identical to that used in medical ultrasound – whereas it is a 2D "en-face" transverse slice in confocal microscopy. Image reformatting software is required to convert between these formats.

(e) The working distances in OCT are much higher than that in confocal microscopy. The working distance (w.d.) of high-NA objectives can be less than 1 mm, whereas OCT can have working distances of several centimetres.

(f) OCT and confocal microscopy operate in complementary regimes of resolution/depth-penetration space (Fig. 1).

2. Materials

The core of an OCT system consists of: (a) the broad-band light source, (b) a fused fibre-optic coupler to act as the beam splitter, (c) a light-detector, and (d) a PC equipped with an analogue-to-digital converter.

1. Light source: The most commonly used light source is known as a "superluminescent diode" or SLD (also SLED in the literature), which possesses the strongly directional light emitting properties of a laser whilst generating emission over a substantial spread in emission wavelengths (lasers are characterised by typically very narrow emission spectra). This, it turns out, is a delicate balancing act (it is much easier to make a laser than an SLD) with the twin goals of broad emission bandwidths and high optical power being especially difficult to achieve together. Improvements in SLD design remain at the forefront of OCT technical developments, and devices operating at various wavelengths from 0.67 to 1.7 μm are commercially available. Key performance figures include the output optical power (typically 0.5–20 mW) and the optical bandwidth (typically 10–90 nm), with high output power generally meaning lower bandwidth and hence reduced resolution. The limiting sensitivity of an OCT scan scales linearly with optical power so that increasing the power by ×10 allows a tenfold reduction in integration time for the same SNR. A variant of OCT known as "swept-source OCT" uses a narrow-bandwidth laser whose wavelength can be rapidly swept over a range comparable with the output bandwidth of an SLD. Here, the figures of merit are the sweep range (typically 110 nm), the output power (typically 10 mW), and the sweep rate (typically 20 kHz for commercial sources and over 300 kHz for research sources).
2. Fibre optics: The fused fibre-optic coupler is a device that is volume-manufactured in millions of units per year within the telecoms industry, where it is used to split and recombine optical signals within fibre-optic communication networks; it typically costs a few hundred dollars or even less when purchased in bulk. Couplers are widely available operating at

0.8 μm and in the telecoms C & L bands (1.3 and 1.55 μm), i.e. in the near-infrared, since signal losses in fibre-optic cables are minimised at these wavelengths (shorter wavelengths undergo stronger light scattering, while longer wavelengths are absorbed by water and oxygen impurities in the glass). These wavelengths are also optimal for imaging the surface 1 mm of biological tissues for similar reasons (water absorption is too strong above 2 μm and light scattering too intense below 0.7 μm, see Note 1).

3. Detectors: Because OCT is a heterodyne technique, the SNR is not generally limited by detector noise, and consequently, the light detectors used are comparatively straightforward. In time-domain OCT and swept-source OCT, a single element linear photodiode is used. To optimise SNR, a "balanced" configuration can be used (2). For spectrometer-based Fourier-domain OCT, a high-speed linear sensor array is used. For wavelengths of 1,060 nm or lower, the sensors are generally silicon, whereas for higher wavelengths, III–V compounds (typically InGaAs) are preferred. Silicon detector technology, particularly linear sensors, is more developed than InGaAs technology so that higher line-scan readout rates can generally be achieved using silicon sensors.

It is also worth noting that OCT is now sufficiently mature that a number of commercial turnkey systems are available. At the time of writing this review, manufacturers include Thorlabs Inc., Michelson Diagnostics Ltd., Bioptigen Inc., Carl Zeiss Meditec, LightLab Imaging Inc., and Volcano Corp. Some of these devices are optimised for lab use, while some are aimed at particular clinical areas such as ophthalmology or intravascular imaging.

3. Methods

3.1. System Selection: Time Domain vs. Fourier Domain

The first step in establishing a useful OCT imaging platform is to decide whether it should operate on the time-domain or the Fourier-domain principle. Time-domain systems are often thought of as "first-generation" because the vast majority of systems constructed between 1991 (when OCT was first introduced) and 2003 were of this type. However, they can also possess distinct advantages in certain applications.

SNR: The majority of systems post-2003 have been Fourier-domain systems because in that year it was confirmed that the SNR is typically 100× better than with time-domain systems so that real-time B-mode imaging became a reality (3). In addition, from 2003 onwards, there has seen rapid development of hardware necessary for the two approaches to Fourier domain, "spectrometer" and "swept-source" OCT (4). Spectrometer-based OCT requires very fast read-out 1D line-scan cameras, and these have evolved from offering a few 1,000 line-scan read-outs per second to several 100,000 today. At the same time, swept-wavelength laser sources have increased in sweep speed by a similar factor. Fourier-domain systems should be chosen where the main requirement is high-speed B or C-mode imaging, and this usually means that sample movement is likely to pose a significant problem.

Lateral resolution: Fourier-domain systems gain their SNR advantage by collecting signal from all depth-resolved voxels in an A-scan in parallel, and this requires a specific optical design for the optical probe. In order to minimise geometric (i.e. confocal) roll-off in the collected signal away from the nominal focus, the probing light beam must have a very long depth of field, and it is a well-known consequence of optical wave propagation that this necessarily involves sacrificing lateral resolution because the numerical aperture (NA) is deliberately made very low. Typical NAs in Fourier-domain OCT are 0.05, and so, lateral resolution at 1.3 μm is also generally fairly low (10–20 μm). Time-domain systems, however, are generally designed to only collect data from one depth-resolved voxel at a time, and so, a natural extension of time-domain OCT is to collect the image in the "en-face" (parallel to the surface) plane using a high NA objective, in which case the technique is generally termed optical coherence microscopy (OCM) (5).

OCM is closely related to confocal microscopy but with specific strengths and weaknesses. OCM offers improved depth penetration over confocal microscopy, first because the coherence gating supplements confocal gating in suppressing multiply scattered light and second because most confocal microscopes operate in the visible, where scattering is stronger. The chief disadvantages of OCM compared with confocal are insensitivity to fluorescence and the presence of image speckle (see Note 1). OCT has received more attention than OCM with the result that, at the time of writing this review, there are no commercial OCM systems on the market. However, OCM should be considered in cases where the most relevant image information is contained in the en-face plane, where conventional confocal imaging struggles to image at the required depth and where high-speed imaging is not a priority. Finally, it is worth noting that OCM can equally well be performed with a Fourier-domain system in which the low-NA probe optics are replaced with high-NA lenses. The SNR will be the same as the time-domain approach, the chief disadvantage being that in the Fourier-domain approach, a fast Fourier transform must be performed to obtain each depth-resolved A-scan, which is computationally inefficient when most of the axial pixels are being discarded by the confocal gating. Recent innovations in swept-source design will soon allow a single en-face plane to be imaged without an FFT being necessary, which will blur the distinction between Fourier and time-domain OCT in this example and may render the latter obsolete. An interesting OCT system that sits somewhere between OCT and OCM is a multi-beam OCT scanner marketed by Michelson Diagnostics Ltd. By combining signals from four moderate-NA beams focussed at different depths into a sample, images with typically 2× higher lateral resolution can be "stitched" together by computer to produce a synthesised image with better lateral resolution than that obtained by a conventional OCT system.

3.2. Wavelength-Selection

OCT systems have been reported that operate at wavelengths from the visible through 1.95 μm. Commercial systems are available at 0.8, 0.98, 1.05, and 1.3 μm, whereas research systems have been developed over a wider range. The chief effect of wavelength selection is in determining the maximum depth to which useful information can be detected in various tissue types (see Note 2). Listed below are the wavelength ranges that have been reported in the literature together with their advantages and disadvantages.

1. 500 nm: Only two research systems have been reported using visible wavelengths (6, 7). The chief disadvantage of visible wavelengths is that depth penetration into biological tissues is likely only to be 100–200 μm due to intense light scattering. The advantages are intrinsically high axial and lateral resolution and the presence of molecular absorbers such as haemoglobin with much higher contrast than in the NIR.
2. 800 nm: The original OCT systems operated in this region, and this is also the most widely used region for ophthalmic applications. The advantages are high resolution, good depth penetration, and excellent penetration through water. There is a wide choice of available light sources, and this is the wavelength range of choice for ultra-high resolution OCT (generally defined as being an axial resolution better than about 5 μm). One limitation is that currently there are no competitive swept-wavelength light sources in this range (see later).
3. 980 nm: There is at least one commercial system operating at this wavelength, from Thorlabs Inc. Water absorption is higher than at 800 nm but is not high enough to pose a significant issue in many applications. Resolution is slightly lower than the best 800 nm systems and depth penetration slightly higher.
4. 1,060 nm: This wavelength is emerging as the new preferred wavelength of ophthalmic imaging (8). The theoretical resolution is typically somewhat lower than for 800 nm; however,

this is compensated for by a higher real-world resolution when imaging the retina because of a reduced sensitivity to blurring due to water dispersion. Also the depth-penetration is improved over 800 nm.

5. 1,300 nm: This wavelength is the most popular for non-ophthalmic applications. Advantages are a very wide choice of light sources and better penetration into biological tissues than either 800 nm or 1,060 nm. However, high water absorption at this wavelength prevents its use through high (greater than several millimetres) depths of interposed clear water, hence its unsuitability for retinal imaging. This remains the case for all longer wavelengths also.
6. 1,550 nm: This wavelength enjoyed early popularity because of the good choice of light sources (it is a popular wavelength used in fibre-optic communications) and reasonably good resolution. However, water absorption is significantly higher than for 1,300 nm, which reduces depth penetration and sensitivity in some applications. The choice of light sources at 1,300 nm is also now greatly improved.
7. 1,700 nm: A handful of research systems have been built at this wavelength (9). Advantages include lower water absorption (similar to 1,300 nm) and improved depth penetration compared with lower wavelengths. Disadvantages include lower resolution and currently a limited choice of commercial light sources.
8. 1,950 nm: There have been a few reports of OCT imaging at this wavelength, which coincides with a strong absorption band of water. Used in conjunction with 1,700 nm imaging, such dual-wavelength measurements could yield depth-resolved measurements of tissue hydration.

3.3. Contrast Method

Like other imaging techniques, OCT offers a variety of contrast mechanisms, and the most appropriate one depends on the application.

1. Structural: The most fundamental source of contrast in OCT is the presence of a strong optical reflection at boundaries between different tissues. In vivo, this gives OCT images to identify, e.g. distinct layers in epithelial tissues such as the retinal nerve fibre layer, retinal pigment epithelium etc. in the retina, the epidermal–dermal junction in the skin and the lamina propria and muscularis mucosa in squamous epithelial tissues of the GI-tract. All commercial OCT systems offer this contrast mechanism. Figure 3 illustrates structural OCT images, which reveal the formation of neo-epidermis in tissue-engineered skin.

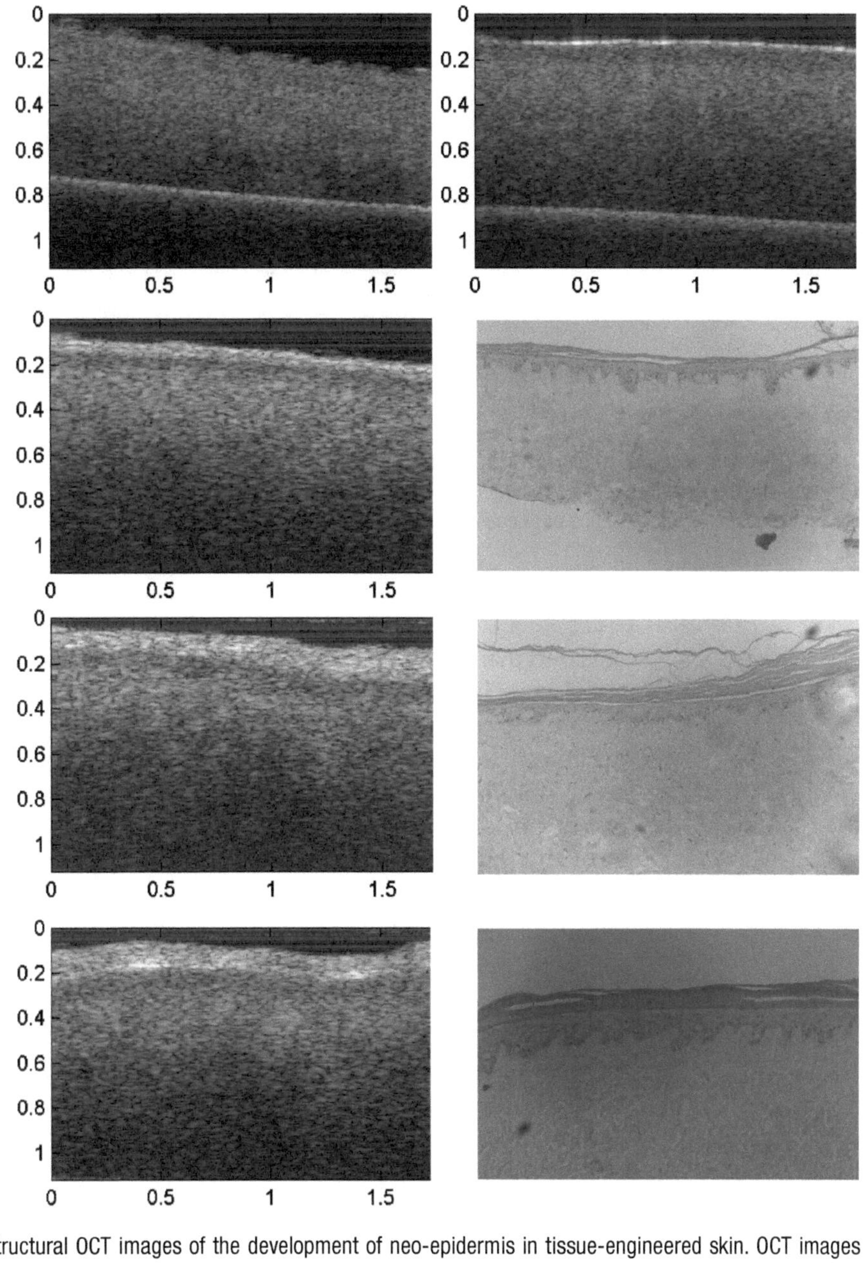

Fig. 3. Structural OCT images of the development of neo-epidermis in tissue-engineered skin. OCT images in the *left column* correspond to time points at Day 1, 8, 15, and 22 and for day 8 onwards are shown next to the corresponding histology. Axis units are in millimetres. Note the gradual appearance of a bright, echo-strong layer at the tissue surface, which histology shows to be a tissue-engineered epidermis formed by keratinocytes and fibroblasts seeded onto the surface of an acellular scaffold (de-epidermized dermis). The metal grid holding the DED at the air–liquid interface is also visible in the OCT images at day 1 and 3 (*top right*), as a bright feature at around 0.8-mm depth. Image courtesy of Dr. L. Smith and Prof. S. MacNeil, University of Sheffield, UK.

2. Polarisation: More advanced OCT systems measure the polarisation state of the reflected light and can thus measure polarimetric properties of the tissue such as birefringence (10). This means that PS-OCT can give information

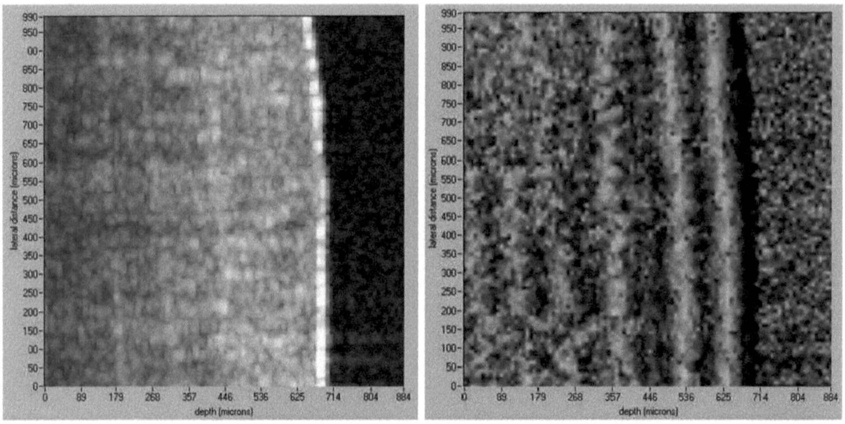

Fig. 4. Structural (*left*) and polarisation-sensitive (*right*) OCT images of annulus fibrosus tissue from the bovine intervertebral disc. Note the strongly banded structure in the polarisation-sensitive image: this arises due to strong optical birefringence and indicates the presence of directionally ordered collagen fibres. Reproduced with permission from Matcher et al. (19).

similar to that obtained on histological sections by polarised-light microscopy but with the advantage that the imaging is reflection mode and therefore non-destructive. Since the chief structural protein of the extracellular matrix – collagen – is highly birefringent when directionally ordered, PS-OCT is potentially a powerful tool to monitor extracellular matrix formation both in vivo and in tissue-engineered constructs.

3. Figure 4 compares structural and polarisation-sensitive OCT images of intervertebral disc; note the characteristic birefringence "banding" on the phase-retardation image, which is a characteristic signature of highly ordered collagen fibres. To the author's knowledge, at the time of writing this review, there is only one commercially available PS-OCT system, from Thorlabs Inc. An additional benefit of a polarization-sensitive system is that it can help to suppress anomalous signals from surface reflections (see Notes 3 and 4).

4. Doppler: Like ultrasound, OCT can be sensitive to the presence of moving scatterers via the Doppler effect (11). Scatterers moving away from the probe light beam back-scatter light, which then has a lower frequency than the original light and this frequency shift is directly proportional to the line-of-sight component of the scatterers 3D velocity vector. Doppler OCT (DOCT) thus shows improved contrast for detecting flowing fluids such as blood in microvessels.

5. Figure 5 compares a structural and Doppler OCT image of the same sample, namely, an elastic tube supporting pulsatile flow of a light-scattering liquid. The significance of such measurements in tissue engineering is that mechanical stimulation

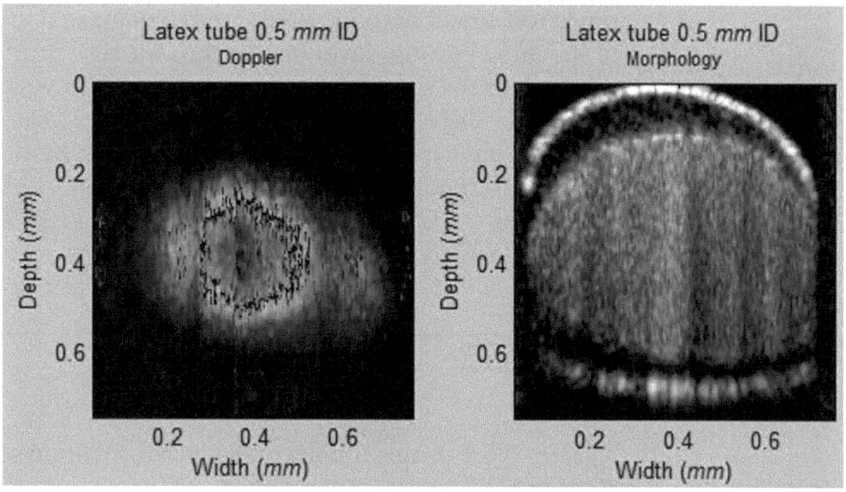

Fig. 5. A comparison of Doppler OCT (*left*) and structural OCT (*right*) images of flowing fluid in a latex rubber microvessel. The *red/blue* colour scale indicates the magnitude of the flow velocity resolved along the line of sight to the detector, as with Doppler ultrasound. Image courtesy of Dr. M. Bonesi, University of Sheffield, UK.

of cells via fluid interfacial shear-stress is emerging as a promising method to upregulate the production of ECM proteins by a variety of cells such as osteoblasts, chondrocytes, and mesenchymal stem cells.

6. Exogenous: As with other optical microscopy modalities, OCT can benefit from the addition of exogenous contrast agents. The main types of contrast agent are those that promote enhanced back-scattering of light and those that promote enhanced absorption. Due to their unusually large scattering and absorption efficiency, biological compatibility, and surface functionalisation, gold nanoparticles (GNP's) have emerged as the favoured optical contrast agent to date. These particles interact with light via the excitation of surface plasmon resonance at the surface of the particle, which means that the scattering and absorption efficiency peak at specific wavelengths that are dependent on the dielectric properties of the particle and of the surrounding medium. Small (<50 nm) GNPs interact predominantly via absorption, and the absorption reaches a maximum in the visible (around 530 nm when in an aqueous environment) which is not ideally matched to the NIR wavelengths used in OCT. Larger GNPs (>100 nm) interact primarily via scattering. Conventional GNPs, that is gold nanospheres, do not attenuate light strongly beyond 0.8 μm; however, more exotic GNP morphologies such as gold nanorods, nanocages, and core–shell nanostructures can have their plasmon resonance peak

tuned to 1 μm or beyond. GNPs can be conjugated to a variety of functional groups including ligands for cell surface receptors such as the sigma and folate receptors (12), which are of interest as these are overexpressed by cancer cells. Recent work has suggested that gold nanorods may be detectable using OCT at concentrations as low as 30 ppm by mass, corresponding to <200 nanorods per 10 μm cell (13). The field is moving rapidly.

4. Notes

1. *Speckle*: OCT images are degraded by a form of noise known as "speckle," which is an optical phenomenon that arises whenever coherent radiation interacts with a medium with properties (especially wave propagation speed) that varies spatially in a random manner (14). As its name suggests, it imparts a "speckled" appearance of randomly varying patches of light and dark on the image which confuses the underlying image structure. These speckles have a size determined by the spatial resolution of the imaging system rather than physical structures within the sample. The speckle effect is generally not seen in conventional or confocal microscopy but will be readily recognisable to anyone familiar with ultrasound images. Generally, speckle is considered to be a nuisance to be suppressed, and a number of methods to accomplish this have been proposed. Spatial filtering of the images using Fourier or wavelet-based approaches has shown promise and is included as a post-processing option in the software of some commercial systems. Generally, such filtering involves a corresponding reduction in spatial resolution, and so, preprocessing methods have also been developed. One approach is to use "diversity" methods to collect multiple versions of the same image with slightly different conditions so as to change the speckle pattern whilst leaving the underlying image largely unchanged. The multiple images can then be averaged. This approach works basically because the speckle structure is much more sensitive to the imaging set-up than the underlying structural image is. "Angular" diversity involves illuminating the sample with light beams directed onto a fixed point at different angles and then averaging the resulting images, whereas "spatial diversity" involves spatially displacing the light beam slightly between the two image acquisitions. "Wavelength" diversity combines images collected with two different light sources of similar but distinct wavelength, whereas "polarisation" diversity uses images collected with different incident polarisation states. While all

these approaches have been implemented by research groups, to the author's knowledge, only polarisation-diversity is available as a standard option on a commercial system (Thorlabs Inc.). Finally, it is worth noting that speckle can be considered a positive benefit as it carries information on the spatial distribution of scatterers on scales below the spatial resolution of the imaging system. The analysis of speckle texture has been shown to predict the size of submicron scatterers in phantom experiments with the implication that cell morphology could be determined indirectly (15). Also, the dynamic displacement of the speckle field is a measure of local elastic strain, which when combined with the application of known stresses allows the determination of local tissue elastic parameters (16).

2. *Depth penetration*: All optical imaging techniques when applied to biological tissues suffer from a comparatively unfavourable resolution vs. depth-penetration trade-off when compared to techniques such as MRI and ultrasound imaging. In practical terms, this means that OCT images can rarely image structures deeper than about 1 mm beneath the surface. In order to optimise the depth penetration of an OCT image, a number of strategies should be considered. *Scattering*: The dominant mechanism by which OCT signal amplitude falls with increasing depth is elastic light scattering: the scattering coefficient of most biological tissues being roughly 100× larger than the absorption coefficient in the NIR. It is well known that the scattering coefficient falls with increasing wavelength in the visible and IR regions of the spectrum, and a rough rule of thumb is that it falls as the square root of the wavelength. Early OCT systems operated at around 0.8 μm wavelength; however, improved depth penetration is generally observed with longer wavelengths with 1.3 μm being popular with several commercial suppliers. Recently, there has been some research activity to develop even longer wavelength systems, e.g. at 1.7 μm; however, these are, at the time of writing this review, not commercially available and also do not currently offer such high depth-resolution as shorter wavelength systems. *Water absorption*: The depth-penetration limit of 1 mm starts from the surface of the biological tissue; however, in some applications a substantial amount of transparent water-rich material lies between this tissue and the optical probe. The key example in OCT is the aqueous and vitreous humour of the eye lying between the retina and the OCT instrument; however, in tissue engineering, examples may arise when tissues are located deep inside perfusion bioreactors. A key advantage of OCT over say confocal microscopy is the much larger working distances that can be used while maintaining high axial resolution, and so, large column

distances of clear medium may frequently be encountered. In this case, it is important to realise that water itself becomes a significant absorber of NIR radiation at wavelengths greater than about 1.2 μm. Hence, for example, in vivo retinal imaging becomes impossible at 1.3 μm. If 5–10 mm of water is likely to be present between the sample and the probe, then shorter wavelengths must be used. Traditionally, this has meant 0.8 μm, but a recent trend has been to use 0.98–1.05 μm. 1.05 μm in particular lies at a minimum of the water absorption spectrum and also lies close to the "zero dispersion" wavelength of water. This brings the advantage of improved resolution at depth as the blurring effects of sample arm dispersion are minimized. Depth penetration is also optimised because of the decreased scattering coefficient c.f. 0.8 μm. There are at least two commercial systems operating in this region. *Optical clearing*: An established procedure in histology has been to reduce the optical opacity of biological samples by immersing them in histological "clearing agents" such as oil of wintergreen (methyl salicylate). Recently, interest in this approach in vivo has been rekindled partly by an improved understanding of the mechanisms by which this "optical transparency" is produced and second by the introduction of so-called "hyperosmotic" clearing agents that are biologically compatible in vivo. Glycerol, glucose, and DMSO have all been used on living skin in animal models and have demonstrated markedly improved depth penetration for confocal, multi-photon, and optical coherence imaging (17). For imaging tissue-engineered samples, it may be fruitful to give consideration not just to the chemical properties of perfusing medium but also to their optical properties, as a careful match between the refractive indices of the medium and the cell and scaffold compartments will lead to a reduced scattering coefficient and hence deeper imaging.

3. *Surface reflections*: One of the major advantages of OCT over ultrasound imaging is that it is completely non-contact, i.e. no coupling medium is needed between the microscope objective and the sample under test. Ultrasound coupling gels are needed to suppress the strong acoustic echoes generated when the incident sound wave strikes the interface between the tissue and the external environment. The physics by which a junction between dissimilar materials gives rise to a wave reflection is well known. In ultrasound, the junction between air and tissue would reflect over 99% of the incident sound energy if not suppressed using the acoustic coupling gel. Optical waves, however, only experience a 4% reflection at an air–tissue interface, and so, coupling agents are generally deemed unnecessary or, in the case of ophthalmic observation, clearly undesirable. Nevertheless, even a 4% reflection

can cause problems because, small though it may seem, it can actually be over 10 billion times stronger than the very deepest signals imaged by OCT. The OCT user should always be aware of the presence of the so-called surface "Fresnel reflection", which presents itself as a very narrow and bright line coincident with the interface between tissue and air, and should not misinterpret this as a real feature. In addition, it should be noted that a depth-resolved cross-section through this surface reflection is actually a measure of the instrument "point spread function" (PSF), so its width is determined by the system resolution and carries no physical significance regarding the sample. Also, the PSF can contain side lobes, which while much weaker than the central peak, can themselves be large compared to the features being imaged. This can result in vertical streaks appearing in an image at points where the surface has a particularly strong Fresnel reflection. There are several ways to attenuate the surface Fresnel reflection. A particularly easy way is simply to tilt the sample so that the light beam enters at an angle rather than at near normal incidence. Due to the law of reflection, in this case, the reflected wave is reflected along a path that does not coincide with the incident beam path, so the reflected light is not collected. This technique, however, may be of limited use if the surface is highly corrugated as there will always be patches that are inclined so as to reflect strongly. Another possibility is to use a water-dipping objective in which case the space between the sample and the glass of the objective is filled with water, and the maximum Fresnel reflection between water and tissue is less than 0.1% as opposed to 4% for air/tissue. However, this option requires either a custom-built system or a close dialogue with the manufacturer because, unlike conventional optical microscopy, water-dipping objectives are not a standard option on current commercial systems. Alternatively, if the Fresnel reflection is troublesome because the region of interest happens to be close to the surface, then, if the sample geometry allows it, a thick film of water can be placed on top of the sample, thus displacing the Fresnel reflection away from the area of interest. Problems with side-lobe artefacts may persist in this case, however. Another option is to use a polarization-sensitive OCT system with a linearly or circularly polarised incident beam and to record only the image in the so-called "cross-polarized" channel. This technique is analogous to suppressing glare on a sunny day by wearing Polaroid sunglasses and works because the surface Fresnel reflection is polarised to the same degree as the incident beam, whereas signal from deeper in the tissue tends to be more randomly polarised. The advantages of this method are efficient suppression of the Fresnel reflection without

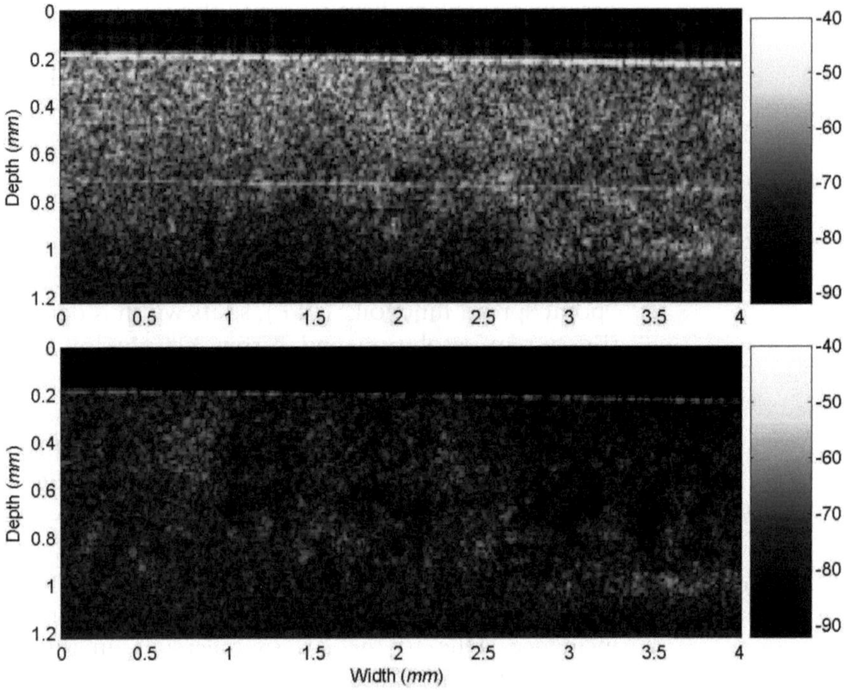

Fig. 6. An OCT image of a sample of dental enamel. The *upper* image is a conventional structural image; note the bright line at the surface which is a Fresnel reflection artefact. The *lower* image is a simultaneous measurement but using cross-polarized detection to reduce this surface artefact. Image courtesy of Dr. M. Bonesi, University of Sheffield, UK.

sample modification or use of immersion media, but this is balanced by increased systems complexity. Figure 6 illustrates the effectiveness of the method when imaging samples of dental enamel.

4. *Image artefacts*: Shadows: OCT suffers from shadow artefacts in the same way that ultrasound does. Because the image is formed by measuring the intensity of wave reflections at a given depth, if the sample contains a structure that heavily attenuates the radiation beam, then all features below this will be illuminated by a weaker beam and will thus be "shadowed" by the overlying structure. Conversely, if a structure is more transparent to the radiation than surrounding tissues, then features below it will appear to be artificially brightened. In ultrasound, shadows are cast by dense structures such as bone, and fluid-filled regions such as cysts can cause brightening of deeper regions. In OCT, regions containing dense concentration of scatterers (e.g. blood vessels) can cause shadows, while fluid-filled voids in porous structures such as tissue-engineered scaffolds can cause brightening artefacts. Figure 7 illustrates some of these effects. *Side lobes*: The mechanism by which OCT generates depth-resolved information is quite different to that by which conventional microscopy achieves

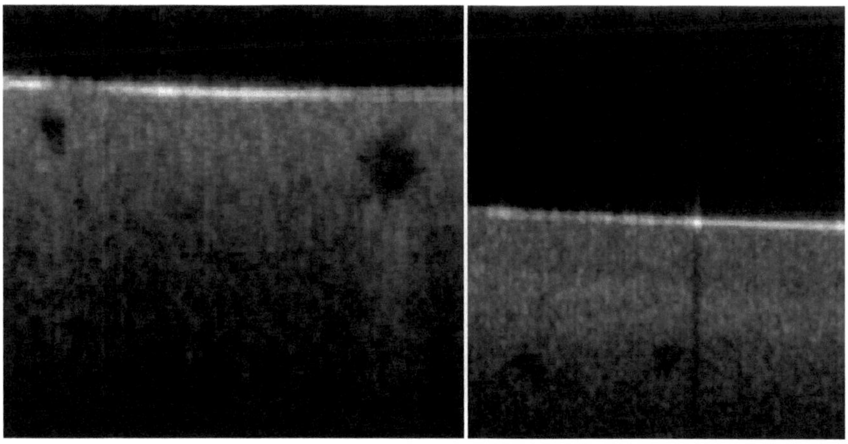

Fig. 7. *Left*: an OCT scan of an agar gel phantom containing air bubbles. Note the characteristic *dark* (echo-weak) bubble and the brightened area immediately below due to reduced beam attenuation. *Right*: an OCT image of a phantom with a highly scattering imperfection on the surface producing a brightened Fresnel reflection and a deeper vertical shadow due to increased beam attenuation. Image courtesy of Dr. Z. Lu, University of Sheffield, UK.

this by virtue of the central role played by the spectrum of the light source. In fact, it can be shown that the axial PSF is just the Fourier Transform of the light source spectrum. The ideal spectrum is a Gaussian distribution, in which case the PSF is also Gaussian. A Gaussian PSF has a smooth decay to zero and the effect of this on the image is to produce a uniform blurring. In reality, however, the light source has a less than ideal spectrum, and a common manifestation of this is that the PSF has a central maximum surrounded by lower amplitude "side lobes." It is the goal of OCT light-source developers to minimise these by optimising the light spectrum to be as near-Gaussian as possible, but in reality, every image will contain side-lobe artefacts to a degree. Usually the side lobes are separated axially from the main peak by a distance comparable to the axial resolution, that is, a few tens of microns. In this case, side lobes appear as weak features immediately above and below a strong signal. The surface Fresnel reflection for example can take on the appearance of a three-layer "sandwich," with the bright central line flanked by weaker lines above and below it and parallel to it. Some light sources, however, can produce side lobes separated from the main peak by a considerable distance (hundreds of microns to millimetres; see Fig. 8 for an example). Given that the entire depth of field of an OCT system might be 1 mm, it is possible to have a side-lobe artefact visible even though the main signal is outside the field of view. In this case, there is a risk that the artefact can be misinterpreted as a real feature. The best strategy to avoid such confusion is to gain familiarity with the

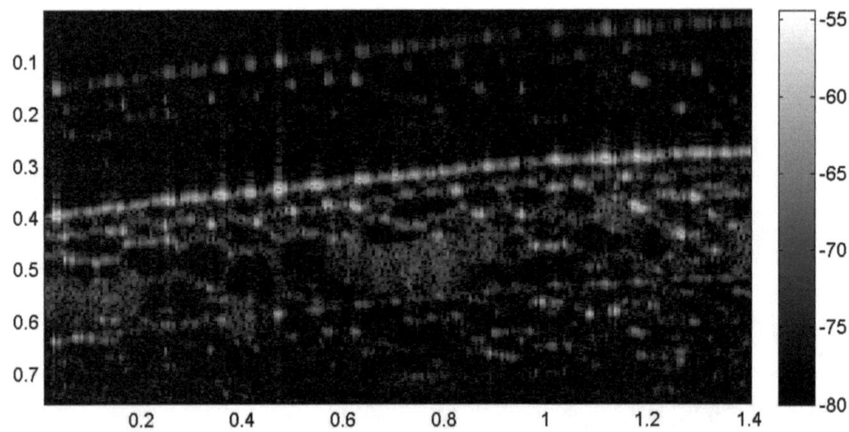

Fig. 8. OCT image of an onion taken using a light source with strong side lobes about 350 μm from the central peak. Note how these side lobes generate "ghost" images (*arrowed*) *above* and *below* the true image. Axis units are in millimetres. Image courtesy of Dr. N. Krstajic of the University of Sheffield (UK) and the FASTDOT consortium.

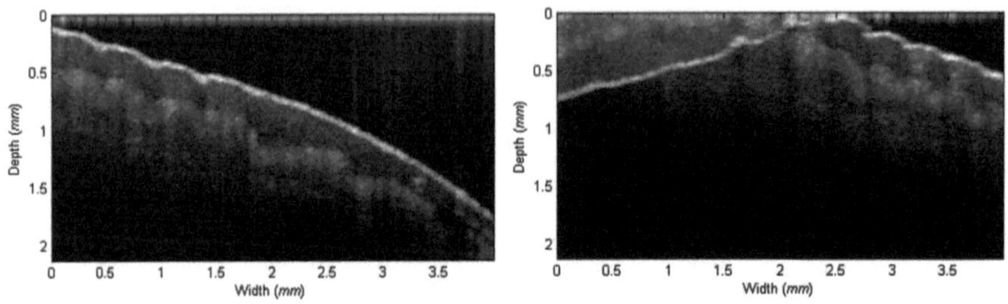

Fig. 9. In vivo Fourier-domain OCT images of palmar skin (fingertip) of a human. *Left*: the object is correctly located within the imaging field of view, yielding structural information on the stratum corneum (superficial *dark layer*) and epidermal layers (deeper *bright layer*). *Right*: the finger tip is too high, and the *left* part of the finger lies beyond the *top* of the field of view. The Fourier-domain mirror artefact causes the image to wrap back into the field of view, yielding the distorted image shown. Image courtesy of Dr. Z. Lu, University of Sheffield, UK.

system when imaging a very simple test object, i.e. one where it is known that only a single reflecting interface is present within the field of view. A glass block whose thickness exceeds the axial field of view of the system is a possible solution, or a transparent block of set gelatine can also suffice. Autocorrelation and mirror artefacts: A peculiarity of Fourier-domain OCT methods compared with time-domain methods is the occurrence of autocorrelation and mirror artefacts. Although both can be removed using more sophisticated hardware and/or data processing algorithms (18), these are not yet universally implemented, and so, it is important to recognise these artefacts. Figure 9 illustrates the Fourier-domain OCT mirror artefact when imaging human palmar skin in vivo, and Fig. 10 illustrates the optical clearing effects of a hyperosmotic agent.

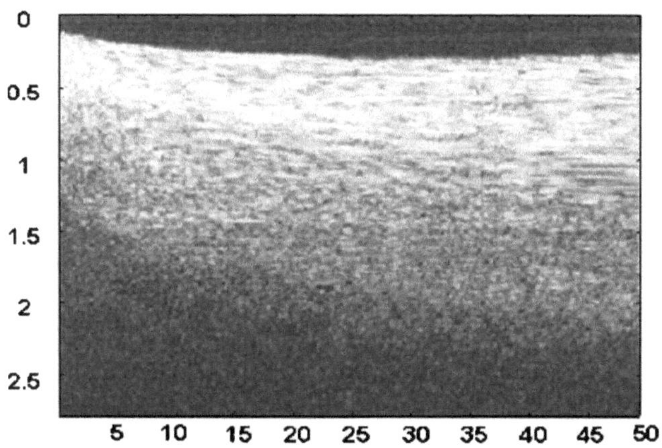

Fig. 10. Illustrating the optical clearing effects of a hyperosmotic agent (glycerol) on the imaging depth of OCT in porcine stomach tissue. The image shows an OCT M-scan (i.e. a sequence of A-scans recorded at a fixed point and displayed using the *x*-axis for time). The *x*-axis units are time in minutes, and the *y*-axis is depth in millimetres. Note the pronounced increase in optical signal from depths below 1.5 mm as the glycerol diffuses into the tissue. Reproduced with permission from (15).

References

1. Tomlins, P.H. and Wang, R.K. (2005) Theory, developments and applications of optical coherence tomography. *J. Phys. D Appl. Phys.* **38(15)**, 2519–2535.
2. Podoleanu, A.G. (2000) Unbalanced versus balanced operation in an optical coherence tomography system. *Appl. Opt.* **39(1)**, 173–182.
3. Leitgeb, R., Hitzenberger, C.K. and Fercher, A.F. (2003) Performance of fourier domain vs. time domain optical coherence tomography. *Opt. Express* **11(8)**, 889–894.
4. Choma, M.A., Sarunic, M.V., Yang, C.H. and Izatt, J.A. (2003) Sensitivity advantage of swept source and Fourier domain optical coherence tomography. *Opt. Express* **11(18)**, 2183–2189.
5. Yang, Y., Dubois, A., Qin, X.P., Li, J., El Haj, A. and Wang, R.K. (2006) Investigation of optical coherence tomography as an imaging modality in tissue engineering. *Phys. Med. Biol.* **51**(7), 1649–1659.
6. Unterhuber, A., Povazay, B., Bizheva, K., Hermann, B., Sattmann, H., Stingl, A., Le, T., Seefeld, M., Menzel, R., Preusser, M., Budka, H., Schubert, C., Reitsamer, H., Ahnelt, P.K., Morgan, J.E., Cowey, A. and Drexler, W. (2004) Advances in broad bandwidth light sources for ultrahigh resolution optical coherence tomography. *Phys. Med. Biol.* **49**(7), 1235–1246.
7. Gangnus, S.V. and Matcher, S.J. (2008) Visible-light OCT spectrometer for microvascular oximetry. *Proc SPIE* **6847**, D8471–D8471.
8. Wang, Y.M., Nelson, J.S., Chen, Z.P., Reiser, B.J., Chuck, R.S. and Windeler, R.S. (2003) Optimal wavelength for ultrahigh-resolution optical coherence tomography. *Opt. Express* **11(12)**, 1411–1417.
9. Sharma, U., Chang, E.W. and Yun, S.H. (2008) Long-wavelength optical coherence tomography at 1.7 μm for enhanced imaging depth 1,700 nm SS-OCT. *Opt. Express* **16(24)**, 19712–19723.
10. de Boer, J.F. and Milner, T.E. (2002) Review of polarization sensitive optical coherence tomography and Stokes vector determination. *J. Biomed. Opt.* **7**(**3**), 359–371.
11. Mason, C., Markusen, J.F., Town, M.A., Dunnill, P. and Wang, R.K. (2004) Doppler optical coherence tomography for measuring flow in engineered tissue. *Biosens. Bioelectron.* **20(3)**, 414–423.
12. Boppart, S.A., Oldenburg, A.L., Xu, C. and Marks, D.L. (2005) Optical probes and techniques for molecular contrast enhancement in coherence imaging. *J. Biomed. Opt.* **10(4)**, 041208.
13. Oldenburg, A.L., Hansen, M.N., Zweifel, D.A., Wei, A. and Boppart, S.A. (2006) Plasmon-resonant gold nanorods as low backscattering albedo contrast agents for optical coherence tomography. *Opt. Express* **14(15)**, 6724–6738.

14. Schmitt, J.M., Xiang, S.H. and Yung, K.M. (1999) Speckle in optical coherence tomography. *J. Biomed. Opt.* **4(1)**, 95–105.
15. Gossage, K.W., Smith, C.M., Kanter, E.M., Hariri, L.P., Stone, A.L., Rodriguez, J.J., Williams, S.K. and Barton, J.K. (2006) Texture analysis of speckle in optical coherence tomography images of tissue phantoms. *Phys. Med. Biol.* **51**, 1563–1575.
16. Ko, H.J., Tan, W., Stack, R., Boppart, S.A. (2006) Optical coherence elastography of engineered and developing tissue. *Tissue Eng.* **12(1)**, 63–73.
17. Tuchin, V.V. (2007) A clear vision for laser diagnostics. *IEEE J. Sel. Top. Quantum Electron.* **13(6)**, 1621–1628.
18. Leitgeb, R.A., Hitzenberger, C.K., Fercher, A.F. and Bajraszewski, T. (2003) Phase-shifting algorithm to achieve high-speed long-depth-range probing by frequency-domain optical coherence tomography. *Opt. Lett.* **28(22)**, 2201–2203.
19. Matcher, S.J., Winlove, C.P. and Gangnus, S.V. (2004) The collagen structure of bovine intervertebral disc studied using polarization sensitive optical coherence tomography. *Phys. Med. Biol.* **49**, 1295–1306.

Chapter 18

Osteogenic Differentiation of Embryonic Stem Cells in 2D and 3D Culture

Lee Buttery, Robert Bielby, Daniel Howard, and Kevin Shakesheff

Abstract

Osteoblasts are the cells that contribute to the formation and function of bone tissue. Knowledge of their biology is important to understanding of the normal processes of bone repair, the development of diseases affecting bone tissue, and to the investigation of approaches to improve bone repair and to treat or prevent bone diseases. Osteoblasts can be readily isolated from bone tissues and grown in culture, and under relatively simple culture conditions, they will recapitulate many aspects of their normal biology. These culture conditions can be also applied to adult stem cells, such as mesenchymal/bone marrow stromal stem cells. More recently, these studies have been extended to include embryonic stem cells. This chapter provides detailed step-by-step protocols to investigate the differentiation of embryonic stem cells into osteoblasts. Several 2D and 3D culture methods are presented and enable comparisons to be made on the efficiency and mechanisms of osteogenic differentiation. Emphasis is also placed on methods to analyse and confirm osteogenic differentiation.

Key words: Osteoblasts, Embryonic stem cells, Cell culture, Differentiation, 3D interactions

1. Introduction

Osteoblasts are the cells that regulate the synthesis of bone tissue. Over the past 40 years or so, various techniques have been developed to study the biology of osteoblasts in vitro using explants or digest cultures of bone tissues and also from aspirates of bone marrow (see (1, 2) for general reviews). Bone marrow contains a population of stem cells (bone marrow stromal stem cells, mesenchymal stem cells), which serve as a reservoir for new osteoblasts (and also other cell types) and thus help to maintain normal bone tissue homeostasis. These marrow-derived stem cells and also osteoblastic cells released from tissues are very amenable to in vitro culture and in response to specific culture conditions and

John W. Haycock (ed.), *3D Cell Culture: Methods and Protocols*, Methods in Molecular Biology, vol. 695,
DOI 10.1007/978-1-60761-984-0_18,

stimuli will readily recapitulate their normal differentiation process. This is characterised by the temporal expression of transcription factors (e.g. cbfa-1/runx2, osterix), cell adhesion molecules, matrix proteins, and enzymes (e.g. cadherin-11, collagen-1, alkaline phosphatase, and osteocalcin). Eventually, after 14–21 days in culture, the cells will form distinct mineralising colonies called bone nodules (see (1, 2) for general reviews).

Various stimuli have been used to promote osteogenic differentiation, and based on a number studies, a relatively simple general osteogenic culture medium has been developed, which is still used extensively today. Basal medium containing 10–15% (batch-tested serum) is supplemented with ascorbic acid and organic phosphate, which help promote osteogenic matrix synthesis. Dexamethasone is also added and through a complex and poorly understood mechanism this synthetic steroid, in the presence of the other biochemical factors, is able to stimulate and "recruit" cells into the osteogenic lineage (1, 2). An array of other molecules and factors have been investigated and shown to promote osteogenesis (e.g. bone morphogenetic proteins) and are important in dissecting and understanding the process of osteogenesis; however, the basic method/medium supplements described above provide a very robust (and cheap) method to investigate osteogenesis (1, 2).

While much of the data on the mechanisms of in vitro osteogenesis have come from studies on explant tissues and marrow stem cells, recent work by us and also others have demonstrated that mouse and human ES cells can be readily differentiated into osteoblasts both in vitro and in vivo (3–7). Given the extensive history on the investigation of osteogenesis using explant tissues and marrow stem cells, issues have been raised over the relevance of investigating osteogenesis using ES cells. However, even though work in this field has been in progress for 40 years or so, many aspects of the process of osteogenesis still remain to be defined and especially the early phases such as the role of marrow stem cells and the mechanism regulating their early differentiation. Based on their capacity for almost unlimited self-renewal and ability to give rise to most somatic cell lineages, ES cells can provide important information on the processes of differentiation and in particular the early mechanisms of lineage commitment and therefore should be considered a very useful cell source to complement studies on osteogenesis using other more established cell types and sources.

The normal function of most cells and tissues is, in addition to soluble factors, dependent on spatial interactions with neighbouring cells and a substratum or extracellular matrix (ECM) (8). Consequently, recapitulating the function of the ECM and 3D cell interactions is an important aspect of investigating mechanisms of differentiation. Moreover, cells may respond very differently to biochemicals or drugs in 2D compared to 3D culture and have significant implications for developing or testing new therapies or medicines. Various natural and synthetic materials have been used to produce 3D scaffolds (9, 10), and through coating with or incorporation of biological molecules, scaffolds can augment adhesion of specific cell types, including osteoblastic cells (11–14). Osteoblasts appear very amenable to culture in 3D environments, and 3D interactions are fundamental to their differentiation. For example, osteogenic differentiation of cells derived from bone tissue explants or bone marrow cells typically involves condensation of cells into adherent discrete 3D clusters also called bone nodules (1, 2) or spheroids in suspension culture (15), mimicking, to some extent, the mechanisms of osteogenesis during development or bone repair.

In this chapter, we focus primarily on osteogenic differentiation of ES cells, but many of the methods can be easily transferred to investigation of other cell types including primary osteoblasts and mesenchymal stem cells. Ideally, it will be useful to use all these cell types in the same experiments to help in quantifying and understanding the mechanisms of osteogenic differentiation.

2. Materials

2.1. General

1. Tissue culture consumables (culture plates, Petri dishes, centrifuge tubes, culture pipettes, syringe filters, etc.) should be gamma-irradiated, single use.
2. Perform all steps using aseptic technique in a suitable microbiological class II cabinet.
3. Once prepared tissue culture media containing supplements should not be kept (even at 4°C) for longer than 2 weeks. Avoid warming the entire stock of supplemented media in a water bath – aliquot enough medium for the procedure into appropriate sterile tubes and warm those prior to beginning cell culture.
4. Trypsin solution should be thawed just prior to use and used on the same day. Do not leave trypsin in 37°C water baths for extended periods of time.
5. All cell cultures should be maintained in a humidified 37°C, 5% CO_2 incubator. Ideally, the incubator should be large enough, with cable access ports to accommodate equipment such as shakers, bioreactors, and magnetic stirrers.
6. Access to orbital shakers, 3D shakers (e.g. gyro twister/ nutating shaker), stirred bioreactors (e.g. CYTOSTIR® Stirred Bioreactor for Microcarrier Cell Culture), and rotary

bioreactors (e.g. Synthecon; http://www.synthecon.com) is required for 3D culture studies.

7. A peristaltic pump or syringe driver is required to help prepare encapsulated cell aggregates.

2.2. Culture Media Reagents and Buffers

1. Dulbecco's Modified Eagles Medium (DMEM), high glucose without sodium pyruvate or
   Minimal Essential Medium Alpha Medium (with ribonucleosides and deoxyribonucleosides (see Note 1).
2. Foetal calf serum (batch tested) (see Note 2).
3. Penicillin/Streptomycin.
4. Trypsin–EDTA.
5. L-Glutamine.
6. L-Ascorbic acid (see Note 3).
7. Beta-glycerophospate (see Note 3).
8. Dexamethasone (see Note 3).
9. Sterile phosphate-buffered saline PBS; 10 mM); To 1 L of distilled water, add 8.79 g of NaCl, 0.272 g of KH_2PO_4, 1.135 g of Na_2HPO_4, mix, and dissolve thoroughly. Adjust to pH 7.2–7.4 with concentrated HCl. Sterilise by autoclaving. PBS is used as a washing buffer and as a diluent for other reagents e.g. ascorbic acid, beta-glycerophosphate, and dexamethasone. It can be made up fresh but will remain stable for several days at normal room temperature. PBS is used in the various washing steps that are important in cell culture, so it is advisable to have several litres of buffer available. Some companies produce ready-prepared PBS tablets, which can be simply added to the appropriate volume of deionised water.
10. Sterile Tris(hydroxymethyl)aminomethane (Tris 200 mM, pH 9.0); dissolve 24.2 g of Tris(hydroxymethyl)aminomethane in 800 mL of distilled water. Adjust to pH 9.0 with concentrated HCl and then adjust volume to 1 L with distilled water. Sterilise by autoclaving and store at room temperate. This buffer is used in the determination of alkaline phosphatase activity. It can be made up fresh each week but will remain stable for several days at normal room temperature.
11. Sterile Tris-buffered saline (25 mM Tris, 150 mM NaCl, 2 mM KCl, pH 7.4); In 800 mL of distilled water, dissolve 8 g of NaCl, 0.2 g of KCl, and 3 g of Tris base. Adjust the pH to 7.4 with HCl. Add distilled water to 1 L. Sterilise by autoclaving. This buffer is used for gelling alginate beads for encapsulation of cells. It can be made up fresh each week but will remain stable for several days at normal room temperature.

2.3. Osteogenic Culture Medium

1. Store the sterile medium at 4°C and use within 1–2 weeks.
2. To DMEM or alpha MEM, add the following: 10–15% v/v foetal calf/bovine serum, 1% v/v penicillin (1,000 U/mL)/streptomycin (10 μg/mL), 2 mM L-glutamine, L-ascorbic acid (final working concentration = 50 μg/mL; see Note 3), beta-glycerophosphate (final working concentration = 10 mM) Note 3, and dexamethasone (final working concentration = 1 μM) (see Note 3).

2.4. Fixatives and Histology Stains

1. Formalin 10% solution (see Note 4); Add 100 mL of 40% formaldehyde to 900 mL of distilled water. Formalin can be stored at room temperature and remains stable for many weeks. Since it is toxic by inhalation and contact with skin, use a fume cabinet.
2. Paraformaldehyde 4% solution (see Note 4); Heat 1 L of 10 mM PBS to 60°C and add 40 g of paraformaldehyde and stir until dissolved thoroughly and cool to room temperature. It may be necessary to readjust pH to 7.4 using concentrated HCl. Paraformaldehye can be prepared fresh or can be stored in aliquots at −20°C. Since it is toxic by inhalation and contact with skin, use a fume cabinet.
3. Alizarin Red S 2% solution: dissolve 2 g of Alizarin Red S powder in 100 mL of PBS and filter (e.g. standard Whatman laboratory filter paper). Adjust to pH 6.2–6.4 with ammonium hydroxide. Store at room temperature, stable for 2–3 weeks (Care: ammonium hydroxide is caustic and a respiratory irritant – use a fume cupboard).

2.5. Immunocytochemistry Reagents (See Note 5)

1. Basic equipment required for immunocytochemistry: metal staining racks or Coplin jars, which are useful for holding glass slides during washing and staining procedures; plastic storage containers (e.g. sandwich box) of ~500 mL capacity, which are useful for washing and staining procedures; plastic Petri-dishes (~20 cm diameter), which are useful for laying slides flat during antibody incubation procedures. If sectioning tissue, access to a cryostat or a wax microtome is required; however, many institutions offer a tissue section cutting service. Access to a photomicroscope (bright field and epi-fluorescence at least, but a confocal microscope is also desirable) is also necessary to observe and record immunostained preparations.
2. For immunocytochemistry, it is not necessary to use sterile buffers, but it is important to ensure that they are prepared fresh, either daily or weekly, as many buffers may precipitate out their salts, altering their effective concentrations, which can affect immunostaining. Buffers like PBS and their containers if left for several weeks or not washed out regularly

with a detergent or disinfectant are also prone to support growth of algae.

3. Diluent for primary antibodies and normal sera: PBS; 0.05% (w/v), bovine serum albumin (BSA); 0.01% (w/v), sodium azide (NaN_3) – (PBS, BSA, and NaN_3). This should be stored at 4°C and be used within 1–2 weeks. The purpose of BSA is to act as an antibody carrier and to help prevent the antibody sticking to the surface of the plastic or glass tube in which it is stored, prepared, or applied. Sodium azide functions as a preservative to prevent microbial contamination and growth. Note: Azide is toxic by skin contact and ingestion.
4. Diluent for secondary antibodies: PBS: 0.05% BSA; this serves the same function as the diluent for primary antibodies, but the azide is omitted as it can interfere with some chromogen reactions used to reveal sites of antibody binding. In the absence of azide, it should be prepared fresh and stored for no longer than a few days at 4°C.
5. Antibodies to investigate different stages of osteogenic differentiation are available from various suppliers. Some may be species-specific i.e. human or mouse, so if working with human or mouse ES cells, confirm cross-reactivity of antibodies before purchasing. Most of the antibodies used in our studies are purchased from R&D systems (see Note 5).
   - Transcription factors (e.g. cbfa-1/runx2, osterix).
   - Cell adhesion/cell surface (e.g. cadherin-11).
   - Cell matrix and mineralisation (e.g. collagen-1, osteocalcin, bone-specific alkaline phosphatase).

2.6. Alkaline Phosphatase Histochemistry

1. Stock naphthol solution: dissolve 60 mg of naphthyl AS phosphate in 1 mL of dimethyl formamide and then slowly add to 200 mL of 200 mM Tris buffer (pH 9.0).
2. Incubating solution: To 10 mL of the stock naphthol solution, add 10 mg Fast Red TR, stir gently to dissolve, pass through a standard Whatman filter, and use immediately. Note: dimethyl formamide is toxic and a suspected carcinogen.

2.7. Alkaline Phosphatase Activity Assay

1. Cell lysis buffer; In PBS, 1% nonyl phenoxylpolyethoxylethanol (NP-40), 0.5% sodium deoxycholate, 0.1% sodium dodecyl sulphate (SDS), 0.1 mg/mL phenylmethylsulphonyl fluoride (PMSF), 30 μg/mL aprotinin, and 1 mM sodium vanadate. Sit until dissolved and use fresh. Note: Most reagents in this buffer but especially SDS, PMSF, and aprotinin are toxic by inhalation and contact with skin.
2. Alkaline assay buffer: 0.1 M solution of 2-amino-2-methyl-1-propanol in distilled water. Adjust to pH 10.5 with HCl and make up to 100 mL. It can be stored at room temperature for several weeks.

3. Substrate:8 mM solution of *p*-nitrophenyl phosphate (PNP) disodium in a 1 mM solution of $MgCl_2$. 1–2 mL should be prepared fresh and can be stored at 2–8°C for up to 24 h. Note: PNP is toxic.
4. PNP standards: 100 mM stock solution of *p*-nitrophenol in 50 mL of 0.25 M NaOH. This can be stored at normal room temperature protected from light for several weeks. Use the stock PNP and 0.25 M NaOH to produce serial dilutions from which a standard curve can be constructed using the following concentrations: 400, 200, 100, 50, 25, 12.5, and 6.25 μM. Absorbance should be read at 405 nm, and a curve should be constructed each time the assay is performed.
5. Total protein quantification: Bradford assay kit or similar is required to normalise total protein concentration with the samples.

2.8. Reagents for Isolation of RNA and RT-PCR Analysis (Table 1)

1. RNA isolation kit with DNAse digestion (e.g. RNeasy and RNase-free DNase, Qiagen).
2. RNAse, DNAse-free microcentrifuge tubes.
3. RNAse, DNAse-free filter pipette tips.
4. Nuclease-free water.
5. Reverse transcription system (e.g. Promega).
6. RNAse, DNAse-free thin-walled PCR tubes.
7. Taq polymerase with hot-start property (e.g. AmpliTaq gold, Applied Biosystems).
8. 10× PCR buffer, Mg-free and 25 mM magnesium chloride solution (e.g. Applied Biosystems). Aliquot and store at −20°C.

Table 1
PCR mix

Reagent	Volume per reaction (μl)	Stock concentration	Final concentration
$MgCl_2$	1.5	25 mM	1.5 mM
10× buffer	2.5	10×	1×
dNTP	0.5	10 mM	200 μM
Fwd primer	0.5–1.25	10 μM	200–500 nM (see Table 2)
Rvs primer	0.5–1.25	10 μM	200–500 nM (see Table 2)
Taq	0.25	5 U/μL	1.25 U per reaction
H_2O	7.75–9.25	–	–

Table 2
Primer sequences for RT-PCR

Gene	Primer concentration	Primer sequences
Runx2 (human & mouse)	500 nM	fwd 5′-CATGGTGGAGATCATCGC-3′ rvs 5′-ACTCTTGCCTCGTCCACTC-3′
Osteocalcin (human)[a]		fwd 5′-ATGAGAGCCCTCACACTCCTC-3′ rvs 5′-GCCGTAGAAGCGCCGATAGGC-3′
Osteocalcin (mouse)	500 nM	fwd 5′-CGGCCCTGAGTCTGACAAA-3′ rvs 5′-ACCTTATTGCCCTCCTGCCTT-3′
Oct4 (human)	200 nM	fwd 5′-GAAGGTATTCAGCCAAAC-3′ rvs 5′-CTTAATCCAAAAACCCTGG-3′
Oct4 (mouse)[b]	500 nM	fwd 5′-GGCGTTCTCTTTGGAAAGGTGTTC-3′ rvs 5′-CTCGAACCACATCCTTCTCT-3′
GAPDH (human)	500 nM	fwd 5′-GAGTCAACGGATTTGGTCG-3′ rvs 5′-CCTTCTCCATGGTGGTGAA-3′
β-Actin (mouse)	200 nM	fwd 5′-GTCGTACCACAGGCATTGTGATGG-3′ rvs 5′-GCAATGCCTGGGTACATGGTGG-3′

a) Moth et al (23)
b) Bourna et al (5)

9. dNTP mix, 10 mM (e.g. Amersham). Aliquot and store at −20°C.
10. Oligonucleotide primers (see Table 2 for sequences). Primers should be reconstituted as 10-μM stocks, aliquoted and stored at −20°C.
11. Agarose, molecular biology grade.
12. 1× TAE buffer (e.g. Sigma-Aldrich).
13. 500 μg/mL ethidium bromide solution (Care! Toxic and carcinogenic).
14. 100 bp DNA ladder (e.g. New England Biolabs).
15. 5× DNA loading buffer (e.g. Bioline).
16. Heat block (capable of temperatures in range of 70–95°C).
17. Thermal cycler.
18. Horizontal gel electrophoresis tank and power pack.
19. UV transilluminator. Care! UV radiation is a hazard. Ensure that local safety regulations are followed and appropriate protective equipment is worn.

2.9. Alginate Encapsulation

1. Low-viscosity alginic acid 2% (w/v) solution: sterile Tris buffered-saline with 0.2% (w/v) bovine or porcine gelatin. Heat to 40–50°C to ensure thorough mixing and then filter-sterilise

through a 0.2-μM filter. Prepare fresh, cool to 37°C, and use immediately.

2. Alginate beads gelling solution: sterile solution of 100 mM $CaCl_2$ in Tris-buffered saline. Autoclave and store at normal room temperature.

2.10. Chemically Engineered Cell Aggregation

1. All reagents should be prepared with sterile buffers and then filter-sterilised through a 0.2-μm filter.
2. Sodium periodate oxidation solution: 1 mM solution of sodium periodate in PBS. Prepare fresh and keep in a darkened tube at 4°C.
3. Biotinylation solution: 5 mM solution of biotin hydrazide in PBS or DMEM/alpha MEM (without serum or other supplements), adjusted to pH 6.5. Prepare fresh and keep at room temperature or warmed to 37°C.
4. Avidin solution: 2 mg/mL solution (w/v) of avidin in PBS or DMEM/alpha MEM (without serum or other supplements). Aliquot and store at −20°C.

3. Methods

Before beginning a study on the differentiation of ES cells, it is important to have a good understanding of their basic biology, and for the purposes of this chapter, knowledge of cell culture and ES cell culture is assumed. Knowledge of the cell type of interest e.g. the osteoblast is also assumed. Possibly, the most crucial element is how the ES cells are cultured and in particular if differentiation is initiated via formation of embryoid bodies (EBs). In our experience, EBs are required to help osteogenic differentiation, but some workers have found that the EB formation step is not necessary (16). There are several methods for promoting formation of EBs, including suspension culture, hanging drops, microfuge pellets and encapsulation in alginate beads, and culture in stirred or rotary bioreactors (17). Together with seeding density, these methods can impact significantly the mechanisms and efficiency of a differentiation protocol. In this chapter, we present several methods including mass suspension with EB size fractionation, microfuge tube pellet culture and alginate encapsulation. We also include a basic method for culture on 3D scaffolds.

We also present several approaches to differentiate the cells in both 2D and 3D culture including simple agitation through use of stirred or rotary bioreactors. We also include various methods for analyzing and quantifying differentiation, which are fundamental to understanding of the process of differentiation and which help in refining methods and approaches to control differentiation.

3.1. Induction of Osteogenic Differentiation via Mass Suspension EB Formation and 2D Culture

1. This protocol involves making EBs and allowing them to form over several days before disrupting them to achieve single cell suspensions, which are then seeded onto tissue-culture-treated plates. This approach works reasonably well in our hands and depending on the nature, timing, and duration of the stimulus used, it can result in significant enrichment with osteoblasts. Note that this is enrichment since, as with most studies on ES cells involving simple biochemical stimulation, other cell types are produced. However, based on quantitation of osteogenesis such as the bone nodule assay, it is possible to achieve levels of 40–60% osteogenic enrichment. Further observations based on this approach were instrumental in identifying specific phases or time points of osteogenic induction, which enabled these cells to be purified close to homogeneity, using cell sorting methods and targeting the cell adhesion molecule cadherin-11 (3–7) (see Figs. 1 and 2).
2. Once ES cells have been dissociated into single cell suspensions and are free from contact with factors used to maintain pluripotency (varies between mouse and human and different cell lines), most cell lines will spontaneously form aggregates or EBs. We recommend seeding cells into a basal osteogenic medium (e.g. DMEM/alpha MEM plus 10–15% FCS) at densities in the region of 5×10^4 to 2.5×10^5 cells per mL. Cells should be cultured in non-tissue-culture-treated plastic Petri dishes to encourage free floating aggregates to form and

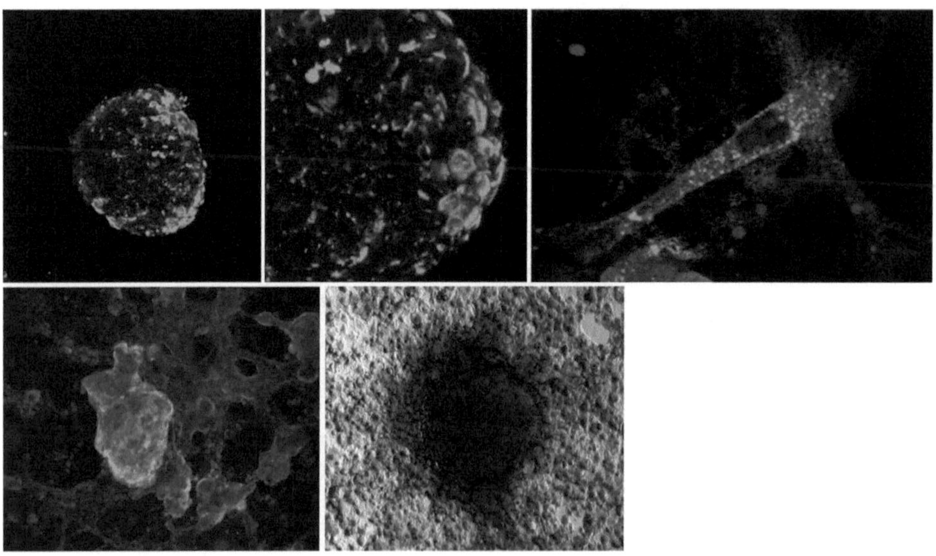

Fig. 1. *Top* three panels show immunostaining for cadherin-11 in a day 5 EB (*top left* and *middle*) and an individual cell 2–3 days after dispersal of EB. *Bottom panels* show discrete bone nodules (21 day cultures) stained with osteocalcin (*left panel*) and alizarin red S (*right panel*). Original magnifications ×50 – *top left* and *bottom panels*; ×80 – *top middle* and *top right panels*.

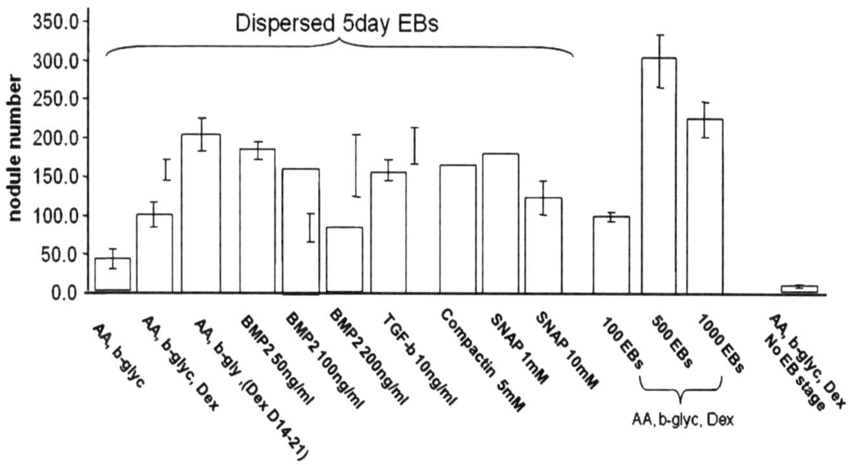

Fig. 2. The effects of various factors on the osteogenic differentiation of mouse ES cells, as assessed by the bone nodule assay. Effects on differentiation vary depending on the type, timing, and concentration of the stimulus, but it is evident from this experiment that intact 3D aggregates support the greatest levels of osteogenic differentiation. *AA* ascorbate, *b-glyc* beta glycerophosphate, *Dex* dexamethasone, *BMP* bone morphogenetic protein, *TGG-b* transforming growth factor beta, Comapctin (a statin drug), *SNAP* *S*-nitrosoacetylpenicillamine (nitric oxide donor), *EBs* embryoid bodies.

to prevent cell or aggregate attachment. Alternatively, EBs can be allowed to form by placing the culture plate on an orbital shaker or gyro twister set at 20 rpm. For 2D osteogenesis, we recommend that EBs be cultured for no longer than 5 days with a medium change on day 3 (see Note 6).

3. To dissociate EBs for seeding onto 2D culture plates, gently draw up EBs from Petri dishes, transfer to sterile centrifuge tubes or universals, and allow settling under gravity at 37°C.

4. Aspirate medium and wash EBs twice for 10 min in sterile PBS at 37°C with gentle agitation (thorough washing is important, particularly for hESC-derived EBs, to ensure efficient trypsinisation and generation of a single cell suspension for seeding).

5. Prior to dissociating EBs, it is often useful to separate EBs into defined size ranges using sterile cell strainers e.g. ~40–100 μm meshes (e.g. BD Falcon™). Embryoid bodies sorted into a ~40–100 μm diameter size range seem to work well in our osteogenesis experiments. Under aseptic conditions, the EBs suspended in PBS are first passed through a 100-μm mesh, collecting the EBs that pass through in a 50-mL conical tube. The EBs are then passed through a 40-μm strainer, which will retain the EBs in the desired range. To recover the EBs, the 40-μm strainer is inverted and placed in a sterile Petri dish, and sterile PBS is added dropwise to dislodge the EBs from the strainer and then processed as normal. The <40 μm and >100 μm fractions can be also used if required.

6. Allow EBs to settle under gravity after completion of washes and aspirate sterile PBS.
7. Add ~2 mL of trypsin–EDTA to the EBs and incubate at 37°C with gentle agitation for 5 min.
8. Add ~4 mL of serum containing medium to inactivate trypsin and pipette cells up and down to obtain a single cell suspension.
9. Centrifuge cells at 300 × *g*, 5 min and re-suspend in osteogenic medium plus biochemical factors/supplements and seed at 5,200/cm^2 into tissue culture treated culture plates or chamber slides, depending upon the intended end-point assay (see Note 7).
10. Feed cells every 2–3 days with fresh medium.
11. Culture for up to 21–28 days depending on intended assay (see Note 8).
12. Differentiation of ES cells produces a wide range of cell types, including beating cardiac myocytes (often apparent after 4–6 days of culture), neurons, epithelium, and mesenchymal lineages. Osteogenic differentiation is well advanced after 21 days in culture and can be assessed by several methods (see below).

3.2. Induction of Osteogenic Differentiation via Intact EBs and "3D Culture"

1. This protocol involves making EBs and keeping them intact over the course of culture period. The EBs can be made by the mass suspension method described above or the microtube method, which produces one EB per tube. While the EBs are kept intact, the approaches to stimulate the cells and analyse them are similar to those described above. The most significant observation that we have made to date is that maintaining 3D interactions within the EB and in the presence of osteogenic stimuli produces greater levels of osteogenic enrichment compared to 2D culture, at least based on the bone nodule assay and also osteocalcin immunostaining. Further work is required to understand the mechanism, but we have noted that cadherin-11, which we and others have shown to be important in osteogenesis, is induced relatively early (e.g. after 3–4 days) in EBs and may be important in coordinating cellular responses and differentiation within the 3D aggregate.
2. Work is still in progress, but we include the microfuge tube, which, although quite labour-intensive, produces EBs that are more uniform in size and shape, and it appears that the physical properties of the aggregates might be important in influencing differentiation – see note above on size ranges of EBs.
3. For the mass suspension method, EBs can be made as described above and sorted into ~40–100 μm size range.

4. For the microfuge tube method, prepare a suspension of ES cells over the density range of 1×10^4 to 5×10^4 cells/mL (we find that 2×10^4 works well for osteogenesis in our experiments). Add 1 mL of the suspension to a single 1.5-mL conical shaped, screw-capped microfuge sample tube. Make up as many aggregates i.e. microfuge tubes as desired. Put the lids on each tube, leaving them loose to allow gas exchange within the incubator. As described above, the EBs are initially formed in DMEM/alpha MEM plus FCS, but no supplements.
5. For both methods, the EBs should be allowed to form for up to 5 days with a medium change on day 3.
6. To induce osteogenesis, EBs formed by either method are cultured in osteogenic medium plus the biochemicals/supplements described above (see Note 2). In our experiments, we tend to use tissue-culture-treated plastic and static culture conditions, allowing the 3D aggregates to adhere. Seed the EBs at densities ranging from 10 to 1,000 EBs cm^2 (see Fig. 3) and culture for up to 12–28 days, with medium changes every 2–3 days.

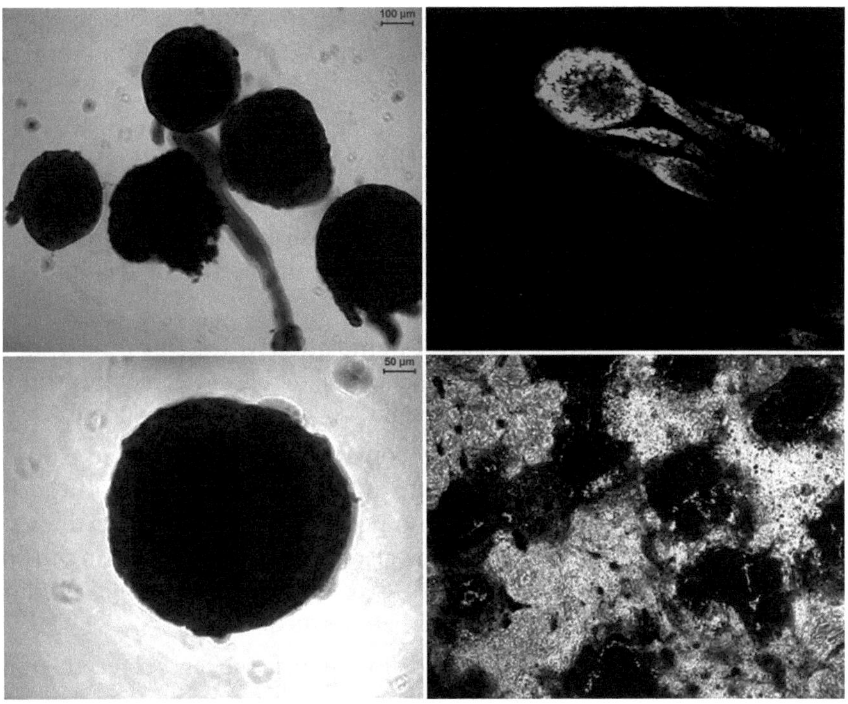

Fig. 3. Examples of alginate encapsulated mouse ES cells cultured in osteogenic medium. *Top left panel* shows staining with toluidine blue to demonstrate presence of cells within the alginate sphere. *Top right panel* shows immunostaining for cadherin-11 in a *sphere* cultured in osteogenic medium for 18 days. *Bottom right panel* shows staining for Alizarin Red S in an alginate *sphere* cultured for 18 days. *Bottom left panel* shows adherent bone nodules released from *alginate spheres* after 18 days and cultured on tissue culture plates for 48 h stained with Alizarin Red S. Original magnifications all = ×20, except *bottom left panel* which = ×50.

7. We have mainly investigated intact EBs under static conditions, but there should be no reason why they cannot be also maintained in osteogenic culture medium under dynamic conditions involving agitation (e.g. gyro twister) or in a stirred bioreactor (see Subheading 3.3).

3.3. Alginate Encapsulation of ES Cells and Osteogenic Differentiation in 3D Cultures

1. Encapsulation of cells using a variety of semi-solid materials such as alginate or agarose is a well-established protocol used routinely as a method for scaling up cell cultures, including ES cells. Here, we use alginate, which as a polysaccharide material gets cross-linked when exposed to Ca^{2+} ions. By forming small droplets of alginate solution and collecting the droplets in buffers containing Ca^{2+} ions, small semi-solid spheres of alginate are formed almost instantaneously. By mixing cells into the alginate solution, they can then be entrapped into the spheres. The method we describe here is a very simple approach that does not require particularly specialised equipment to perform the encapsulation process and produces spheres in the range of 1–2 mm in diameter. The method and the spheres produced here work perfectly well in our experiments; however, it is possible to use more specialised methods and equipment to produce spheres with more defined size ranges e.g. from 50 to 500 μm (see Fig. 3).
2. Prepare ~5 mL of a single cell suspension of ES cells at density of 2.5×10^5 cells/mL to 1×10^6 cell/mL in Tris-buffered saline and then pellet by centrifugation at $300 \times g$.
3. Aspirate the buffer and re-suspend the pellet in the same volume (~5 mL) of sterile 2% alginic acid in Tris-buffered saline, pre-warmed to 37°C.
4. Using a peristaltic pump fitted with sterile tubing (or a syringe driver) fitted with a 25 gauge blunt needle flow through sterile Tris-buffered saline to achieve a flow rate of ~30 mL/h.
5. Prepare a sterile container such as sterilised 200-mL glass beaker containing a sterile solution of 100 mM $CaCl_2$ in Tris-buffered saline and position the tip of the needle so that it is ~3 cm above the surface of the liquid. Changing the height of the needle and flow rate will obviously change the dimensions of the spheres that are formed.
6. Once the desired flow rate has been achieved, transfer the tubing to the cell suspension, ensuring that they are evenly dispersed in the alginate solution prior to collection. This can be done by gentle agitation or stirring. Flow through the cell mixture and collect the spheres in the $CaCl_2$ buffer. Allow the spheres to remain in the $CaCl_2$ buffer for ~5 min to ensure the sphere structure has stabilised before removing.

7. Wash the spheres in several changes of Tris-buffered saline and sample a few spheres to assess cell encapsulation. This can be done by visualising using phase contrast or differential interference contrast microscopy or by histological staining such as with a 1% solution of toluidine blue in 50% (v/v) isopropanol.
8. Re-suspend the spheres in osteogenic medium plus biochemical factors/supplements as described in Subheading 3.1. The alginate spheres are amenable to dynamic culture using simple agitation (e.g. gyro twister set at 20 rpm). We have also used both stirred bioreactors and rotary bioreactors (set at 20–30 rpm). All methods seem to work equally well in our hands. Cultures can be maintained for up to 21–28 days, and medium should be exchanged every 2–3 days (see Note 2).
9. Cells/aggregates can be released from the alginate spheres by placing them in sterile solution of 30 mM sodium citrate in distilled water. The released cells/aggregates can be then pelleted by gravity sedimentation, washed with PBS, and then re-suspended in culture medium and re-plated onto tissue culture plates.

3.4. Chemically Engineered Cell Aggregation

1. Here, we describe a method of promoting 3D cell–cell interactions by engineering cells to incorporate a "chemical handle" into their extracellular surfaces and use this to "actively" co-ordinate cellular interactions. One such approach is engineering of cell surface sugar residues and in particular sialic acid, which is present on the surfaces of virtually all cell types. Studies by Mahal et al. (1997) (18) showed that a ketone group, which is not normally found in any cell surface biological molecule, could be added to sialic acid and subsequently the ketone group could be used for ligation of specific probes and molecules, enabling manipulation of cell interactions and targeting. We have extended these studies, and rather than the metabolic labelling described above, our approach is based on performing a periodate oxidation of the cells in suspension to form an unnatural, reactive aldehyde on a proportion of cell surface sialic acid groups. This has significant advantages over metabolic labelling in that it can be achieved in a matter of minutes, compared to days. Cell aggregation was subsequently controlled by biotinylation of the cell surface via a biotin hydrazide reaction and cross-linking by introduction of the protein avidin (18). We have reported application of this method to a variety of cell types (19) including mouse ES cells (20).
2. This method enables cell aggregates to form more rapidly and interestingly in the case of ES cells can augment induction of mesoderm formation and subsequently can enhance osteogenic differentiation (see Fig. 4).

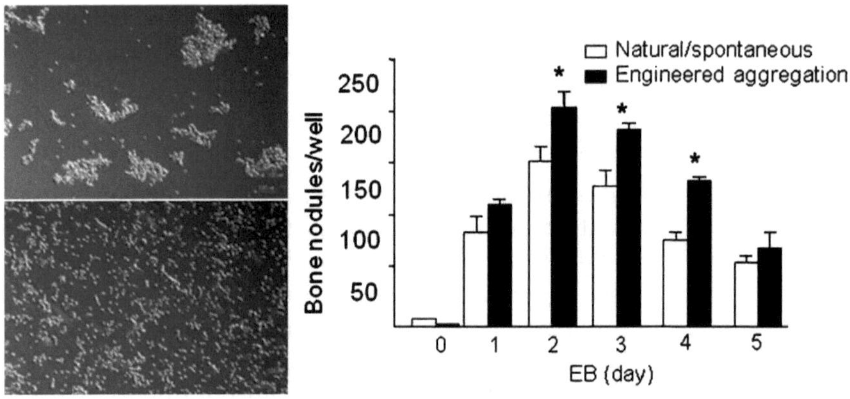

Fig. 4. Example of engineered versus natural spontaneous aggregation in mouse ES cells. The photomicrographs demonstrate formation of distinct aggregates after hours in culture using the chemically engineered method (*top panel*) compared to non-engineered cells over the same time course. The *graph* shows quantitation of bone nodules formed by the engineered and spontaneous aggregation relative to length of time EBs were allowed to form before being cultured in osteogenic medium.

3. Suspend ES cells at a density of 5×10^4 cells/mL or up to 2.5×10^5 cells/mL in sterile PBS or DMEM/alpha MEM.
4. Pellet cells by centrifugation at $300 \times g$, remove supernatant, and re-suspend pellet in sodium periodate oxidation solution at 4°C. Incubate in the dark at 4°C for 10 min.
5. Pellet cells by centrifugation at $300 \times g$, remove supernatant, and wash with 2–3 changes of biotinylation buffer.
6. Pellet cells by centrifugation at $300 \times g$, remove supernatant, and re-suspend in biotinylation solution and incubate at room temperature with gentle shaking e.g. gyro twister at 20 rpm.
7. Pellet cells and wash with biotinylation buffer, then pellet cells by centrifugation at $300 \times g$ and re-suspend in sterile PBS or DMEM/alpha MEM. Cells can be then plated at the required density (e.g. 5×10^4 to 2.5×10 cells/mL) in any non-adherent vessel.
8. Add avidin solution to achieve a final working concentration of 10 μg/mL and incubate at 37°C with gentle agitation until aggregates form (can range from a few hours to overnight).
9. Once formed, aggregates can be manipulated as described above to investigate osteogenic differentiation.

3.5. Osteogenic Differentiation on 3D Scaffolds

1. There is an extensive body of literature on the use of scaffolds for 3D culture and osteogenic differentiation. Scaffolds can vary in their intrinsic chemistry (e.g. natural or synthetic), their porosity and roughness, whether they are engineered to be biodegradable and also whether they include growth factors or other bioactive molecules that get either adsorbed

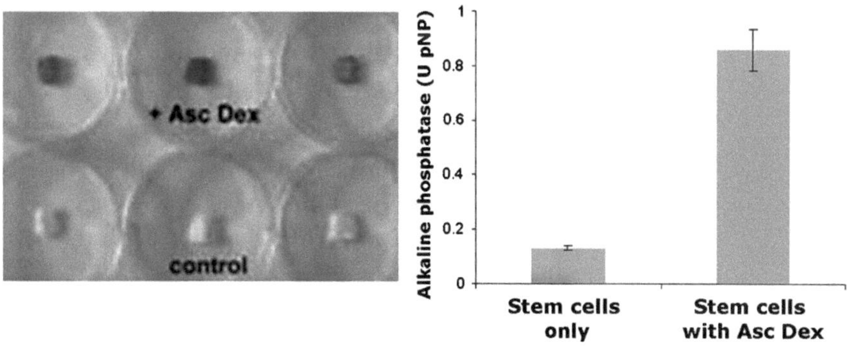

Fig. 5. Example of poly lactide scaffolds in which human embryonic stem cells have been cultured in the presence or absence of osteogenic medium. The photomicrograph shows an example of alkaline phosphatase staining (red/dark areas) after 21-days in culture. The *graph* shows an example of alkaline phosphatase activity of human embryonic stem cells cultured on the scaffolds in the presence or absence osteogenic medium.

onto their surfaces or encapsulated within their structure. It is not the intention to recommend any specific configuration of scaffold. Here, we present some basic approaches that we have used to investigate osteogenesis on porous 3D polylactide scaffolds (see Fig. 5).

2. Prepare a single cell suspension of cells – this can be of ES cells, but we recommend inducing differentiation by formation of EBs, using any of the approaches described above.
3. Place the scaffold into the well of a culture plate so that the scaffold just fits. For example, we used polylactide scaffold (5 mm^3) and a standard 24-well culture plate – ideally a non-tissue-culture-treated plate to minimise attachment of cells to the plate. Ensure the scaffold is wetted with osteogenic culture medium (without supplements) by soaking overnight in a cell culture incubator at 37°C and 5% CO_2.
4. Remove the culture medium and dynamically seed the cells onto the scaffold by taping the plate to an orbital shaker with 2 cm radius at 60 rpm overnight. In our experiments, we typically seed a minimum of 1×10^5 cells per scaffold and up to 1×10^6 cells per scaffold – but it depends on size of scaffold! (see Note 9).
5. Transfer the scaffolds to fresh non-tissue-culture-treated plates and culture the scaffolds in osteogenic medium plus biochemical factors/supplements for up to 21–28 days, with agitation at 30–40 rpm.

3.6. Assessing Mineralisation with Alizarin Red staining – Bone Nodule Assay

1. Demonstration of calcium salts is a standard histological method for identifying biomineralisation. Alizarin Red is a calcium binding dye commonly used for this purpose. The von Kossa method whereby silver is substituted for calcium and then reduced to metallic silver is also suitable.

Increased differentiation of cells along the osteogenic lineage in response to growth factors is indicated by increased number of mineralised bone nodules (see Fig. 1).

2. Aspirate culture medium and wash cells briefly in PBS.
3. Fix cells with 10% formalin for 10–20 min at room temperature in fume cupboard and rinse with several changes of PBS – discard of all waste fixative and washes in the appropriate chemical waste.
4. Stain cells for ~5 min with 2% Alizarin Red S solution at room temperature.
5. Tip off staining solution and de-stain in running tap water for 20–30 min.
6. Tip off water and allow to air-dry.
7. Mineralised nodules can be identified as cell clusters stained a deep violet/red. Count the total number of mineralised nodules. This can be carried out either manually on a light box or using digital imaging and colony counting software or image analysis software (see Note 10).

3.7. Immunocytochemistry

1. A brief protocol based on immunofluorescence is presented here. Immunofluorescence is recommended over immunoenzyme methods as the process of in vitro osteogenic differentiation is usually marked by the formation of 3D mineralising bone nodules. Immunofluorescence staining of such nodules offers the possibility of using confocal microscopy to optically slice through the nodules and visualise localisation of proteins/markers within the structure (see Fig. 1). We strongly recommend using immunostaining as a method to help verify osteogenic differentiation and especially to complement the bone nodule assay. Ideally, we recommend using antibodies to the transcription factors cbfa-1/runx2 or osterix, the cell adhesion molecule cadherin-11, the matrix proteins collagen-1, and osteocalcin. There are, of course, many others that can be used including alkaline phosphatase, but this should be the bone-specific isoform.

Immunostaining can be performed on fixed cells or aggregates grown in any culture dish, but larger plates require larger volumes of antibody, making it very expensive; hence, it is recommended to perform immunostaining on cultures grown on coverslips or cell culture chamber (4–8 wells) slides (e.g. Nunc). Free-floating aggregates or encapsulated cells can be readily immunostained as intact structures and visualised by confocal microscopy to produce optical slices through the structure or sectioned using a microtome to produce physical slices through the structure.

2. Aspirate culture medium and wash with several changes of PBS.

3. Fix cells with a 4% solution of paraformaldehyde for 20–30 min and then wash with several changes of PBS. Dispose of paraformaldehyde and post-fixative washes in the appropriate chemical waste.
4. Incubate with PBS: 0.2% Triton X-100 for ~30 min to permeabilise the cells and then wash with several changes of PBS.
5. Incubate with 3% (v/v) normal serum from same species whichever the secondary antibody to be used was raised for ~20 min.
6. Aspirate normal serum and without washing incubate with primary antibody overnight at 4°C. "Optimal" dilution of antibody should be determined empirically, using the manufacturer's antibody data sheet as a guide.
7. Tip off antibody and wash with several changes of PBS.
8. Incubate in the appropriate fluorophore-conjugated secondary antibody for ~1 h at room temperature (use the manufacturer's antibody data sheet as a guide for dilution).
9. After washing with PBS, slides, coverslips, or plates can be mounted with PBS, glycerol, or Vectashield (plus or minus counterstains like DAPI or propidium iodide).

3.8. Assessing Osteogenesis by Alkaline Phosphaste Staining

1. Alkaline phosphatase is a membrane-bound enzyme which catalyses the hydrolysis of phosphate monoesters. It is present in many tissues but is particularly abundant in tissues like the bone, liver, kidney, placenta, and also in embryonic tissues including ES cells. The location and also activity of the enzyme can be determined based on use of substrate which is hydrolysed by the alkaline phosphatase and results in a colour change of the substrate. It should be noted that although bone tissue (osteoblasts) typically have some of the highest levels of this enzyme, it is not specific and should be verified by additional methods like the bone nodule assay and immunostaining. In the enzyme histochemistry method, this reaction takes place in situ, and the substrate is normally coupled to a diazonium salt which results in deposition of coloured insoluble dye (can be red or blue depending on which salt is used) at the site of the cells containing the enzyme. This method can be performed on 2D cultures, 3D aggregates, and scaffolds.
2. Wash the cells, aggregates, or scaffolds to be stained in several changes of Tris buffer and then fix by immersion in 10% formalin for 20–60 min, depending on ion size of tissue/object. Wash in several changes of Tris buffer.
3. Aspirate Tris buffer and incubate by immersion in alkaline phosphate incubation solution (10 mL of the stock naphthol solution containing dissolved 10 mg Fast Red TR) for

5–30 min. Development of the colour reaction can be watched and stopped as required.

4. Wash in several changes of distilled water and mount in a suitable mounting medium (e.g. 50:50 distilled water and glycerol). This is required if cells are going to be visualised microscopically. For scaffold, they can be stored in distilled water or air-dried.

3.9. Assessing Osteogenesis by Alkaline Phosphatase Activity

In the enzyme activity method, the cells are usually lysed to release the alkaline phosphatase, which is then mixed with the substrate usually on 96-well plate, and the colour change is quantified colorimetrically and plotted against defined standards (see Fig. 5). This method can be performed on 2D cultures, 3D aggregates, and scaffolds.

1. Wash cells, aggregates, or scaffolds in several changes of Tris buffer.
2. Homogenise cells in lysis buffer. For monolayers, this can be done by scraping the cells. Aggregates can be mechanically disrupted using a scalpel or scissors. Scaffolds can be mechanically disrupted using a scalpel or scissors (if amenable) or crushed with a mortar andpestle. Homogenates should be collected into a centrifuge tube and incubated with lysis buffer with continuous agitation at 4°C for 60 min. To improve protein extraction yield, the homogenates can be rapidly free-thawed three times by immersion in liquid nitrogen. Centrifuge at 1,000 × *g* and collect supernatant. Store at −80°C until required.
3. Prepare a 96-well plate and add 50 μL of sample together with 50 μL of alkaline assay buffer and 50 μL of substrate into each well. The plate should also include a similar volume of PNP standards plus separate sample, buffer, and substrate blanks/controls.
4. Protect plate from light and incubate plate at 37°C for ~30 min. Add 0.25 M NaOH to stop the reaction and read absorbance at 405 nm.
5. Alkaline phosphatase activity is expressed in activity Units as calculated from the PNP standards. One Unit of activity is defined by the amount of enzyme activity that liberates 1 μmol of PNP per hour. The results should be normalised for total protein concentration determined by for example Bradford assay.

3.10. Assessing Osteoblast Differentiation by gene Expression Analysis

1. RT-PCR analysis can be used to demonstrate the expression of genes important to the differentiation and mineralisation of osteoblasts. Various gene markers have been described in the literature as indicative of osteogenic differentiation. Two genes commonly used are cbfa-1/runx2, a member of the

runt family of transcription factors, which is necessary and sufficient for osteoblast differentiation, and osteocalcin (also known as bone Gla protein or BGP), a major non-collagenous matrix protein of bone, which is produced by mature osteoblasts. Loss of ES pluripotency markers such as Oct-4 and Nanog can also be assessed. A housekeeping gene such as GAPDH, β-tubulin, or β-actin should also be included. We have found GAPDH to be a suitable housekeeping gene for differentiating human ES cells, but not in differentiating mouse ES cells, where β-actin expression is more stable. Step 2–7 describe RNA isolation; 7–11 reverse transcription; 12–19 PCR.

2. Prepare RNeasy reagents as directed in kit manual.
3. Aspirate culture medium and wash cells briefly twice with D-PBS.
4. Lyse cells in buffer RLT and homogenise the sample. Homogenised lysates may be stored for several months at −70°C.
5. Isolate RNA as directed by kit instructions, including on-column DNAse treatment step.
6. Elute RNA from column in 30 μL of nuclease-free water and store at −70°C.
7. Prepare a 1-μg aliquot of RNA and heat it to 70°C for 10 min, then centrifuge briefly and place on ice.
8. Prepare a reverse transcription master mix as described in the kit protocol, using random hexamer primers.
9. Prepare reaction mixture (20 μL volume) by mixing RNA and master mix. Allow to incubate for 10 min at room temperature and then for 45 min at 42°C.
10. Heat reaction at 95°C for 5 min and then at 0–5°C for 5 min.
11. Dilute with 80 μL of nuclease-free water and store cDNA at −20°C.
12. Prepare PCR master mix (25 μL reaction volume).
13. Prepare reaction mix by adding 10 μL of target cDNA plus 15 μL of master mix in thin-walled PCR tubes. Include a negative control with nuclease-free water substituted for cDNA. If DNase treatment was omitted during RNA isolation, then it is also good practice to do RT-ve controls using original RNA samples.
14. Put reaction tubes into cycler and run on settings as shown in Table 3. PCR products can be stored at 4°C before electrophoresis if desired.
15. Prepare a 1.5% agarose gel in TAE buffer, including 10 μl of ethidium bromide solution.

Table 3
Thermal cycling conditions for RT-PCR

Gene	Cycling conditions	Expected product size (bp)	Cycles
Runx2	94° 30 s, 58.5° 30 s, 72° 45 s	286	32
Osteocalcin (human)	94° 60 s, 60° 60 s, 72° 60 s	293	34
Osteocalcin (mouse)	94° 30 s, 61° 30 s, 72° 60 s	193	30
Oct4 (human)	94° 30 s, 59° 30 s, 72° 45 s	649	30
Oct4 (mouse)	94° 30 s, 61° 30 s, 72° 60 s	312	30
GAPDH (human)	94° 30 s, 58° 30 s, 72° 30 s	303	35
β-Actin (mouse)	94° 15 s, 60° 60 s, 72° 45 s	492	30

16. Mix 9.4 μL of PCR product with 2.4 μL of 5× loading buffer.
17. Mix 3 μL of 100 bp ladder with 2.4 μL of 5× loading buffer and 6.6 μL of nuclease-free water.
18. Load 10 μL of PCR product or ladder per lane and run gel for 60 min at 90 V.
19. Visualise products on a UV transilluminator (care – UV radiation can be hazardous!) Check for specific product at anticipated size, presence of non-specific products, and primer dimer.

3.11. Conventional Bright-Field and Epi-fluorescence Microscopy

1. Samples can be readily viewed by conventional bright-field microscopy. In addition, techniques like differential interference microscopy and polarised light microscopy can be applied to improve image contrast and also provide information on the orientation of extracellular matrix tissues like collagen and lipids. A stereo-microscope is also useful for macroscopic imaging of culture plates or scaffolds. Immunofluorescently stained samples require an epi-fluorescence microscope fitted with appropriate excitation and emission filters (always check the configuration before using a given fluorophore) or a confocal microscope.

3.12. Scanning Electron Microscopy

1. By scanning an electron beam across the surface of a sample, Scanning Electron Microscopy (SEM) allows high-resolution images of the topography of the sample and also any cells and tissues growing on that scaffold to be obtained. Most research institutions will have access to both SEM and TEM microscope facilities and services.

2. For SEM (and transmission electron microscopy; TEM), tissues and scaffolds are usually fixed with a 3% (v/v) solution of glutaraldehyde in PBS. Glutaraldehyde is preferred to formaldehyde as its reaction with tissues is more rapid and ultimately produces better preservation of tissue micro-architecture, which works well with SEM and TEM applications. The down side of glutaraldehyde is that unlike formaldehyde it can cause marked denaturation of protein and changes in the conformation of proteins and DNA – this could affect conventional histological, histochemical, and immunochemical staining. Osmium tetroxide is often used as a secondary fixative and contrast agent in electron microscopy. It has relatively poor tissue penetration but cross-links phospholipids. Being a heavy metal, it is electron-dense and produces strong contrast with particular cellular structures such as lipid membranes.
3. Scaffolds should be fixed in glutaraldehyde overnight at 4°C, before washing with several changes of sodium phosphate buffer – dispose of waste fixative and washings in the appropriate chemical waste. Scaffolds are then post-fixed with 1% solution (v/v) of osmium tetroxide in sodium phosphate buffer for 1 h at room temperature. Glutaraldehyde and osmium tetroxide are highly toxic and should be handled wearing gloves, in a fume cabinet and disposed of in the appropriate chemical waste.
4. Samples are then dehydrated through an increasing series of ethanol (25% [v/v], 50% [v/v], 70% [v/v], 90% [v/v], 95% [v/v], and 100% [v/v]) and dried using hexamethyldisilaxane (HMDS). Scaffolds are then cut down to an appropriate size using a scalpel blade for mounting on carbon-coated electron microscope stubs and sputter coating with gold for ~4 min. Samples are then imaged with the scanning electron microscope.

3.13. Micro-Computed Tomography

1. This is an imaging method combining X-ray transmission and tomographical reconstruction algorithms, to create 2D and 3D images of objects. It has been used extensively to investigate the architecture of bone tissue and more recently has been applied to imaging the structure of porous scaffolds. Obviously, this is quite a specialised piece of equipment, but bench-top versions are now available, and it is likely that most research institutions would have access to such equipment. It is a non-destructive method, and the samples require relatively processing. However, we have found that by fixing and staining scaffolds with a 1% solution of osminum tetroxide, it is possible to obtain images of the distribution of cells within a scaffold. The osmium serves as a contrast agent binding specifically to cell membranes and yields low-resolution

images of the cells. The software can enable full 3D rendering or 2D sections through the scaffold. It can also enable the image to be quantified to determine properties such as percentage porosity of the scaffold or percentage cell/tissue occupancy.

4. Notes

1. Basal culture medium can influence differentiation, and in many protocols, use of alpha MEM with ribonucleosides and deoxyribonucleosides is recommended. This medium is usually expensive and may be 5–6 times the price of other media. We have, therefore, also used DMEM in our osteogenesis experiments and found it to work well.
2. Batch testing of FCS is essential, and it is recommended to obtain a number of different batches and perform a basic osteogenic differentiation assay – e.g. culture cells for 21 days in the presence of osteogenic supplements (acorbic acid, etc.) and then perform a bone nodule assay (Alizarin Red S staining) to determine which serum best supports differentiation. Prior to making up culture medium, we routinely filter the serum through a 0.2-μm filter – it is not essential to do this as the serum should be sterile, but if FBS is prepared and aliquoted as a general laboratory consumable, it is worthwhile filter-sterilising to be sure.
3. The osteogenic supplements ascorbic acid, beta-glycerophosphate, and dexamethasone should be all prepared as aliquots, filter-sterilised through 0.2-μm filter, and stored frozen. They should be added fresh to the culture medium and only when the cells are fed/stimulated. Ascorbic acid and beta-glycerophosphate are water-soluble and can be prepared in sterile PBS or medium (without FBS) and filter-sterilised into aliquots. Dexamethasone is usually water-insoluble (although some batches are water-soluble) and should be dissolved in DMSO or ethanol and then diluted to the appropriate working concentration with PBS/medium. It may be helpful to prepare aliquots at 10–1,000× stock concentrations and dilute as required. It is possible to use a range of other factors to stimulate osteogenesis (see Fig. 2), but as many of these factors are expensive (e.g. cytokines), the method described here can be useful in defining a set of basic conditions or parameters or time course prior to use of cytokines.
4. Formalin is a good all-round fixative, but if samples are to be also used for other investigations such as immunocytochemistry, formalin can sometimes be too "harsh" and

can potentially mask antigenic sites, requiring antigen retrieval step. If immunocytochemistry is to be used, it is worth considering a more gentle fixative such as a 4% (v/v) solution of paraformaldehyde in PBS. Samples can be fixed and processed in exactly the same way, but being a more gentle fixative, tissue micro-architecture may not be as well preserved – but it is still good!

5. Basic protocols and reagents for immunocytochemistry are provided and as described work well in our hands, providing a useful investigative measure of the extent of osteogenic differentiation. Many other antibodies and probes are available, and it is strongly advised that these, along with ones we have included, should be evaluated in your own laboratories and experiments. When using an antibody for the first time or testing on an unknown cell or tissue, the antibody should be titrated to determine an optimal dilution. Ideally, a known positive control tissue should be used – bone tissues from rodents usually work well. Immunofluorescence staining of thick tissue sections or aggregates (50–500 μm) is performed with the section or aggregate "free-floating" in a small Petri dish, and all solutions are added to and decanted from the section. Overall, however, the same routine is followed except incubation times (depending on section thickness) have to be increased; Subheading 3.7, step 3 incubate for up to 1–2 h; Subheading 3.7, step 4 incubate for up to 60 min; Subheading 3.7, step 6 incubate for up to 48 h; Subheading 3.7, step 8 incubate for up to 3 h. All washing stages should be increased to ~30 min. When staining scaffolds, it might be better to consider using immunoenzyme and chromogen methods as many scaffolds have intrinsic autofluorescence, which can mask immunofluorescence signals. These approaches could be also extended to include Western blotting.

6. As discussed above, there are many ways for forming embryoid bodies. We routinely use suspension culture, which in our hands works well. However, it is advisable to try other methods such as the microfuge pellet method, which is also described. Others have used centrifugation of cells in 96-well (V bottom) plates to form EBs (21). We have also found that it is useful to prepare embryoid bodies in a basal osteogenic medium, which comprises DMEM or alpha MEM supplemented with 10–15% FBS (batch tested for osteogenesis). We find that forming the embryoid bodies in this medium can help "acclimatise" the cells. If EBs were made in basic medium used for ES cell culture and then dispersed as single cells into osteogenic medium, then many cells often failed to adhere to the culture plates. We have also found that the time period over which EBs are allowed to form can also influence

osteogenic differentiation. In our initial ES studies, we used 5-day-old EBs, but subsequently, we have found that osteogenesis can be improved if younger 2–3-day-old EBs are used (see also Notes 7 and 8).

7. Cell plating density is known to have a significant effect on osteogenic differentiation and the density recommended works well, although again it is advisable to try different densities in your own experiments. We have also investigated plating out intact embryoid bodies, and this also seems to influence osteogenesis (Fig. 2). To encourage either dispersed single cell suspensions or EBs to stick to the culture plate, it is advisable to briefly coat the plate with a 0.1% sterile solution of bovine or porcine gelatin for ~20 min.
8. The duration of the culture can be varied to investigate specific phases of the osteogenic process, and it is advisable to set up several replicates so that detailed time courses can be performed. Typically, osteogenesis is divided into three phases of 7 days, with days 1–7 associated with induction of early genes/markers like cbfa-1/runx2, osterix, and cadherin-11. Days 7–12 are associated with collagen-1 and alkaline phosphatase, and days 12–21 are associated osteocalcin and mineralisation.
9. It is important to assess seeding efficiency of scaffolds using methods such as Alamar Blue assay and the Hoescht assay to compare number of cells retained on the scaffolds compared to number seeded (22). This involves sacrificing samples, so ensure that a suitable number of replicates are available. At this point, scaffolds can be either cultured as described or used in bioreactor studies or in animal models (with the appropriate Home Office licences).
10. The bone nodule assay is a quantitative assay as each bone nodule is believed to be the product of one osteoprogenitor. Thus, by counting individual bone nodules, it is possible to determine the numbers of cells, osteogenic potential within a sample, and to assess the influence of culture conditions (e.g. density, biochemical stimuli etc.) on those cells. It might be helpful to divide the plate into grids by overlying the bottom of the plate with an acetate sheet scribed with grids of known dimensions. Some image analysis software packages are able to count the nodules or measure the stained areas. Using this approach, we have found that timing for delivering a particular stimulus like dexamethasone can significantly influence osteogenesis and for both mouse and human ES cells adding dexamethasone to culture ~14 days after plating resulted in a several fold increase in the numbers of bone nodules. Such studies are useful in understanding the process of differentiation

within stem cell cultures and phases/time points when there might be enrichment for particular cell types. This approach also enabled us to apply methods to purify osteogenic progenitors at specific time points using an antibody to cadherin-11 and magnetic-activated cell sorting (MACS).

References

1. Aubin, J.E., Triffitt, J.T. (2002) Mesenchymal stem cells and osteoblast differentiation. In: Bilezikian, J.P., Raisz, L.G., Roadan, G.A., Editors. Principles of Bone Biology. Academic Press, USA, pp. 59–82.
2. Bruder, S.P., Fink, D.J., Caplan, A.I. (1994) Mesenchymal stem cells in bone development, bone repair, and skeletal regeneration therapy. *J. Cell. Biochem.* **56**, 283–294.
3. Buttery, L.D.K., Bourne, S., Xynos, J.D., Wood, H., Hughes, F.J., Hughes, S.P.F., Episkopou, V., Polak, J.M. (2001) Differentiation of osteoblasts and in vitro bone formation from murine embryonic stem cells. *Tissue Eng.* 7, 89–99.
4. Bielby, R., Polak, J.M., Buttery, L.D.K. (2004) In vitro differentiation and in vivo mineralization of osteogenic cells derived from human embryonic stem. *Tissue Eng.* **9/10**, 1518–1525.
5. Bourne, S., Polak, J.M., Hughes, S.P.F., Buttery, L.D.K. (2004) Osteogenic differentiation of mouse ES cells: differential gene expression analysis by cDNA microarray and purification of osteoblasts by cadherin-11 magnetic activated cell sorting. *Tissue Eng.* **10**, 796–806.
6. Sottile, V., Thomson, A., McWhir, J. (2003) *In vitro* osteogenic differentiation of human ES cells. *Cloning Stem Cells* **5**, 149–155.
7. Randle, W.L., Cha, J.M., Hwang, Y.S., Chan, K.L., Kazarian, S.G., Polak, J.M., Mantalaris, A. (2007) Integrated 3-dimensional expansion and osteogenic differentiation of murine embryonic stem cells. *Tissue Eng.* **13**, 2957–2970.
8. Yamada, K.M., Clark, K. (2002) Cell biology: survival in three dimensions. *Nature* **419**, 790–791.
9. Mueller-Klieser, W. (1997) Three-dimensional cell cultures: from molecular mechanisms to clinical applications. *Am. J. Physiol.* **273**, C1109–C1123.
10. Freed, L.E., Guilak, F., Guo, X.E., Gray, M.L., Tranquillo, R., Holmes, J.W., Radisic, M., Sefton, M.V., Kaplan, D., Vunjak-Novakovic, G. (2006) Advanced tools for tissue engineering: scaffolds, bioreactors, and signaling. *Tissue Eng.* **12**, 3285–2305.
11. Howard, D., Partridge, K., Yang, X., Clarke, N.M., Okubo, Y., Bessho, K., Howdle, S.M., Shakesheff, K.M., Oreffo, R.O. (2002) Immunoselection and adenoviral genetic modulation of human osteoprogenitors: in vivo bone formation on PLA scaffold. *Biochem. Biophys. Res. Commun.* **299**, 208–215.
12. Karageorgiou, V., Kaplan, D. (2005) Porosity of 3-D biomaterial scaffolds and osteogenesis. *Biomaterials* **26**, 5474–5491.
13. Liu, X., Ma, PX. (2004) Polymeric scaffolds for bone tissue engineering. *Ann. Biomed. Eng.* **32**, 477–486.
14. Kanczler, J.M., Oreffo, R.O. (2008) Osteogenesis and angiogenesis: the potential for engineering bone. *Eur. Cell. Mater.* **15**, 100–114.
15. Kale, S., Biermann, S., Edwards, C., Tarnowski, C., Morris, M., Long, M.W. (2000) Three-dimensional cellular development is essential for ex vivo formation of human bone. *Nat. Biotechnol.* **18**, 954–958.
16. Karp, J.M., Ferreira, L.S., Khademhosseini, A., Kwon, A.H., Yeh, J., Langer, R.S. (2006) Cultivation of human embryonic stem cells without the embryoid body step enhances osteogenesis *in vitro*. *Stem Cells* **24**, 835–843.
17. Kurosawa, H. (2007) Methods for inducing embryoid body formation: *in vitro* differentiation system of embryonic stem cells. *J. Biosci. Bioeng.* **103**, 389–398.
18. Mahal, L.K., Yarema, K.J., Bertozzi, C.R. (1997) Engineering chemical reactivity on cell surfaces through oligosaccharide biosynthesis. *Science* **276**, 1125–1128.
19. De Bank, P.A., Kellam, B., Kendall, D.A., Shakesheff, K.M. (2003) Surface engineering of living myoblasts via selective periodate oxidation. *Biotechnol. Bioeng.* **81**, 800–808.
20. De Bank, P.A., Hou, Q., Warner, R.M., Wood, I.V., Ali, B.E., Macneil, S., Kendall, D.A., Kellam, B., Shakesheff, K.M., Buttery, L.D. (2007) Accelerated formation of multicellular

3-D structures by cell-to-cell cross-linking. *Biotechnol. Bioeng.* **97**, 1617–1625.

21. Burridge, P.W., Anderson, D., Priddle, H., Barbadillo Muñoz, M.D., Chamberlain, S., Allegrucci, C., Young, L.E., Denning, C. (2007) Improved human embryonic stem cell embryoid body homogeneity and cardiomyocyte differentiation from a novel V-96 plate aggregation system highlights interline variability. *Stem Cells* **25**, 929–938.
22. Unsworth, J.M., Rose, F.R., Wright, E., Scotchford, C.A., Shakesheff, K.M. (2003) Seeding cells into needled felt scaffolds for tissue engineering applications. *J. Biomed. Mater. Res. A* **66**, 425–431.
23. Noth U, Osyczka AM, Tuli R, Hickok NJ, Danielson KK, Tulan KS (2002) Multilineage Mesenchymal differentiation potential of human molecular bone-derived cells. J. *Orthop. Res* 20, 1060–1069.

Chapter 19

3D Structuring of Biocompatible and Biodegradable Polymers Via Stereolithography

Andrew A. Gill and Frederik Claeyssens

Abstract

The production of user-defined 3D microstructures from biocompatible and biodegradable materials via free-form fabrication is an important step to create off-the-shelf technologies to be used as tissue engineering scaffolds. One method of achieving this is the microstereolithography of block copolymers, allowing high resolution microstructuring of materials with tuneable physical properties. A versatile protocol for the production and photofunctionalisation of pre-polymers for microstereolithography is presented along with a discussion of the possible microstereolithography set-ups and previous work in the field.

Key words: 3D cell culture, Tissue engineering scaffolds, Biocompatible/biodegradable scaffolds, Block copolymers, Biomaterials, Ring-opening polymerisation

1. Introduction

Biodegradable and biocompatible polymers are attracting growing interest in the field of scaffold based tissue engineering (1). The tuneable mechanical and biological properties of polymeric materials (2) and recent advances in microstructuring (3) make these materials ideal for simulating the range of biological environments encountered in the human body, both for soft and hard tissue engineering (4–6).

1.1. Polymer Synthesis

Biodegradable polymers such as polycaprolactone (PCL) (7), polylactide (PLA) (8), polyglycolide (PGL) (9), trimethylene carbonate (TMC) (10), poly-4-hydroxybutyrate (11), δ-valerolactone (δ-VL) (12), and β-propiolactone (13) are attractive materials for tissue engineering due to their steady degradation in vivo to produce relatively nontoxic, readily excreted products. Many of the

John W. Haycock (ed.), *3D Cell Culture: Methods and Protocols*, Methods in Molecular Biology, vol. 695,
DOI 10.1007/978-1-60761-984-0_19,

individual polymers exhibit both attractive and unattractive physical properties for use as biomaterial, e.g. PCL displays remarkable elasticity and thermal properties, but cellular attachment is poor due to its hydrophobicity (14). PLA, on the other hand, presents good biodegradability and processability, but poor mechanical properties and thermal stability (3). The aforementioned polymers can be copolymerised to combine the properties of two individual polymers. Since these biocompatible polymers have a very similar synthesis route (see Fig. 1), this copolymerisation procedure can be easily carried out in a one-pot synthesis (to produce random copolymers) or in a consecutive manner (to produce block-copolymers) (15). By changing the monomeric mixing ratios of these polymers, the materials properties of the resulting copolymers can be easily tuned, making these materials interesting to be used as scaffolds for tissue engineering. Indeed, these materials have been used in applications such as nerve tract guidance channels (16), drug delivery systems (17), experimental bladder tissue scaffolds (18), and in general tissue-engineering applications (4).

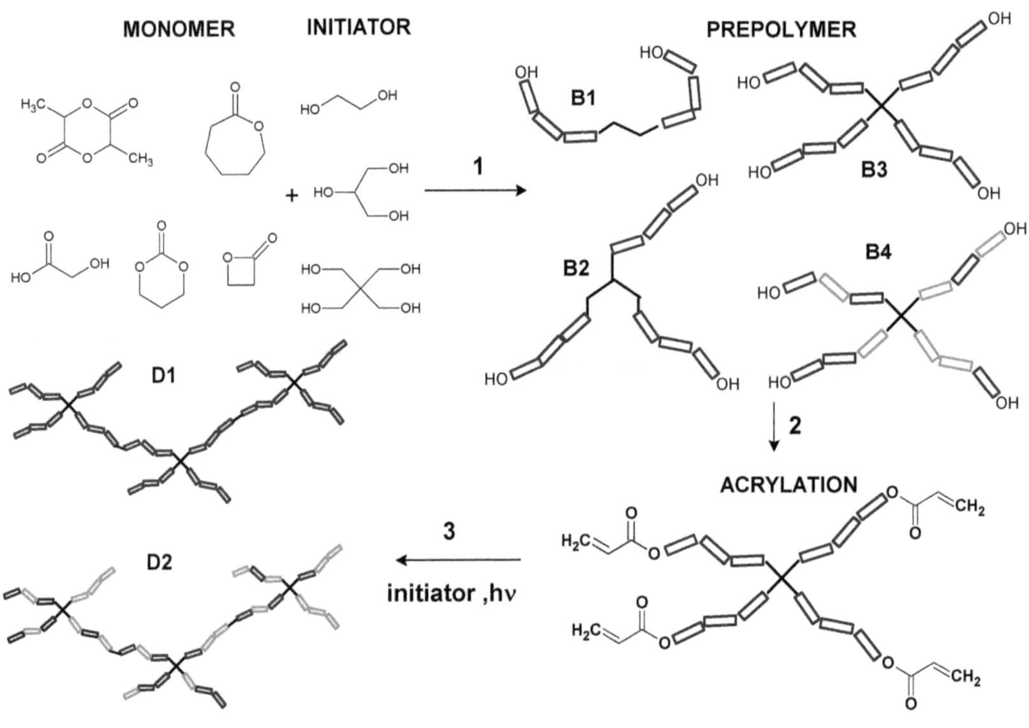

Fig. 1. Polymerisation scheme for producing photocurable biodegradable polymers. Reaction (1) shows a schematic reaction route for making low molecular weight pre-polymers with hydroxyl ending, (B1) two-armed, (B2) three-armed, (B3) four-armed (homo)polymers, and (B4) random copolymers can be produced via this route; reaction (2) illustrates the acrylation of the pre-polymers and these can be used as further building blocks for UV cross-linking as exemplified in reaction (3).

The reaction route in Fig. 1 can be divided in three reaction steps: (1) the production of a low-molecular weight pre-polymer by combining a selected monomer and pre-polymer initiator, (2) acrylation of the hydroxyl end-groups to produce a photocurable pre-polymer and (3) mixing with a photoinitiator and UV irradiation to produce the polymer.

By varying the monomeric units, the chain length and the pre-polymer initiator in reaction step 1 (Fig. 1), block copolymers with varying properties can be produced. For example, the synthesis of four-armed pre-polymers, with pentaerythritol and caprolactone, followed by further polymerisation with l-lactide to give PCL–PLA block-copolymers with well-defined chain structure has been shown to produce materials with impressive biodegradability, miscibility, toughness, and mechanical properties (15). Pre-polymers produced via this reaction can be photosensitised and photocured in consequent reaction steps (steps 2 and 3 in Fig. 1). This reaction route also allows for 3D (micro)structuring of these polymers via (micro)stereolithography or (μ)SL. This is a solid free form (SFF) fabrication technique which allows fabrication of user defined 3D structures via computer aided design/computer aided manufacturing (CAD/CAM) (19).

It has been shown that 3D and 2D cell cultures behave fundamentally different (20). Indeed, the gene expression and biological activity of 3D cell cultures reflects more closely that of living 3D tissues. 3D cell culture scaffolds allow close simulation of these in vivo conditions. Additionally, controlled microstructuring of biomaterials is important for constructing defined microporous structures to be used as tissue replacements in tissue engineering (21). Production of specific 3D topologies has been shown to aid guided cell growth and cell proliferation via mimicking the biological (extracellular matrix) environment (22, 23). CAD/CAM techniques, such as μSL, are well suited to produce these microstructured biomaterials.

A number of biocompatible polymers with interesting physical properties have been synthesised via the reaction route described in Fig. 1, i.e. (1) shape memory polymers (24) and (2) polymers with tuneable degradation rates (25). Biodegradable shape memory polymers form the basis for shape memory surgical sutures, developed by Lendein and Langer (26). These sutures are able to tighten and seal the wound when the shape memory effect is activated. Shape memory behaviour is also useful in implants as they can be inserted/injected into the body in compressed form and then activated within the body to expand to their full functional state, thus providing a route for introducing a scaffold with well-defined structure with a minimal amount of surgical intervention. Additionally, the degradation rate of biodegradable medical implants in vivo is an area of considerable interest (27, 28). A tissue scaffold needs to provide initial rigidity and support as the tissue grows around it. As the tissue starts building its own extra-cellular matrix to retain its structure and function, it is advantageous that the scaffold degrades before it inhibits further tissue growth. Fine tuning of the polymer microstructure allows the production of polymers (e.g. ε-caprolactone–trimethylenecarbonate copolymers (25)), which degrade on a similar timescale to tissue growth.

1.2. Analysis Techniques

1.2.1. NMR Spectroscopy

NMR spectroscopy is used for the analysis of the products. In Fig. 2, we show a spectrum of a *co*-[*ran*(PCL/TMC)-PEO] polymer polymerized via the aforementioned synthesis route. The spectrum provides a measure of an average molecular weight (M_w) via end group analysis. The end group signal is located and used to calculate the integral signal for one proton. The sum of the integrals of the monomeric hydrogens is then divided by the number of monomeric hydrogens and this figure is divided by the integral for one proton as obtained earlier to give the number of repeating monomer units, thus giving the molecular weight. In copolymers, the monomeric ratios can be determined by comparison of selected peak integrations. For example, in the case of

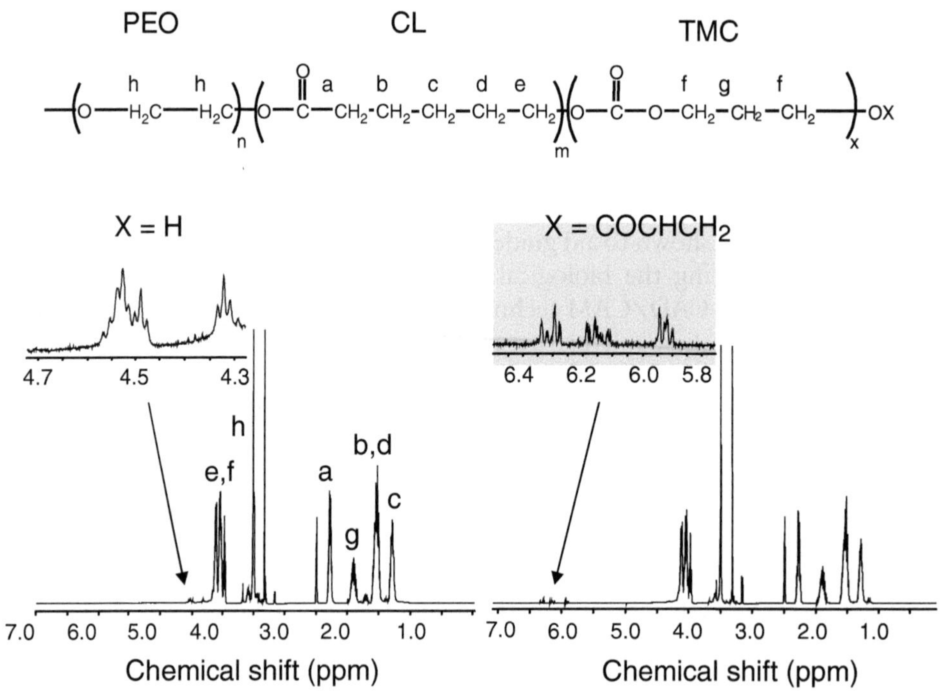

Fig. 2. NMR spectra of a *co*-[*ran*-(PCL/TMC)-PEO] copolymer, as depicted in the chemical formula, with hydroxyl (*left* side) or acrylate (*right* side) groups as end groups. The assignment of the peaks is given in the *left* side spectra and corresponds to the letter code assignment given in the chemical formula. After acrylation the ^{1}H NMR signal of the hydroxyl hydrogens (~4.5 ppm) disappear and a feature corresponding to the ^{1}H NMR signals corresponding to the acrylate groups (~6.2 ppm).

a butyrolactone PCL copolymer the integration of a well-defined methylene peak from the ε-caprolactone monomer can be compared to that of a proton on that of a well-defined tertiary hydrogen peak from the butyrolactone to give monomeric ratios (29). Additionally, a spectrum before and after the acrylation reaction is shown in Fig. 2 indicating that, in this case, this reaction has gone to completion, as confirmed by the disappearance of the signal corresponding to the hydroxyl group proton and the appearance of the signal corresponding to the acrylate protons.

1.2.2. Differential Scanning Calorimetry

By monitoring the amount of heat absorbed by a sample to raise it to a given temperature compared to a standard, a thermogram can be plotted which shows the phase transition temperatures of the sample. For example, when an amorphous polymer reaches its crystallisation temperature less heat is required from the calorimeter as crystallisation is an exothermic process, whereas when the polymer reaches its melting point more heat will be required as melting is an endothermic process. The enthalpies of transition can be determined from the thermogram using a simple equation linking the calorimetric constant and the area under the curve. Differential scanning calorimetry (DSC) is useful for following polymer degradation as the melting point is molecular weight dependent. Blend/copolymer composition can also be investigated with DSC; for example, the copolymerisation of small amounts of β-butyrolactone with PCL significantly reduces PCL crystallinity and can be used to tune the physical properties of PCL polymers (29).

1.2.3. Size Exclusion Chromatography

Size exclusion chromatography (SEC) can be used to attain a measure of both the polydispersity of a polymer, and when coupled with a technique such as light scattering or viscometry, gives absolute molecular weights. SEC involves the sample being carried through a stationary phase (typically a gel such as agarose under low pressure for biomolecules, or a porous solid media such as silica under high pressure for polymers). Filtration is dependent upon the hydrodynamic volume and unrelated to interactions between substrate and stationary phase. When an aqueous solvent is used, the technique is referred to as gel filtration chromatography. When an organic solvent is used, the technique is known as gel permeation chromatography (GPC). For example, in ref. 30, the molecular weight of the segmented glycolide–TMC copolymer (Maxon™) was determined using SEC coupled to a diffractometer with a poly(methyl methacrylate) standard for comparison. As well as giving the molecular weight, SEC was used to study the thermal degradation of the polymer and showed that the polymer rapidly degrades into smaller molecular fragments at 240°C, characterised by the sharp drop in molecular weight and large increase in polydispersity index.

1.3. Microstereolithography

Microstereolithography (μSL) generally consists of three basic components: (1) a layer-by-layer CAD computer model, (2) a scanning laser, and (3) a control/feedback system. A schematic of two alternative set-ups for μSL is shown in Fig. 3; (a) a scanning μSL and (b) a projection μSL set-up. In a scanning μSL set-up the focal point of the light/laser beam is moved within the photopolymerisable resin via a 2D galvanometric scanner fabricating an object directly from the coordinates of the CAD model (as shown in Fig. 3a). In a projection μSL experiment, the beam is projected through either a static or dynamic (spatial light modulator, SLM) photomask (as shown in Fig. 3b) (3). The sample is typically mounted on a high-precision linear *xyz*-translation stage; stepping in the *z*-direction enables the layer-by-layer building of the 3D structure, while *xy*-translation provides the opportunity to produce larger 3D structures than achievable by galvanic scanning/masking alone.

The wavelength of the light source used for μSL is critically dependent on the photoinitiator used to start the polymerisation. Photoinitiators such as Irgacure 369 and 4,4′-bis(diethylamino) benzophenone have an absorption maximum in the UV (300–350 and 350–410 nm, respectively), so mercury lamps and UV lasers (e.g. 355 nm, frequency tripled output of a Nd:YAG laser) can be used for photocuring. With UV light, the material within the focal spot polymerises via a single-photon polymerisation route. One of the drawbacks of conventional stereolithography using UV light is that the beam is absorbed within the first 10–50 μm of the resin. This limits SL to a planar process occurring

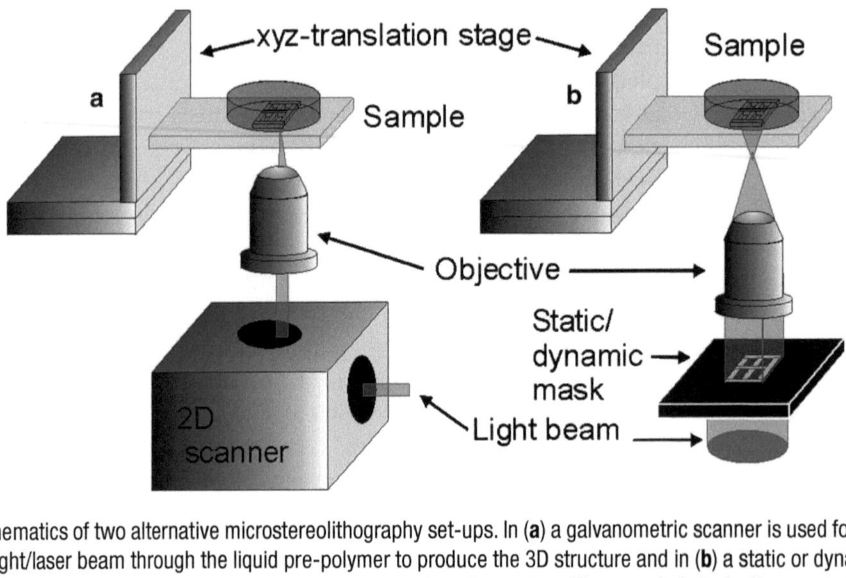

Fig. 3. Schematics of two alternative microstereolithography set-ups. In (**a**) a galvanometric scanner is used for 2D scanning the light/laser beam through the liquid pre-polymer to produce the 3D structure and in (**b**) a static or dynamic mask is used to project a 2D pattern onto the sample to produce the 3D structure. The sample is typically mounted on a high precision *xyz*-translation stage to achieve maximum versatility during direct write. The light can be focused from below through the transparent support (typically a microscope slide) as shown in this figure, or from the top.

only at the surface of the resin. Additionally, radical polymerisation is quenched by oxygen, so reactive species at the surface are inhibited. To overcome this, SL must be performed in an inert gas atmosphere.

The technique of two photon polymerisation (2PP) overcomes these drawbacks as it allows polymerisation to take place within the volume of the resin. In 2PP, photons with double the wavelength of the absorption maximum of the photoinitiator are used (for UV absorbing initiators typically ~800 nm). This means that the photoinitiator needs to absorb two photons to generate the reactive species. The photoinitiator is promoted first to an intermediate state by the absorption of one photon, this intermediate state can be either real (sequential 2PP), with a lifetime of 10^{-4}–10^{-9} s or virtual (simultaneous 2PP) with a lifetime of 10^{-15} s. A second photon must then be absorbed to generate the reactive species. Due to the short lifetimes involved, these 2PP initiations are rare and require a high photon flux, and in practice, 2PP is performed with femtosecond lasers, such as Ti:sapphire lasers.

This two-photon absorption process depends on two photons interacting with the absorbing molecule nearly simultaneously, and results on a quadratic dependence on the intensity of the incident light, rather than a linear dependence in conventional stereolithography. If we approximate the intensity (I) distribution near the focus as a Gaussian distribution instead of the more complex exact Lommel functions (31), then we can estimate the (radial/axial) I and I^2 distribution as shown in Fig. 4a–c for a Gaussian spatial profile laser beam focussed by a high numerical aperture (NA = 1.3) objective lens. Since 2PP cross-section is dependent on I^2, the 3D spot size of the focal point is smaller and more confined within the focal plane than in the case of 1PP (for a more detailed discussion, *see* ref. 3). This allows the production of user defined (sub)micrometre definition structures with these techniques. Indeed, ~100 nm resolution structures have been produced via 2PP (31).

Both 1PP and 2PP techniques have been used to fabricate micro-scale high resolution structures from various photocurable media (13, 32–35). More specifically, 3D microstructuring of the photocurable biodegradable pre-polymers based on TMC/PCL (with PEG as pre-polymer initiator) fabricated via the protocol described in the methods section has been achieved in a number of studies. These TMC/PCL polymers are primarily interesting for nerve conduits or neural tissue scaffolds (36, 37). Mizutani et al. (25) assessed the in vivo biocompatibility/biodegradability of this class of materials via implantation of polymers films into the subcutaneous tissues of rats for up to 5 months. A minimal immune response was observed over the implantation period and the degradation rate depended strongly on the pre-polymer

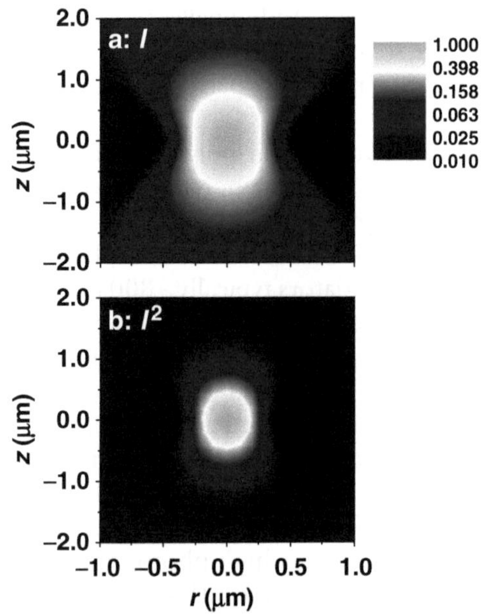

c

$$w(z) = w_0\sqrt{1+\left(\frac{z}{z_0}\right)^2}$$

$$I(r,z) = I_0\,\frac{w_0^2}{w(z)^2}\,exp\left(\frac{-2r^2}{w(z)^2}\right)$$

where
z_0 = Raleigh range
w_0 = beam waste
(0.61λ/NA for diffraction limited focus)

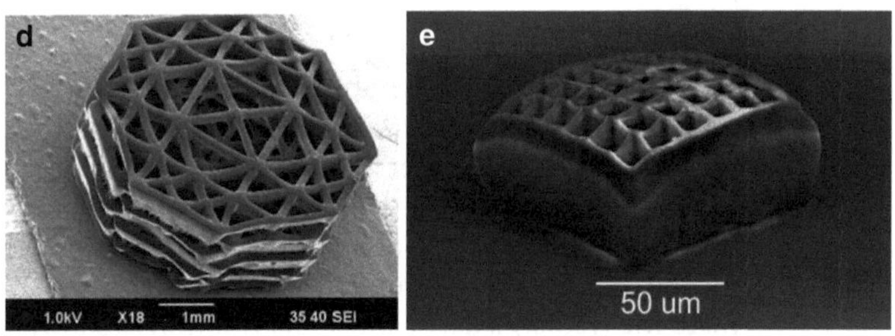

Fig. 4. (**a**) 2D intensity profile close to the focal spot and (**b**) intensity squared profile close to the focal spot for a 740 nm wavelength laser beam focussed by a 1.3 numerical aperture objective, as calculated via the equations given in (**c**). (**d**, **e**) Polymeric structures formed by 1PP (*lower left*) and 2PP (*lower right*). Note the difference in scale bars between the two structures. Figures 4(d) and (e) are reproduced with permission from reference (33) and (38), respectively.

initiator used. Additionally, Kwon et al. (2) explored the fabrication of complex 3D structures of these materials with stereolithography, and a plethora of microstructures of interest for biomedical applications (microbanks, microneedles, and multi-microtunnels) are exemplified. Lee et al. (33) demonstrated the use of similar scaffolds for cartilage regeneration (e.g. see Figure 4d). Interestingly, this study also shows that the specific 3D structure of the scaffolds strongly influences chondrocyte adhesion and proliferation. Higher resolution structures (3–4 μm) have been also achieved via 2PP and have been reported in ref. 38 (see also Fig. 4e).

In this protocol review, we have illustrated a versatile and powerful technique to produce 3D scaffolds for tissue engineering and 3D cell culturing, via combining the synthesis of biocompatible photocurable polymers with stereolithography. The physical properties (degradation rate, elasticity, and shape memory properties) of these μSL photo-polymerised constructs can be tuned. This makes these microstructured materials well-suited for a range of biomedical applications, such as drug delivery vehicles, implants, and tissue engineering.

In this protocol review, we discuss the synthesis route of this class of photocurable, biocompatible, and biodegradable polymers, specifically highlighting the synthesis of PCL/TMC (co) polymers.

2. Materials

For the materials section, we summarise the full spectrum of pre-polymer initiators (two-branched to four-branched) and monomeric units and radical initiators for the first reaction step (see Fig. 1). This first reaction can be used for the synthesis of a large number of pre-polymers via combination of different pre-polymer initiator with different monomers to provide a wide range of materials with tuneable physical properties.

2.1. Pre-polymer Initiator

Two-branched:

1. Polyethylene glycol: M_w: 200–1,000.
2. Glycol: anhydrous, 99.8%.

Three-branched:

3. Glycerol: 99%.

Four-branched:

4. Pentaerythritol: ≥98%.
5. Diglycerol poly-(oxyethylene glycol ether) (b-PEG).

2.2. Monomeric Units

1. ε-caprolactone: purum, ≥99.0%.
2. TMC.
3. L-Lactide: 98%.

2.3. Radical Initiator

1. Stannous 2-ethylhexanoate.

2.4. Acrylation Reaction

1. Methacrylic anhydride.
2. Triethylamine: ≥99%.

2.5. Photoinitiators

1. Irgacure 369.
2. 4,4′-bis(diethylamino)benzophenone.

2.6. Solvents

1. Toluene: anhydrous, 99.8%.
2. Dichloromethane: ≥99.5%, 50 ppm amylene stabiliser.

3. Diethylether: anhydrous, 99.8%.
4. Tetrahydrofuran: anhydrous, ≥99.9%, inhibitor free.

3. Methods

3.1. Polymer Synthesis

3.1.1. Pre-polymer Manufacture

1. The initiator is transferred to a two-necked round bottomed flask and dried for 1 h via a Dean–Stark set-up at 180°C using toluene as the azeotropic solvent. The amount of toluene depends on the particular Dean–Stark set-up. In our typical Dean–Stark set-up (250 mL round bottom flask and 25 mL Dean–Stark Trap), we use ~150 mL of solvent.
2. The monomer is added under a nitrogen atmosphere. The pre-polymer initiator/monomer ratio is an important parameter for tuning the amount of monomeric units per pre-polymer and thus the M_w of the pre-polymer product – e.g. in our typical set-up for PCL-based polymers, we use a 0.025/0.2 mol pre-polymer initiator/monomer ratio – will result in a pre-polymer with, on average, eight monomeric units per pre-polymer. For a typical random copolymer (e.g. a TMC–PCL copolymer) we use 0.025/0.1/0.1 mol initiator/TMC/PCL.
3. 1 μmol of tin(II) ethylhexanoate catalyst is added to the mixture and the reaction is stirred at 180°C under nitrogen for 8 h.
4. After 8 h the reaction is allowed to cool. On cooling, the mixture forms two layers, the polymer being the lower layer. This can then be separated, filtered, and dried with via rotary evaporation (see Note 1).

3.1.2. Acrylation

1. The pre-polymer is dried in a vacuum dessicator for 8 h. All glassware is dried in a vacuum oven at 120°C overnight prior to use (see Note 2).
2. The pre-polymer is dissolved in dry dichloromethane (typically 150–200 mL for 5–10 g of pre-polymer) and added to a two necked round bottom flask.
3. The mixture is stirred vigorously under nitrogen for 1 h.
4. Triethylamine is added to the flask, followed by the slow dropwise addition of methacrylic anhydride using an addition funnel in the ratio of 1 mol (–OH):4 mol methacrylic anhydride:8 mol TEA. A small quantity (~500 ppm) of hydroquinone is added to prevent premature polymerisation. Since the reaction is exothermic the reaction is cooled with ice water in the initial stages (30 min).

5. The reaction is stirred at room temperature for 24 h under nitrogen, before washing with an equal volume of saturated sodium bicarbonate (2×), brine (1×), and then dried with magnesium sulphate which is removed by gravity filtration (see Note 3).
6. The product is dried by rotary evaporation.

4. Notes

1. For solid polymers, a solvent such as isopropanol or methanol can be added to the reaction mixture to induce precipitation of the product, which can be collected by filtration and dried in a vacuum dessicator.
2. Dryness of glassware and reagents is absolutely essential for this reaction, as is the inert atmosphere.
3. Patience is needed with the washing as the phases are slow to separate. Alternative purification methods involve dialysis or using a flash column packed with an appropriate medium such as alumina or silica.

Acknowledgements

FC thanks EPSRC for an EPSRC Life Science Interface Fellowship (Grant No. EP/C532066/1).

References

1. Martina, M., Hutmacher, D.W. (2007) Biodegradable polymers applied in tissue engineering research: a review. *Polym. Int.* **56**(2), 45–57.
2. Kwon, I.K., Matsuda, T. (2005) Photo-polymerized microarchitectural constructs prepared by microstereolithography (μSL) using liquid acrylate-end-capped trimethylene carbonate-based prepolymers. *Biomaterials* **26**(14), 1675–1684.
3. Lee, K.S., Kim, R.H., Yang, D.Y., Park, S.H. (2008) Advances in 3D nano/microfabrication using two-photon initiated polymerization. *Prog. Polym. Sci.* **33**(6), 631–681.
4. Amsden, B. (2007) Curable, biodegradable elastomers: emerging biomaterials for drug delivery and tissue engineering. *Soft Matter* **3**(11), 1335–1348.
5. Declercq, H.A., Cornelissen, M.J., Gorskiy, T.L., Schacht, E.H. (2006) Osteoblast behaviour on in situ photopolymerizable three-dimensional scaffolds based on d, l-lactide, ε-caprolactone and trimethylene carbonate. *J. Mater. Sci. Mater. Med.* **17**(2), 113–122.
6. Meretoja, V.V., Helminen, A.O., Korventausta, J.J., Haapa-aho, V., Seppala. J.V., Narhi, T.O. (2006) Crosslinked poly(ε-caprolactone/d, l-lactide)/bioactive glass composite scaffolds for bone tissue engineering. *J. Biomed. Mater. Res. A* **77A**(2), 261–268.
7. Porter, J.R., Henson, A., Popat, K.C. (2009) Biodegradable poly([ε]-caprolactone) nanowires for bone tissue engineering applications. *Biomaterials* **30**(5), 780–788.
8. Griffith, L.G. (2000) Polymeric biomaterials. *Acta Mater.* **48**(1), 263–277.

9. Ashammakhi, N., Makela, E.A., Tormala, P., Waris, T., Rokkanen, P. (2000) Effect of self-reinforced polyglycolide membrane on osteogenesis: an experimental study in rats. *Eur. J. Plast. Surg.* **23**(8), 423–428.
10. Hou, Q., Grijpma, D.W., Feijen, J. (2009) Creep-resistant elastomeric networks prepared by photocrosslinking fumaric acid monoethyl ester-functionalized poly(trimethylene carbonate) oligomers. *Acta Biomater.* **5**(5), 1543–1551.
11. Martin, D.P., Williams, S.F. (2003) Medical applications of poly-4-hydroxybutyrate: a strong flexible absorbable biomaterial. *Biochem. Eng. J.* **16**(2), 97–105.
12. Zeng, F., Lee, H., Allen, C. (2006) Epidermal growth factor-conjugated poly(ethylene glycol)-block-poly(δ-valerolactone) copolymer micelles for targeted delivery of chemotherapeutics. *Bioconjug. Chem.* **17**(2), 399–409.
13. Han, L.H., Mapili, G., Chen, S., Roy, K. (2008) Projection microfabrication of three-dimensional scaffolds for tissue engineering. *J. Manufact. Sci. Eng. Trans. ASME* **130**(2), 4 021005–1–4.
14. Lv, N.J., Meng, S., Guo, Z. et al. (2008) Improving the biocompatibility of poly(ε-caprolactone) by surface immobilization of chitosan and heparin. *E-Polymers*, **139**, 1–13.
15. Liu, F., Zhao, Z., Yang, J., Wei, J., Li, S. (2008) Enzyme-catalyzed degradation of poly(l-lactide)/poly(ε-caprolactone) diblock, triblock and four-armed copolymers. *Polym. Degrad. Stab.* **2**(94), 227–233.
16. Wen, X., Tresco, P.A. (2006) Fabrication and characterization of permeable degradable poly(dl-lactide-co-glycolide) (PLGA) hollow fiber phase inversion membranes for use as nerve tract guidance channels. *Biomaterials* **27**(20), 3800–3809.
17. Kenawy, E-R., Abdel-Hay, F.I., El-Newehy, M.H., Wnek, G.E. (2009) Processing of polymer nanofibers through electrospinning as drug delivery systems. *Mater. Chem. Phys.* **113**(1), 296–302.
18. Baker, S.C., Rohman, G., Southgate, J., Cameron, N.R. (2009) The relationship between the mechanical properties and cell behaviour on PLGA and PCL scaffolds for bladder tissue engineering. *Biomaterials* **30**(7), 1321–1328.
19. Hollister, S.J. (2005) Porous scaffold design for tissue engineering. *Nat. Mater.* **4**, 518–524.
20. Abbott, A. (2003) Cell culture: biology's new dimension. *Nature* **424**, 870–872.
21. Hutmacher, D.W., Sittinger, M., Risbud, M.V. (2004) Scaffold-based tissue engineering: rationale for computer-aided design and solid free-form fabrication systems. *Trend Biotechnol.* **22**(7), 354–362.
22. Roach, P., Eglin, D., Rohde, K., Perry, C.C. (2007) Modern biomaterials: a review – bulk properties and implications of surface modifications. *J. Mater. Sci. Mater. Med.* **18**(7), 1263–1277.
23. Engelmayr, G.C., Cheng, M.Y., Bettinger, C.J., Borenstein, J.T., Langer, R., Freed, L.E. (2008) Accordion-like honeycombs for tissue engineering of cardiac anisotropy. *Nat. Mater.* 7, 1003–1010.
24. Nagata, M., Kitazima, I. (2006) Photocurable biodegradable poly(Îμ-caprolactone)/poly(ethylene glycol) multiblock copolymers showing shape-memory properties. *Colloid Polym. Sci.* **284**(4), 380–386.
25. Mizutani, M., Matsuda, T. (2002) Liquid photocurable biodegradable copolymers: in vivo degradation of photocured poly(ε-caprolactone-co-trimethylene carbonate). *J. Biomed. Mater. Res.* **61**(1), 53–60.
26. Lendlein, A., Langer, R. (2002) Biodegradable, elastic shape-memory polymers for potential biomedical applications. *Science* **296**(5573), 1673–1676.
27. Nicodemus, G.D., Bryant, S.J. (2008) Cell encapsulation in biodegradable hydrogels for tissue engineering applications. *Tissue Eng. Part B* **14**(2), 149–165.
28. Jeon, O., Bouhadir, K.H., Mansour, J.M., Alsberg, E. (2009) Photocrosslinked alginate hydrogels with tunable biodegradation rates and mechanical properties. *Biomaterials* **30**(14), 2724–2734.
29. Li, S.M., Pignol, M., Gasc, F., Vert, M. (2004) Synthesis, characterization, and enzymatic degradation of copolymers prepared from ε-caprolactone and beta-butyrolactone. *Macromolecules* **37**(26), 9798–9803.
30. Zurita, R., Franco, L., Puiggalí, J., Rodríguez-Galán, A. (2007) The hydrolytic degradation of a segmented glycolide–trimethylene carbonate copolymer (Maxon™). *Polym. Degrad. Stab.* **92**(6), 975–985.
31. Serbin, J., Egbert, A., Ostendorf, A, et al. (2003) Femtosecond laser-induced two-photon polymerization of inorganic–organic hybrid materials for applications in photonics. *Opt. Lett.* **28**(5), 301–303.
32. Ovsianikov, A., Ostendorf, A., Chichkov, B.N. (2007) Three-dimensional photofabrication with femtosecond lasers for applications in photonics and biomedicine. *Appl. Surf. Sci.* **253**(15), 6599–6602.
33. Lee, S.J., Kang, H.W., Park, J.K., Rhie, J.W., Hahn, S.K., Cho, D.W. (2008) Application of microstereolithography in the development of

three-dimensional cartilage regeneration scaffolds. *Biomed. Microdev.* **10**(2), 233–241.

34. Wu, S.H., Serbin, J., Gu, M. (2006) Two-photon polymerisation for three-dimensional micro-fabrication. *J. Photochem. Photobiol. A Chem.* **181**(1), 1–11.
35. Doraiswamy, A., Jin, C., Narayan, R.J., et al. (2006) Two photon induced polymerization of organic–inorganic hybrid biomaterials for microstructured medical devices. *Acta. Biomater.* **2**(3), 267–275.
36. Pego, A.P., Poot, A.A., Grijpma, D.W., Feijen, J. (2001) Copolymers of trimethylene carbonate and e-caprolactone for porous nerve guides: synthesis and properties. *J. Biomater. Sci. Polym. Ed.* **12**(1), 35–53.
37. Pego, A.P., Vleggeert-Lankamp, C.L.A.M., Deenen, M., et al. (2003) Adhesion and growth of human schwann cells on trimethylene carbonate (co)polymers. *J. Biomed. Mater. Res. A* **67**(3), 876–885.
38. Claeyssens, F., Hasan, E.A., Gaidukeviciute, A, et al. (2009) Three-dimensional biodegradable structures fabricated by two-photon polymerization. *Langmuir* **25**(5), 3219–3223.

Chapter 20

Alvetex®: Polystyrene Scaffold Technology for Routine Three Dimensional Cell Culture

Eleanor Knight, Bridgid Murray, Ross Carnachan, and Stefan Przyborski

Abstract

A broad range of technologies have been developed to enable three dimensional (3D) cell culture. Few if any however are adaptable for routine everyday use in a straightforward and cost effective manner. Alvetex® is a rigid highly porous polystyrene scaffold designed specifically to enable routine 3D cell culture. The scaffold is engineered into thin membranes that fit into conventional cell culture plasticware. The material is inert and offers a polystyrene substrate familiar to cell biologists worldwide. The 3D geometry of the scaffold provides the environment in which cells grow, differentiate, and function to form close relationships with adjacent cells thus creating the equivalent of a thin tissue layer in vitro. This chapter introduces the features required by a technology that enables routine 3D cell culture. Using Alvetex® as a product that satisfies these requirements, its application is demonstrated for the growth of a recognised cell line. Procedures detailing the use of Alvetex® for 3D cell culture are provided. This is followed by a series of detailed methods describing ways to analyse such cultures including histological techniques, immunocytochemistry, and scanning electron microscopy. Examples of data generated from these methods are shown in the corresponding figures. Additional notes are also included where further information about certain procedures is required. The use of Alvetex® in combination with these methods will enable investigators to routinely produce complex 3D cultures to research the growth, differentiation, and function of cells in new ways.

Key words: Alvetex®, Three dimensional, 3D, Cell culture, In vitro, Routine, Polystyrene scaffold, Scaffold membrane, Plasticware, Histology, Tissue processing, Immunocytochemistry, Scanning microscopy, Biochemical assay

1. Introduction

Cell-based in vitro assays are an important component of basic research, drug discovery, disease modelling, and toxicity testing. The availability of accurate, informative in vitro models is an increasingly important challenge facing researchers worldwide. Rising cost-to-delivery ratios, limitations of models representative

John W. Haycock (ed.), *3D Cell Culture: Methods and Protocols*, Methods in Molecular Biology, vol. 695, DOI 10.1007/978-1-60761-984-0_20, of real-life tissues, and the poor predictive value of existing in vitro tests place great emphasis on the development of more realistic cell culture systems. In vivo, cells possess a natural three-dimensional (3D) geometry and are supported by a complex 3D extracellular matrix (ECM) which facilitates cell–cell communication via direct contact and through the secretion of signalling molecules. In contrast, cells grown in culture are generally confined in two-dimensional (2D) monolayers without many of the physical and chemical cues, which underlie their identity and function in vivo. Consequently, cells grown on standard flat plasticware behave very differently such that the growth and function of cells as multi-cellular 3D structures is significantly different to their growth as conventional 2D monolayer cultures (1) (The reference list relates to the development of Alvetex® and its usage for biological applications). As a result of such improved cell performance there has been a drive to develop materials for 3D cell culture with the objective of creating a 3D spatial environment that will overcome some of the restrictions associated with traditional 2D culture.

A range of alternative materials have been developed to support 3D cell growth in vitro. These include naturally occurring substances as well as products fabricated from naturally-derived and synthetic polymers. Natural substrates such as alginate, a seaweed-derived material, have been used to support cell growth in a number of ways including cell encapsulation (2). More recently, this material has been used as a macroporous scaffold to support the development of 3D spheroids and the terminal differentiation of hepatocytes in vitro (3). Similar such technology has been released commercially, and Algimatrix™ (Invitrogen) is marketed as an animal-free scaffold available for the development of high-fidelity cell culture models that are more predictive of disease states and drug responses. While this technology may enable a degree of 3D cell growth, growing cells as individual spherical masses are not suitable for all requirements; the distribution of cells throughout the material is not entirely even, and there are issues about mass transfer given the thickness of the scaffold under static growth conditions. Furthermore, it is not clear whether cells will respond differently to alginate compared to conventional plasticware which is familiar and has been employed for many years.

Using biodegradable polymers such as poly(glycolic acid), poly(lactic acid) and their co-polymers (poly(lactic-*co*-glycolic acid)) has been one of the more successful approaches for 3D cell culture (4). Their degradation over time aids the integration of cells with host tissues during transplantation. However, biodegradability is not necessarily a feature for routine in vitro cell culture, where there are shelf life issues, and improper storage of such biopolymers can render such products useless due to degradation leading to changes in their properties, quality, and variation of results. Moreover, biodegradation during an in vitro experiment introduces another variable that will influence how cells behave.

Similarly, injectable hydrogels have developed as a system with proven success for tissue repair (5). Hydrogels comprise a cross-linked natural base material such as agarose, collagen, fibrin, or hyaluronic acid with high water content. They can be engineered to support preferential cell growth and function. Hydrogels in essence trap cells in an artificial ECM environment and may be modified to incorporate biologically active molecules. However, their use for routine 3D cell culture is also restricted by various practical issues including expense, shelf life, gel preparation, storage, and inconsistency.

The development of inert non-degradable scaffolds made from synthetic polymers overcomes several of the limitations these other technologies experience. Such scaffolds consist of voids into which cells can grow, and which are joined by interconnecting pores. Methods of fabrication for porous materials include emulsion templating (6, 7), leachable particles (8), and gas foaming technology (9). Gas-in-liquid foam templating has been used as a method to create porous scaffolds for cell culture applications (10). However, gas bubbles can coalesce leading to a broad range of scaffold porosities hence it is difficult to control the consistency of the material and subsequent growth of cells. Electrospinning methods have given rise to electrospun synthetic fibres woven into mats to support cell growth (11). Ultra-Web™ (Corning) was developed as a commercial polyamide electrospun nanofibre mat for cell culture. Cells grow as monolayers on the uneven topography created by the nanoscale Ultra-Web™ fibre mat rather than in 3D within the body of the material.

1.1. Technology for Routine 3D Cell Culture

Although promising, these and many other examples of new scaffold technologies designed to support cell growth have not been successfully developed into commercial products that enable routine 3D cell culture in a straightforward and cost-effective manner. A polystyrene scaffold, however, has the potential to achieve this need. Polystyrene is an attractive substrate to support 3D cell culture since it is familiar, readily accessible, and easily transferred to cell biology laboratories worldwide. Polystyrene is also chemically inert, stable, and is a material consistent with conventional 2D tissue culture plasticware. Polystyrene scaffolds are also advantageous given that they are generally simple and inexpensive to mass produce, and they are designed as a consumable product with a long shelf life. These attributes make polystyrene-based substrates well suited for routine 3D cell culture.

Alvetex® is a novel polystyrene scaffold in which the porosity of the material is carefully controlled and tailored to support 3D cell culture (6, 7). The scaffold is formed by polymerisation in a biphasic emulsion, consisting of an aqueous and a non-aqueous monomer/surfactant phase, termed a high internal phase emulsion, or HIPE (6, 7). The resulting polymer (poly-HIPE) consists of a relatively homogeneous porous network of voids, linked by interconnecting pores (Fig. 1). Alvetex® has been developed as a solution for routine 3D cell culture and is designed for incorporation into existing cell culture products, such as multi-welled plates or well inserts (Figs. 2 and 3). Careful consideration has been given

Fig. 1. Structure of Alvetex®, a novel substrate for routine 3D cell culture consisting of a highly porous polystyrene scaffold. Scanning electron micrographs show structural characteristics of Alvetex® at high (**a**), low (**b**) and medium (**c**) magnifications (see scale bar insert). Panel (**c**) shows a transverse section of the polystyrene membrane at approximately 200 μm thick.

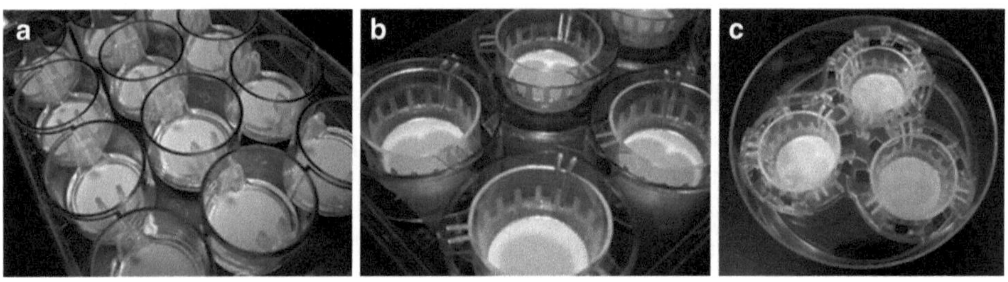

Fig. 2. Alvetex® is supplied in a variety of formats including multi-welled plates ((**a**); 12-well plate example shown) and well inserts ((**b**); six-well insert shown). A cradle housing three 6-well inserts has been developed to fit into a 90 mm Petri dish (**c**), in order to accommodate larger medium volumes that can support long term 3D cell culture.

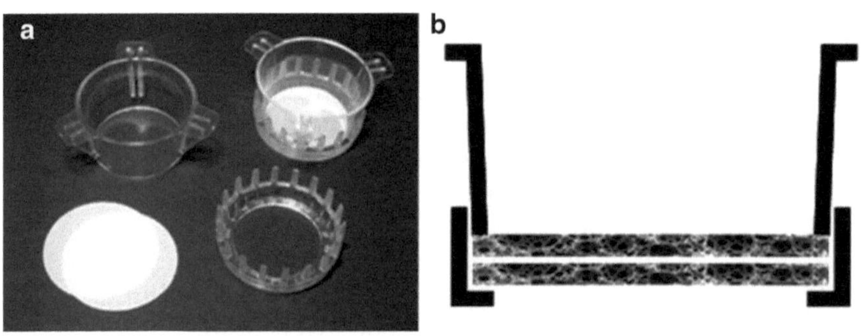

Fig. 3. The well insert consists of a two part assembly that fits together to clamp the scaffold in position (**a**). The clamping arrangement has been designed such that different thicknesses of scaffold can be used if needed. This provides the user with the flexibility to study the interaction of two or more 3D cultures as illustrated in the schematic diagram (**b**).

to the way Alvetex® is presented and used during cell culture. For example: a polystyrene clip has been specifically designed to hold Alvetex® in place during transportation of the product and during its use in cell culture (Fig. 2a); a cradle designed to hold three 6-well inserts containing Alvetex®, fits into a standard 100 × 20 mm Petri dish, enabling incubation with larger medium volumes to support the greater metabolic requirements of 3D cultures (Fig. 2c). Alvetex® has been engineered into a thin 200 μm thick membrane to address the issue of mass transfer, enabling cells to enter the material and allowing for sufficient mass exchange of gases, nutrients, and waste products during static culture. The polystyrene that forms Alvetex® is cross-linked, which gives the material improved structural stability and strength even when presented as a thin membrane. Alvetex® is compatible with standard cell culture plasma treatment, gamma sterilisation methods, and if required can be coated using standard cell culture reagents such as collagen, fibronectin, etc.

For routine 3D cell culture, the development of any new technology must consider issues such as cost, ease of use, application, and reproducibility, especially when the application is for high throughput research. A technology that is expensive, difficult to use, or is inconsistent in some manner, will not satisfy these demands and will fail to be accepted as an industrial standard. Importantly, the ability of Alvetex® to support true 3D cell culture has been vigorously exemplified and validated over a range of different cell types (6, 12, 13) (e.g. Figs. 4 and 6).

This chapter describes the use of Alvetex® for the growth of keratinocytes as an example application of this generic technology for 3D cell culture. Subsequent to the culture of cells in Alvetex®, the resulting 3D culture can be analysed using a broad range of standard techniques commonly used in general cell and molecular

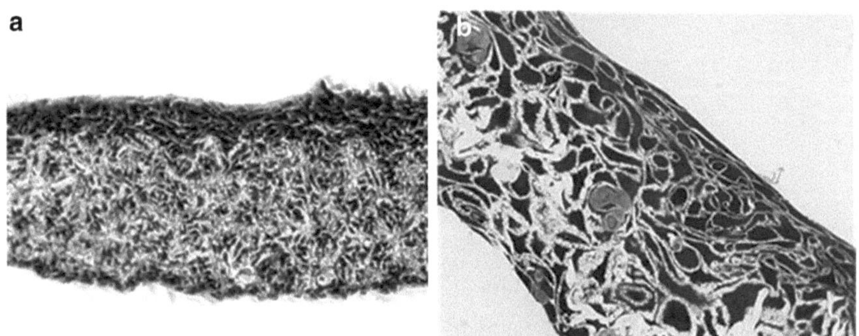

Fig. 4. Cells grown in Alvetex® can be fixed and processed for histological analysis using standard methods. Two examples of cultured human keratinocytes grown in Alvetex® membranes that have been sectioned in the transverse plane are shown: (**a**) Culture fixed in Bouins, embedded in paraffin wax, processed and counter stained with H&E; (**b**) Sample fixed in Karnovsky's, embedded in LR White resin, processed and counter stained with Toluidine Blue. Scale bars: 100 μm.

biology. Several examples of these methods have been provided including histological analysis, immunocytochemistry, and electron microscopy. This is not in any way exhaustive and there is plenty of scope to include additional methods of analysis as the user sees appropriate.

2. Materials

Table 1 provides a list of the primary reagents required, including the supplier and corresponding catalogue number.

Table 1
Suppliers of primary chemicals

Name	Supplier	Catalogue number
Saturated picric acid	Sigma	P6744-1GA
Formaldehyde	Fisher	BP531-500
Glacial acetic acid	Fisher	A/0400/PB17
Paraformaldehyde powder	Sigma	P6148
Sodium hydroxide	Fisher	BPE 357-500
Disodium monohydrogen phosphate	Fisher	BPE166-500
Sodium dihydrogen phosphate	Fisher	BPE329-500
Sodium cacodylate	Sigma	C0250
glutaraldehyde	Fluka	49362
Osmium tetroxide	Electron Microscopy Services	19900
Triton X100	Sigma	T8787
Bovine serum albumin	Sigma	A1653
Normal goat serum	Sigma	G6767
Toluidine blue	Sigma	T3260
LR White resin embedding kit	Electron Microscopy Services	14380
Polymount	Polysciences	08381-120
Histoclear	National Diagnostics	HS-200
DPX mountant	Sigma	44581
Vectashield/Dapi	Vector Laboratories	H-1200

2.1. 3D Cell Culture Using Alvetex®

1. Example cell line to demonstrate 3D culture system: human HaCaT keratinocytes (American Type Culture Collection (ATCC), Manassas VC, US).
2. Growth medium for HaCaT lineage: (a) Dulbecco's modified Eagle medium (DMEM) supplemented with 10% (v/v) foetal calf serum, 100 μg/mL penicillin, and 10 μg/mL streptomycin; (b) Quantum 153 (PAA) specialised medium that supports keratinocyte growth and differentiation.
3. General cell culture consumables (culture flasks, serological pipettes, etc.), reagents (70% ethanol, phosphate buffered saline (PBS; calcium-free), and trypsin), and access to equipment for standard 2D culture to expand cell populations (laminar flow hoods, CO_2 incubators, etc.).
4. Alvetex® well inserts (six-well format) (Fig. 2b) and Petri dish cradle (Fig. 2c) to enable height adjustment to the air/liquid interface and large medium reservoir for long-term culture. Each well insert contains one 200-μm thick disc of Alvetex® (Reinnervate Limited, Sedgefield, UK).

2.2. Preparation of 3D Cultures for Histological Analysis Using Wax Embedding

1. Bouins fixative (saturated picric acid 750 mL, formaldehyde 250 mL, and glacial acetic acid 50 mL. Measure out the fluids in a fume hood. Mix well in a 1 L Duran bottle. Fixative is stable for 12 months. Safety: this fixative is carcinogenic, irritant, and is toxic).
2. Dehydration ethanols (30, 50, 70, 80, 90, 95, and 100%).
3. Basic equipment: spade forceps to handle 3D Alvetex® cultures, scalpel (to trim/cut scaffold if required), and disposable pipettes.
4. Materials and equipment needed for embedding in paraffin wax: absolute ethanol, Histoclear, paraffin wax, Falcon tubes or glass vials, oven set at 60°C, embedding moulds, and tissue-processing cassettes.
5. Materials and equipment needed for tissue sectioning, mounting slides, and cell counterstaining: microtome, microscope slides (e.g. Histobond), graded series of ethanol solutions, Histoclear, counterstain, staining jars, coverslips, and DPX mountant.

2.3. Preparation of 3D Cultures for PFA Fixation and Immunocytochemistry

1. Four percent paraformaldehyde (PFA) fixative (PBS 1,000 mL; PFA powder 40 g; and sodium hydroxide solution, 1 N. Prepare 1× solution of PBS before use. Working in the fume hood, heat PBS up to 60°C (maximum) on a hotplate stirrer. Dissolve PFA powder in warm PBS using a heated magnetic stirrer. Slowly add drops of 1 N sodium hydroxide until the solution is clear. Filter the fixative and adjust to pH 7.3–7.4. Allow fixative solution to cool to room

temperature. Bottle and label with date, owner, and hazard rating. Store at 4°C. Safety: PFA is toxic).

2. Dehydration ethanols (30, 50, 70, 80, 90, 95, and 100%).
3. Basic equipment: spade forceps to handle 3D Alvetex® cultures, scalpel (to trim/cut scaffold if required), disposable pipettes.
4. Materials for wax embedding as listed in Subheading 2.2 above.
5. Materials and equipment needed for sectioning and preparation of slides for immunocytochemistry: microtome, microscope slides (e.g. Histobond), graded series of ethanol solutions (50, 70, 90, 95%, and absolute ethanol).
6. Materials and equipment needed for immunocytochemistry: 1× citrate buffer, pH 6.0; 1× PBS; blocking buffer (5% normal goat serum, 1% bovine serum albumin, 0.2% Triton X100 in PBS); permeabilisation solution (1% (w/v) Triton X100 in PBS); primary and secondary antibodies (specific to experimental analysis); microwave oven; humidified chamber; Vectashield/DAPI mountant; nail varnish; and cover slips.

2.4. Preparation of 3D Cultures for Histological Analysis Using Resin Embedding

1. Karnovsky's fixative (2% PFA; 2.5% glutaraldehyde in 0.1 M phosphate buffer pH 7.4. Work in the fume hood. Mix well and make up to 100 mL with dH_2O. Karnovsky's fixative should be made up and used fresh).
2. Dehydration ethanols (30, 50, 75, 95, and 100%).
3. Basic equipment: spade forceps to handle 3D Alvetex® cultures; scalpel (to trim/cut scaffold if required); disposable pipettes.
4. Materials and equipment needed for embedding in resin for subsequent toluidine blue staining: LR White resin embedding kit, gelatin embedding capsules or PTFE embedding moulds.
5. Materials and equipment needed for tissue sectioning, mounting slides, and toluidine blue staining: ultra-microtome, glass knives, microscope slides (e.g. Histobond), hotplate, 1% toluidine blue in 1% sodium borate, dH_2O wash bottle, 50% ethanol wash bottle, Polymount.

2.5. Preparation of 3D Cultures for Scanning Electron Microscopy

1. Karnovsky's fixative as listed in Subheading 2.4 above.
2. Dehydration ethanols (30, 50, 75, 95, and 100%).
3. 1% buffered osmium tetroxide (osmium tetroxide solution (2%); 0.2 M phosphate buffer (pH 7.4). Working in a fume hood throughout, combine equal amounts of the 2% osmium tetroxide solution and the 0.2 M phosphate buffer in a clean 15 mL centrifuge tube. Vortex gently. Store in a fume hood

at room temperature. Final concentrations: 1% osmium tetroxide in 0.1 M phosphate buffer. Cacodylate buffer may be used to replace the phosphate buffer. Safety: osmium tetroxide is highly toxic and volatile, and should be used and stored only in a chemical fume hood. Wear gloves, lab coat, and goggles).

4. Basic equipment: spade forceps to handle 3D cultures; scalpel (to trim/cut Alvetex® cultures if required); disposable pipettes.
5. Specialised equipment to prepare samples for scanning electron microscopy (SEM): critical point dryer (CPD) and sputter coater.

3. Methods

The following methods detail the use of Alvetex® for 3D cell culture. The growth of a well known human keratinocyte cell line is used as an example to demonstrate this application. A series of detailed methods describing ways to analyse such cultures are then provided and examples of data generated from these methods are shown in the corresponding figures. Additional notes are provided where further information about certain procedures is required.

3.1. 3D Cell Culture Using Alvetex®

1. Expand populations of HaCaT cells in DMEM-based medium as monolayer cultures in conventional 2D plasticware according to standard procedures. Prepare single cell suspension of HaCaTs in PBS using enzymatic methods (e.g. trypsinisation) as standard. Determine cell number and calculate medium volume and cell number needed to seed 1×10^6 HaCaT cells in 150 μL Quantum 153 medium per six-well insert.
2. To prepare Alvetex® for use, first, carefully remove product from wrapping within the culture hood. Note that Alvetex® is delicate and should be handled carefully at all times (see Note 1).
3. Alvetex® is supplied either tissue culture treated (using oxygen plasma) or non-treated. Cell suspensions can be seeded directly onto dry treated Alvetex® that rapidly "wick" into the scaffold by capillary action. Alternatively, cells are seeded onto Alvetex® that has been hydrated (see Note 2). For certain applications, Alvetex® may also be supplied in a non-sterile format. For sterilisation of Alvetex® see Note 2.
4. Prepare Petri dish and cradle to receive well inserts containing either plasma-treated or hydrated Alvetex® (see Fig. 2c). Place inserts into cradle and set them to their lowest position. Do not add medium to the Petri dish at this time.

5. For each six-well insert, 1×10^6 HaCaT cells are seeded per Alvetex® disc in 150 μL of Quantum 153 medium. Seed cells centrally within each insert. This volume of cell suspension will move into the body of the scaffold when added. Place cultures in cell culture incubator (37°C and 5% CO_2) for 15 min to allow cells to settle.
6. Slowly and carefully add approximately 50 mL Quantum 153 medium into the Petri dish. As the level rises, this will flood the well inserts and submerge the Alvetex® containing the cultured cells.
7. Maintain the cells in the incubator for 3–5 days to enable the culture to become established. A medium change may not be necessary during this period.
8. After 3–5 days, refresh medium and raise the inserts within the cradle such that the level of the medium is in line with the level of the Alvetex®. This creates the air–liquid interface which induces epithelial differentiation and initiates the formation of the stratum corneum.
9. Continue to grow cells for required period, changing the culture medium as needed. Once complete, prepare cells for cell and molecular analysis as highlighted below.
10. See Note 3 about further optimization of 3D cell cultures using Alvetex®.

3.2. Preparation of 3D Cultures for Histological Analysis Using Wax Embedding

1. Aspirate off the medium and carefully wash the 3D culture in PBS twice.
2. Remove well inserts from cradle, place in a conventional six-well plate and add 5 mL Bouins fixative. Fix specimens overnight (samples may remain in fixative up to 3 days at room temperature).
3. Aspirate off the fixative, wash the specimen using 5 mL dH_2O or PBS three times to thoroughly remove excess fixative, discarding the waste liquid. The general location of cells can be readily observed by the distribution of yellow staining on Alvetex® left by the Bouins fixative.
4. Remove Alvetex® from the well insert either by separating the two parts of the device or using a scapel to cut out the scaffold. At this time, samples can be transferred to a tissue-processing cassette to minimise direct handling and damage to the 3D culture.
5. Aspirate off the water/PBS and add 5 mL of 30% ethanol. Leave to equilibrate for at least 15 min. Aspirate off the ethanol and discard.
6. Repeat with 50, 70, 80, 90% and then 95% ethanol. A gradual dehydration of the sample will result in less tissue shrinkage.

Material can be stored in 95% ethanol prior to paraffin embedding.

7. Fully dehydrate specimens stored in 95% ethanol by replacing with 100% absolute ethanol for at least 30 min, transferring them to Falcon tubes or glass vials as appropriate.
8. Aspirate off the ethanol and replace with 15 mL Histoclear for at least 30 min.
9. Replace the Histoclear with a 50:50 Histoclear and molten paraffin wax (60°C) mix and incubate in a convection oven at 60°C for 30 min.
10. Replace the Histoclear:wax mix with 100% molten wax and incubate at 60°C for a further 60 min. Replace wax once and repeat.
11. Transfer the polymer to plastic embedding moulds and orientate the polymer into its required embedding position with plane of section in mind.
12. Allow wax to cool and set at room temperature for 1–2 h or overnight.
13. Once the wax has hardened, remove the wax embedded block from the plastic mould. The sample is now ready for sectioning on a suitable microtome (e.g. Leica RM2125).
14. Following the microtome manufacturers' instructions throughout, align the block correctly with the microtome blade and proceed to cut 5–10 μm sections of the sample block.
15. Transfer sections to a slide water bath (40°C), floating them on the surface of the water to enable them to flatten out.
16. Transfer selected sections to slides by flotation. Histobond slides are recommended, since sections adhere to these slides well.
17. Place on a slide drier and leave overnight. The sections should now be ready for histological counterstaining (see Note 5).
18. Subsequent to counterstaining, mount the sections in DPX, coverslip, and dry. Figure 4a shows an example of a paraffin wax embedded 3D Alvetex® culture, sectioned and counterstained with haematoxylin and eosin (H&E).

3.3. Preparation of 3D Cultures for PFA Fixation and Immunocytochemistry

1. Aspirate off the medium and carefully wash the 3D culture in PBS twice.
2. Remove well inserts from cradle, place in a conventional six-well plate and add 5mL 4% PFA fixative at 4°C. Fix specimens at 4°C for a minimum of 12 h, but not longer than 24 h.
3. Repeat steps 3–17 in Subheading 3.2 above. The sections are now ready for histological counterstaining (see Note 5) or immunocytochemistry (as follows).

4. Deparaffinise sections in Histoclear for 10 min. Handle samples carefully to avoid loss of section from slide.
5. Hydrate specimen through a graded series of ethanols (100, 90, and 70%) with 5 min incubation in each solution.
6. To assist in retrieval of antigens, place slides in 1× citrate buffer (pH 6) and microwave (800 W) for 6 min.
7. Leave to stand outside microwave for 1 min and microwave for a further 3 min at 800 W.
8. Allow to cool for 20 min and wash in 1× PBS for 10 min. Repeat 3×.
9. Treat cells with permeabilisation solution for 15 min, followed by a further 15 min treatment with blocking buffer.
10. Incubate with primary antibody diluted in blocking buffer (this will be specific to the antigen chosen and will be used at a pre-determined concentration) overnight at 4°C in a humidified chamber.
11. Wash 3× 10 min with 1× PBS.
12. Incubate with secondary antibody diluted in blocking buffer (this will be specific to the antigen chosen and will be used at a pre-determined concentration) for 2 h in the dark at room temperature.
13. Wash 3× 10 min with 1× PBS.
14. Mount in Vectashield/DAPI. This solution simultaneously mounts the specimen and stains cell nuclei.
15. Seal the perimeter of the cover slip with nail varnish and dry in the dark.
16. Store slides in the dark at 4°C until ready for inspection using a fluorescence microscope equipped with the appropriate filters. Figure 5 shows an example of a 3D Alvetex® culture fixed with 4% PFA, embedded in paraffin wax, sectioned, and probed with Ki67 antibody using immunocytochemistry and standard fluorescence microscopy.

3.4. Preparation of 3D Cultures for Histological Analysis Using Resin Embedding

1. Aspirate off the medium and carefully wash the 3D culture with PBS twice.
2. Remove well inserts from cradle and carefully cut small samples (2–3 mm square) from the scaffold disc.
3. Immerse samples in Karnovsky's fixative at 4°C for 90 min. The ratio of fixative volume to specimen size should be approximately 20:1.
4. After fixation, aspirate the fixative into a waste container for disposal.
5. Wash the specimen twice for 2 min in 3 mL 0.1 M phosphate to remove excess fixative, discarding waste liquid. Use phosphate

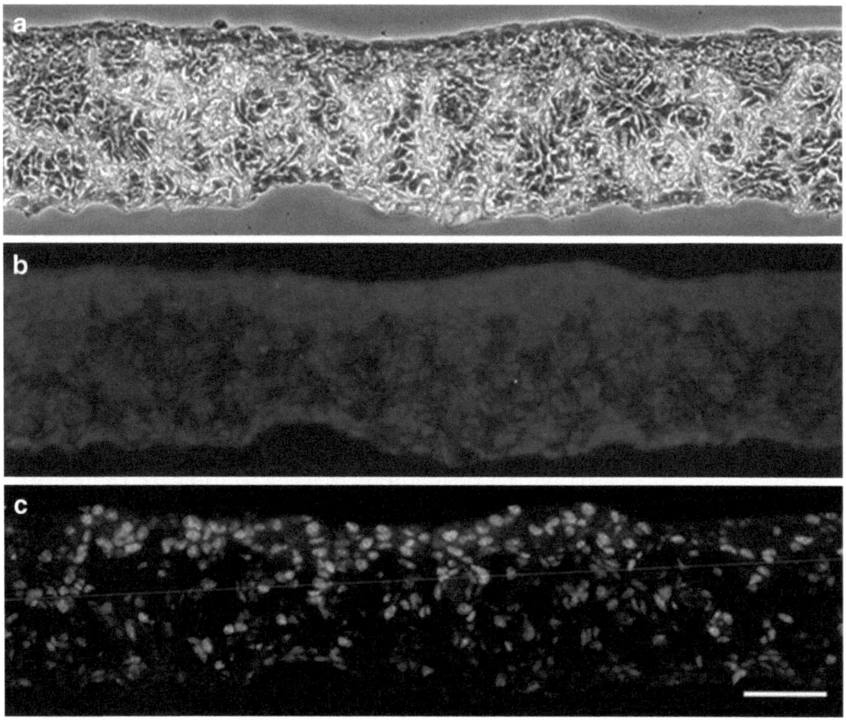

Fig. 5. Protein expression in 3D cell cultures can be localised using immunocytochemistry and fluorescence microscopy. Three corresponding images from the same region of Alvetex® show: (**a**) cell morphology by phase microscopy; (**b**) expression of cell nucleus marker DAPI; (**c**) expression of cell proliferation marker Ki67. Scale bar: 100 μm.

buffer if planning to stain with toluidine blue, but not cacodylate buffer (either can be used for SEM).

6. Add 20× the specimen volume of 30% ethanol. Leave to equilibrate for 5 min preferably on a rotating wheel. Pour off the ethanol and discard. Repeat twice.
7. Repeat step 6 with 50, 75, 95%, and then absolute ethanol. The sample can be stored in absolute ethanol prior to resin embedding.
8. Prepare the resin/catalyst mix according to the manufacturer's instructions.
9. Immerse samples in a 1:1 mix of absolute ethanol and LR White resin (20× the specimen volume) and leave on a rotor for 1 h at room temperature.
10. Replace the resin/ethanol mix with three changes of resin alone for at least 60 min each. Samples can also be left in resin at room temperature overnight.
11. Embed the specimen according the manufacturer's instructions. Specimens may be embedded in gelatin capsules or flat embedding moulds made of PTFE.

12. It is possible to cut LR White resin blocks on a standard microtome with a steel knife blade. The method of choice, however, is to use a motorised ultra-microtome and glass knife. For toluidine blue staining, 0.5–1.0 μm sections should be cut.
13. Place a drop of distilled water approximately 1 cm in diameter on a clean glass slide. Pick the cut sections up with a wire loop and invert the loop to place 2–4 sections into the drop of water on the slide.
14. Place the slide with the water droplet containing sections on a hot plate adjusted to approximately 60–65°C and wait until all the water evaporates.
15. Locate the sections on the slide and circle the bottom of the slide with a permanent marker to indicate their position.
16. Place 1–2 drops of 1% toluidine blue in 1% sodium borate onto the sections and place the slide on the hot plate. Wait until the edges of the drop of stain begin to turn golden (approximately 20–30 s).
17. Remove the slide and rinse the sections with a gentle stream of distilled water from a wash bottle to wash off the excess stain.
18. Differentiate the sections by rinsing in a gentle stream of 50% ethanol.
19. Replace the washed slide on the hot plate after gently drying the bottom of the slide with a piece of paper towel and wait for the residual ethanol to evaporate from the slide surface.
20. Remove the slide from the hot plate and add one to two drops of Polymount mounting medium to the top of the sections adhering to the slide. Gently lower a coverslip onto the droplet of Polymount over the sections.
21. Allow to dry and store in a dust free location. Figure 4b shows an example of a resin-embedded 3D Alvetex® culture, sectioned and stained with toluidine blue.

3.5. Preparation of 3D Cultures for Scanning Electron Microscopy

1. Follow steps 1–5 of Subheading 3.4.
2. Post fix samples in 1% buffered osmium tetroxide for 90 min at 4°C then remove the osmium tetroxide solution.
3. Add 20× the specimen volume of 30% ethanol. Leave to equilibrate for 5 min preferably on a rotating wheel. Pour off the ethanol and discard. Repeat twice.
4. Repeat step 6 with 50, 75, 95%, and then absolute ethanol. The sample can be stored in excess absolute ethanol.
5. Follow the manufacturer's instructions to completely dry the specimen using a CPD.

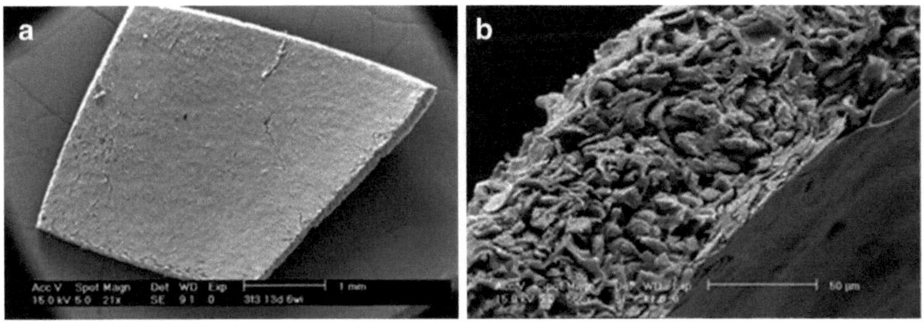

Fig. 6. Detailed structure of 3D cell cultures can be visualised using scanning electron microscopy. Inspection of pieces of Alvetex® at low magnification shows homogeneous coverage by cultured cells (**a**). Higher magnification imaging in this transverse section reveals cells growing throughout the scaffold (**b**). See scale bar inserts.

6. Follow the manufacturer's instructions to gold plate the sample using a sputter coater. The gold coating will increase the sample's conductivity and consequently improve image quality.
7. The samples are now ready for imaging using a conventional scanning electron microscope. Figure 6 shows an example of 3D Alvetex® culture imaged using the SEM technique.

4. Notes

1. *Handling Alvetex®*. Wear gloves when handling the scaffold to avoid damage and contamination. The scaffold is reasonably fragile. Therefore handle the material carefully when performing any manipulation including medium changes, transfer of substrate to an alternative culture vessel, embedding, etc. When using forceps, exercise care as manipulating the scaffold can damage its structure. Try to handle the scaffold only around the edges. When dispensing medium or other components over the scaffold, place the end of the pipette tip towards the wall of the culture vessel and avoid touching the scaffold. All procedures concerning handling of the scaffold should be performed according to standard aseptic methods required for cell culture. It is recommended that *all* handling is performed in a tissue culture hood to minimise contamination and dust.
2. *Preparing non-plasma treated and/or non-sterile Alvetex® for 3D cell culture*. Perform this procedure in a sterile environment (i.e. cell culture hood). The scaffold disc can be hydrated and sterilised using a solution of 70% ethanol. Immersion in 70% ethanol will sterilise the scaffold within a minute and also pre-wet the material in preparation for incubation in aqueous

solutions (e.g. PBS, culture medium). Sterilisation can be performed in the plate provided (ensure walls of wells are sterilised), or in a fresh sterile vessel after transfer of the scaffold. Carefully aspirate the 70% ethanol solution and immediately wash the scaffold in sterile PBS for a minute. Repeat this procedure twice. Carefully aspirate PBS and replace with appropriate cell culture medium. Additional wash(es) with culture medium can be performed prior to cell seeding to further prepare the scaffold for receipt of cells. Alternatively, the above procedure can be performed by passing the insert through the appropriate solutions in separate vessels (e.g. Petri dishes) using forceps. This avoids aspirating the fluids between many wells.

3. *Optimising use of Alvetex® for 3D cell culture.* For best results enabling 3D cell growth on Alvetex®, cell seeding density and cell culture medium may require optimisation in line with the specific requirements of the cell-type cultured and the nature of the assay(s) to be performed. As a rough guide, seed approximately fivefold to tenfold more cells per well than for a standard 2D culture to initiate 3D cell growth. The vertical axis of the third dimension essentially enables additional layers of cells to form complex 3D interactions with adjacent cells simulating the structure of a tissue. At optimal conditions, cells will grow throughout the scaffold effectively forming a 3D slab of tissue as shown in the examples (Figs. 4 and 6). It may be necessary to change the cell culture medium frequently due to the larger numbers of cells growing per unit volume of medium within a 3D scaffold compared to the equivalent 2D culture. Prior to cell seeding, Alvetex® can be pre-coated with standard cell culture reagents such as collagen, fibronectin, laminin, poly-d-lysine, and poly-l-ornithine to encourage cell adhesion, differentiation, and function. Subsequent to sterilisation and wetting of the scaffold, aqueous solutions of such coatings can be applied to the scaffold following standard procedures for coating culture plastic. Depending on the application, it is recommended to consider the concentration of ECM proteins that deposit as filamentous networks (e.g. collagen) so not to completely block the porous structure of the scaffold and inhibit cell entry. Alternatively, users can choose to seed cells in a solution of ECM protein(s) simultaneously into the scaffold.

4. *Fixation of 3D cell cultures.* "Fixation" is a process of stabilisation important for anatomical study of any biological tissue. To achieve this, decomposition caused by tissue enzymes and decay must be arrested. During the process, tissue is hardened for convenient handling. There are two basic approaches; rapid heating/quick-freezing and chemical methods. The mechanisms of action by which fixatives work are poorly

understood, but generally involve denaturation of protein. Survival of tissue antigens for immunochemical staining therefore depends on the type and concentration of fixative, on fixation time, and on the size of the tissue specimen to be fixed. Given that the 3D Alvetex® culture is relatively thin, fixation of cells is rapid and efficient. The following summarises the features of the three fixatives highlighted herein:

(a) *Bouins* – excellent fixative for use when samples are to be embedded in paraffin wax, sectioned, and stained for general histology, especially for trichrome stains. Bouins fixation is poor for immuno-detection.

(b) *PFA* (4% PFA) – fixative for paraffin wax embedding and general sectioning. These samples can be stained for general histology, but the degree of fixation is less vigourous than bouins fixation so the quality of the morphology is not quite as good. PFA fixation is compatible with subsequent immuno-detection of certain antigens and should be used therefore when the objective is to study morphology and protein expression simultaneously.

(c) *Karnovsky's* – this is a mixture of PFA and glutaraldehyde, with post fixation in osmium tetroxide and is suitable for use when preparing samples for resin sectioning and electron microscopy. Resin embedded samples subsequently stained with toluidine blue have improved tissue morphology compared to standard wax embedded samples.

5. *Histological counter staining*. Traditionally sections of tissue are counterstained with various dyes and chemicals to bring out the inherent contrast within a tissue specimen. The same methods can be applied to cells grown in 3D culture when using Alvetex® technology. Sections generated from either Bouins or PFA fixed, paraffin wax-embedded cultures can be counterstained using a variety of conventional methods. These include: Mayer's haematoxylin (haemalum) and eosin (H&E) – staining nuclei blue and cytoplasm pink (Fig. 4a); Masson's trichrome – nuclei purplish brown to black, collagen blue, cytoplasm of muscle and erythrocytes, red. These and other counterstains used in histology are well known and the details of these procedures are widely available.

6. *General notes on molecular/cellular assays*. Intact, viable cells can be retrieved from the scaffold subsequent to their culture. This can be achieved by dissociation of 3D culture using enzymatic treatments such as trypsin in addition to gentle mechanical disruption by rocking or shaking. Typically 60–80% of the cells can be removed from the scaffold. Isolated viable cells may be passaged for further culturing. Intact cells can also be prepared for flow cytometry using standard methods,

or cytospun onto glass slides for standard histological or immunocytochemical procedures. Cell viability can be assessed directly on 3D Alvetex® cultures using standard biochemical assays such as MTS or MTT. These tests are widely available as commercially as kits with full manufacturer's recommended instructions provided. Cells can be lysed directly within Alvetex® using standard approaches with reagents that do not react with polystyrene for subsequently preparing samples of protein and/or nucleic acid for subsequent expression profiling.

References

1. Schmeichel, K. L., and Bissell, M. J. (2003) Modeling tissue-specific signaling and organ function in three dimensions. *J. Cell Sci.* **116**, 2377–2388.
2. Zimmermann, H., Shirley, S. G., and Zimmermann, U. (2007) Alginate-based encapsulation of cells: past, present, and future. *Curr. Diab. Rep.* 7, 314–320.
3. Dvir-Ginzberg, M., Elkayam, T., and Cohen, S. (2008) Induced differentiation and maturation of newborn liver cells into functional hepatic tissue in macroporous alginate scaffolds. *FASEB J.* **22**, 1440–1449.
4. Mikos, A. G., Sarakinos, G., Leite, S. M., Vacanti, J. P., and Langer, R. (1993) Laminated three-dimensional biodegradable foams for use in tissue engineering. *Biomaterials.* **14**, 323–330.
5. Blackshaw, S. E., Arkison, S., Cameron, C., and Davies, J. A. (1997) Promotion of regeneration and axon growth following injury in an invertebrate nervous system by the use of three-dimensional collagen gels. *Proc. Biol. Sci.* **264**, 657–661.
6. Bokhari, M., Carnachan, R. J., Przyborski, S. A., and Cameron, N. R. (2007) Emulsion-templated porous polymers as scaffolds for three dimensional cell culture: effect of synthesis parameters on scaffold formation and homogeneity. *J. Mater. Chem.* **17**, 4088–4094.
7. Carnachan, R. J., Bokhari, M., Przyborski, S. A., and Cameron, N. R. (2006) Tailoring the morphology of emulsion-templated porous polymers. *Soft Matter.* **2**, 608–616.
8. Aydin, H. M., El Haj, A. J., Piskin, E., and Yang, Y. (2009) Improving pore interconnectivity in polymeric scaffolds for tissue engineering. *J. Tissue Eng. Regen. Med.* **3**, 470–476.
9. Salerno, A., Oliviero, M., Di Maio, E., Iannace, S., and Netti, P. A. (2009) Design of porous polymeric scaffolds by gas foaming of heterogeneous blends. *J. Mater. Sci. Mater. Med.* **20**, 2043–2051.
10. Barbetta, A., Gumiero, A., Pecci, R., Bedini, R., and Dentini, M. (2009) Gas-in-liquid foam templating as a method for the production of highly porous scaffolds. *Biomacromolecules.* **10**, 3188–3192.
11. Sun, T., Norton, D., McKean, R. J., Haycock, J. W., Ryan, A. J., and MacNeil, S. (2007) Development of a 3D cell culture system for investigating cell interactions with electrospun fibers. *Biotechnol. Bioeng.* **97**, 1318–1328.
12. Bokhari, M., Carnachan, R. J., Cameron, N. R., and Przyborski, S. A. (2007) Culture of HepG2 liver cells on three dimensional polystyrene scaffolds enhances cell structure and function during toxicological challenge. *J. Anat.* **211**, 567–576.
13. Bokhari, M., Carnachan, R. J., Cameron, N. R., and Przyborski, S. A. (2007) Novel cell culture device enabling three-dimensional cell growth and improved cell function. *Biochem. Biophys. Res. Commun.* **354**, 1095–1100.

Index

John W. Haycock (ed.), *3D Cell Culture: Methods and Protocols*, Methods in Molecular Biology, vol. 695, DOI 10.1007/978-1-60761-984-0, © Springer Science+Business Media, LLC 2011

FSC
www.fsc.org
MIX
Papier aus verantwortungsvollen Quellen
Paper from responsible sources
FSC® C105338

If you have any concerns about our products,
you can contact us on
ProductSafety@springernature.com

In case Publisher is established outside the EU,
the EU authorized representative is:
Springer Nature Customer Service Center GmbH
Europaplatz 3, 69115 Heidelberg, Germany

Printed by Libri Plureos GmbH
in Hamburg, Germany